HORMONES, BEHAVIOR, AND PSYCHOPATHOLOGY

HORMONES, BEHAVIOR, AND PSYCHOPATHOLOGY

Edited by

Edward J. Sachar, M.D.
Department of Psychiatry
Albert Einstein College of Medicine
Bronx, New York

Raven Press, Publishers ▪ New York

Made in the United States of America

International Standard Book Number 0-89004-062-1
Library of Congress Catalog Card Number 75-16658

Preface

The chapters in the volume were presented at the 65th Annual Meeting of the American Psychopathological Association in March, 1975. The theme of the conference reflected a consensus among the members that the study of interrelations between psychiatry and endocrinology—an ancient concept in western medicine—had made spectacular progress in the past few years, which deserved recognition and discussion.

The field has drawn in a creative way on new techniques—in psychologic assessment, endocrine biochemistry, hormonal and neuropharmacologic manipulation, and neurochemical analysis—to clarify old problems, as well as to discover previously unsuspected relationships. The findings have important implications for psychobiological research, for psychiatric diagnosis, for psychosomatic and internal medicine, and for the strategy of therapeutic interventions.

Hormones exert significant influences on brain and behavior; and the brain, in turn, sensitively modulates pituitary function. For purposes of grouping the chapters, we have separated these two types of psychoendocrine relationships, although physiologically they are probably nearly always integrated.

Studies of behavioral responses to hormones have been enormously aided by the isolation, purification, and, in some cases, partial synthesis of pituitary, hypothalamic, and target gland hormones. Dr. de Wied's work is a remarkable illustration of such recent research. Trophic hormones of the pituitary are shown to exert significant effects on memory and motivational processes in animals, quite independent of their endocrine effects. Indeed, specific fragments of the ACTH and melanocyte stimulating hormone (MSH) molecules, totally devoid of endocrine properties, produce the same behavioral changes. Other investigators are now exploring the therapeutic value of these substances in humans, particularly those suffering from memory disturbances associated with senility or electroconvulsive therapy.

Not only pituitary peptide hormones but also hypothalamic releasing hormones may be psychoactive. Dr. Lipton and his co-workers present data that thyrotropin releasing hormone (TRH) is such a psychoactive agent in animals and humans. This view is supported by Dr. Itil's electroencephalographic findings in human subjects. He reports that the EEG effects of TRH closely resemble those of stimulant drugs, whereas MSH inhibiting factor (MIF) has a different profile. Other investigators are exploring the behavioral effects of such hypothalamic hormones as luteinizing hormone releasing factor (LRF), and somatostatin.

Dr. Prange and his colleagues describe ways in which thyroid hormones enhance the clinical response of depressed women to tricyclic antidepressants by a mechanism which they believe to be mediated by central neuroreceptors; they also report unusual adverse side effects when imipramine is combined either with estrogens in women or with testosterone in men. An especially striking example of the therapeutic application of synthetic hormones in psychiatry is the chapter by Dr. Money and his colleagues, describing the successful use of androgen-suppressing hormones in the rehabilitation of sex offenders; Dr. Rose's comments deal thoughtfully and constructively with the ethical issues involved in using such biological aids in the treatment of patients with combined psychiatric and legal problems.

In addition to the experimental techniques employed by these investigators, experiments of nature have also provided important information about hormonal influences on brain. Dr. Reinisch reviews the extensive animal literature, indicating that brief periods of prenatal hormonal exposure can exert significant enduring effects on subsequent psychosexual development and behavior patterns; she well summarizes the provocative human data that suggest that prenatal androgenization may exert similar effects in girls. Dr. Bardwick's review highlights the fact that women indeed experience a significant fluctuation in mood, mental content, and outlook during the menstrual cycle; she speculates that the mechanism may involve the influence of gonadal hormones on the enzymes involved in catecholamine metabolism, particularly monoamine oxidase—an intriguing hypothesis that awaits further verification. Drs. Whybrow and Hurwitz undertook the task of reviewing perhaps the most ancient set of psychoendocrinological observations, involving serious psychiatric disturbances in patients suffering from endocrinopathies. These investigators doubt that there are specific psychiatric syndromes associated with endocrine disease, but they also highlight a paradox: this particular area has not been restudied by the modern methods of psychiatric and endocrine assessment illustrated in the other chapters, and it is very hard to draw inferences from the literature. The observation of serious depression associated with reserpine therapy led to an international psychopharmacological research effort; the high incidence of depression associated with Cushing's syndrome has, curiously, not attracted similar interest from psychobiologists.

The latter half of this volume focuses on exciting developments in the study of brain and behavioral influences on the endocrine system. A major research breakthrough in this area has been the discovery of the role of monoaminergic neurons in modulating secretion of releasing and inhibiting factors from hypothalamic neuroendocrine cells. Drs. Wurtman and Fernstrom describe some of the ingenious neuropharmacologic techniques used to work out the neurotransmitter regulation of neuroendocrine function, using the regulation of prolactin secretion by dopamine as a paradigm. In this instance, the investigators offer the provocative suggestion that the secretion of prolactin, in contrast to all other pituitary hormones, is directly inhibited by a neurotransmitter (dopamine) without an intervening hypothalamic hormone. Dr. Olney reports on an innovative method of chemically lesioning a brain area critically involved in neuroendocrine regulation, the arcuate nucleus of the hypothalamus, using monosodium glutamate (MSG). Although he does not emphasize it in his chapter, his work also has profound public health implications, in that MSG is a common food additive, only recently withdrawn from baby food.

Just as neuropharmacological techniques have been used to clarify the role of neurotransmitters in hormone regulation, neuroendocrine techniques can now be employed to solve questions in neuropharmacology. Thus, Dr. Sachar and colleagues report that d- and L-amphetamine equally stimulate growth hormone and equally suppress cortisol, responses which are not affected by blockade of dopamine receptors; they conclude that these neuroendocrine effects are noradrenergically mediated and that the two amphetamine isomers must be equipotent on noradrenergic neurons in the hypothalamus. Similarly, they reason that because dopamine inhibits prolactin secretion, measurement of plasma prolactin responses can provide a simple method for assessing the dopamine-blocking action of antipsychotic drugs in humans. Their data suggest that this approach can be used clinically to monitor the brain responses to these drugs in patients, to evaluate such pharmacological features of the drugs as their relative potency and duration of action, and to test the hypothesis that all effective antipsychotic drugs are dopamine blockers. Dr. Meltzer and his co-workers applied this technique in following the clinical course of psychotic inpatients receiving thioridazine therapy. These investigators call attention to some fascinating findings: the prolactin (brain) responses long precede the clinical responses; those cases with the lowest prolactin responses seemed to have required the largest doses of medication; and prolactin levels return to normal within only 48 hr after drug withdrawal.

Neuroendocrine techniques have also been applied to the study of the psychobiology of depressive illnesses, particularly as a way of documenting apparent hypothalamic dysfunction in these disorders. Drs. Carroll and Mendels review the extensive evidence for abnormalities in growth hormone and ACTH and cortisol secretion in severe endogenous depressions. They and Dr. Stokes describe in detail the hypersecretion of cortisol and its resistance to suppression by dexamethasone in these patients. Drs. Carroll and Mendels present striking data showing for the first time that this dexamethasone resistance is most apparent in evening rather than in morning hours. Their observation may provide important clues to the understanding of the biological bases of the peculiar circadian fluctuation in depressive symptomatology.

Normally, cortisol secretion ceases during the evening hours, probably because of CNS inhibition of ACTH secretion; Dr. Curtis attempted to determine if anxiety could stimulate a cortisol stress response during this period. He exposed phobic patients to their specific phobic stimulus at this time, and, although the patients appeared greatly anxious, there was no cortisol response. Dr. Curtis suggests that the pituitary-adrenocortical system is refractory to stress at this time but, since many other investigators, using other stimuli, have produced in humans substantial cortisol stress responses during this period, Dr. Curtis data may be more informative about special characteristics of phobic behavior than about the adrenal cortex.

Returning to abnormalities of human growth hormone (HGH) secretion in depression, it appears likely that the HGH response to hypoglycemia is noradrenergically mediated. The deficient HGH responses to this stimulus in depressed patients may thus reflect a diminution in hypothalamic noradrenergic activity. Dr. Garver's preliminary report suggests that correlation of HGH responses with urinary excretion of the brain noradrenalin metabolite MHPG in these patients may be a way to confirm this hypothesis.

Anorexia nervosa and psychosocial dwarfism are two serious diseases in which psychological and endocrine dysfunction are intimately entwined. Dr. Katz and his colleagues present remarkable findings that in late adolescent and young adult women suffering from anorexia nervosa, the pattern of secretion of the pituitary gonadotrophin LH reverts to a pattern characteristic of early puberty. They raise the possibility that this illness may be primarily a hypothalamic disorder, with secondary psychological manifestations. Dr. Halmi presents a unique histopathological study of the pituitary of a deceased anorexia nervosa patient, involving advanced staining techniques; lesions in the gonadotropin secreting cells are obvious and striking.

Dr. Brown and Dr. Money and his colleagues offer a thorough review and analysis of endocrinological and psychiatric aspects of psychosocial dwarfism. These unfortunate children fail to grow, and also fail to secrete HGH physiologically and in response to stimulation tests. Dr. Money's group, through their family studies, amply confirm that these children have been subjected to severe neglect and, often, horrendous abuse in their homes. They show that correcting the pathological psychosocial situation results in a prompt resumption of growth and of HGH secretion.

In a fitting conclusion, Dr. Shagass' Presidential Address puts the chapters and the subject of this volume into a larger context: that of the Medical Model as applied to psychiatry. Dr. Shagass eloquently argues against the notion that the Medical Model implies an antihumanistic bias. Analyzing the various dimensions of the model, Dr. Shagass concludes that it represents the highest order of a physicianly approach to the understanding and treatment of patients, including an awareness of the ethical context of the relationship, and of the effort to appreciate the biological factors contributing to those problems in functioning we define as psychiatric.

In closing, I would like to express my appreciation to Arthur Prange, Jr., M.D., co-chairman of the symposium published here, and to the members of the Council of the American Psychopathological Association, who offered much helpful advice on the selection of topics and authors whose work would best illustrate some of the many facets of this rapidly advancing discipline.

Edward J. Sachar
November 1975

Contents

Contributors

N. Altman
Department of Psychiatry
Albert Einstein College of Medicine
Bronx, New York 10461

C. Annecillo
Department of Psychiatry and Behavioral
    Sciences and Department of Pediatrics
The Johns Hopkins University School of
    Medicine and Hospital
Baltimore, Maryland 21205

J. M. Bardwick
Department of Psychology
University of Michigan
Ann Arbor, Michigan 48104

B. Bohus
Rudolf Magnus Institute for Pharma-
    cology
Medical Faculty
University of Utrecht
Vondellaan 6, Utrecht, The Netherlands

R. M. Boyar
Department of Oncology and the Institute
    for Steroid Research
Montefiore Hospital and Medical Center
Bronx, New York 10467

G. R. Breese
Biological Sciences Research Center
University of North Carolina
School of Medicine
Chapel Hill, North Carolina 27514

G. M. Brown
Neuroendocrinology Research Section
Clark Institute of Psychiatry
and
Department of Psychiatry
University of Toronto
Toronto, Ontario, Canada

B. J. Carroll
Department of Psychiatry
University of Michigan
Ann Arbor, Michigan 48104

B. R. Cooper
Biological Sciences Research Center
University of North Carolina

School of Medicine
Chapel Hill, North Carolina 27514

G. C. Curtis
Neuropsychiatric Institute
University of Michigan Medical Center
University Hospital
Ann Arbor, Michigan 48104

J. M. Davis
Illinois State Psychiatric Institute
1601 West Taylor Street
Chicago, Illinois 60612

L. S. Davis
Department of Psychiatry
Yale University
and
The Connecticut Mental Health Center
34 Park Street
New Haven, Connecticut 06519

H. Dekirmenjian
Illinois State Psychiatric Institute
1601 West Taylor Street
Chicago, Illinois 60620

D. de Wied
Rudolf Magnus Institute for Pharma-
    cology
Medical Faculty
University of Utrecht
Vondellaan 6, Utrecht, The Netherlands

V. S. Fang
Department of Medicine
University of Chicago Pritzker School of
    Medicine
Chicago, Illinois 60637

J. D. Fernstrom
Laboratory of Brain and Metabolism
Department of Nutrition and Food
    Science
Massachusetts Institute of Technology
Cambridge, Massachusetts 02139

A. G. Frantz
Department of Medicine
Columbia University
College of Physicians and Surgeons
New York, New York 10032

D. Gain
Department of Psychiatry and Behavioral Sciences and Department of Pediatrics
The Johns Hopkins University and Hospital
Baltimore, Maryland 21205

D. L. Garver
Illinois State Psychiatric Institute
1601 West Taylor Street
Chicago, Illinois 60612

W. H. Gispen
Rudolf Magnus Institute for Pharmacology
Medical Faculty
University of Utrecht
Vondellaan 6, Utrecht, The Netherlands

G. Gorzynski
Department of Psychiatry
Montefiore Hospital and Medical Center
Bronx, New York 10467

H. A. Gross
Department of Mental Hygiene
New York State Psychiatric Institute
722 West 168th Street
New York, New York 10032

P. H. Gruen
Department of Psychiatry
Albert Einstein College of Medicine
Bronx, New York 10461

K. A. Halmi
Department of Psychiatry
University of Iowa College of Medicine
Iowa City, Iowa 52242

F. S. Halpern
Department of Psychiatry
Albert Einstein College of Medicine
Bronx, New York 10461

L. Hellman
Department of Oncology and the Institute for Steroid Research
Montefiore Hospital and Medical Center
Bronx, New York 10467

G. R. Heninger
Department of Psychiatry
Yale University
and
The Connecticut Mental Health Center
34 Park Street
New Haven, Connecticut 06519

T. Hurwitz
Department of Psychiatry
Dartmouth Medical School
Hanover, New Hampshire 03755

T. M. Itil
Division of Biological Psychiatry
Department of Psychiatry
New York Medical College
New York, New York 10029

J. L. Katz
Department of Psychiatry
Montefiore Hospital and Medical Center
Bronx, New York 10467

M. A. Lipton
Biological Sciences Research Center
University of North Carolina
School of Medicine
Chapel Hill, North Carolina 27514

M. R. Mattson
Payne Whitney Clinic
New York Hospital
Cornell Medical Center
New York, New York 10021

H. Y. Meltzer
Department of Psychiatry
University of Chicago Pritzker School of Medicine
and
Illinois State Psychiatric Institute
1601 West Taylor Street
Chicago, Illinois 60612

J. Mendels
Department of Psychiatry
University of Pennsylvania
and
Depression Research Unit
Veterans Administration Hospital
Philadelphia, Pennsylvania 19104

J. Money
Department of Psychiatry and Behavioral Sciences and Department of Pediatrics
The Johns Hopkins University and Hospital
Baltimore, Maryland 21205

P. S. Mueller
Department of Psychiatry
Rutgers University
New Brunswick, New Jersey 08902

J. W. Olney
Department of Psychiatry

Washington University School of
Medicine
St. Louis, Missouri 63110

G. N. Pandey
Illinois State Psychiatric Institute
1601 West Taylor Street
Chicago, Illinois 60612

A. J. Prange, Jr.
Biological Sciences Research Center
University of North Carolina
School of Medicine
Chapel Hill, North Carolina 27514

J. M. Reinisch
Department of Developmental Psychology
Columbia University
Teachers College
New York, New York 10027

V. Rhee
Department of Psychiatry
Washington University School of Medicine
St. Louis, Missouri 63110

H. Roffwarg
Department of Psychiatry
Montefiore Hospital and Medical Center
Bronx, New York 10467

R. M. Rose
Department of Psychosomatic Medicine
Boston University School of Medicine
Boston, Massachusetts 02118

E. J. Sachar
Department of Psychiatry
Albert Einstein College of Medicine
Bronx, New York 10461

B. Schainker
Department of Psychiatry
Washington University School of Medicine
St. Louis, Missouri 63110

C. Shagass
Temple University Medical School
and
Eastern Pennsylvania Psychiatric Institute
Philadelphia, Pennsylvania 19129

R. N. Sollod
Payne Whitney Clinic

New York Hospital
Cornell Medical Center
New York, New York 10021

P. E. Stokes
Payne Whitney Clinic
New York Hospital
Cornell Medical Center
New York, New York 10021

P. M. Stoll
Payne Whitney Clinic
New York Hospital
Cornell Medical Center
New York, New York 10021

I. Urban
Rudolf Magnus Institute for Pharmacology
Medical Faculty
University of Utrecht
Vondellaan 6, Utrecht, The Netherlands

Tj. B. van Wimersma Greidanus
Rudolf Magnus Institute for Pharmacology
Medical Faculty
University of Utrecht
Vondellaan 6, Utrecht, The Netherlands

P. Walker
Department of Psychiatry and Behavioral
Sciences and Department of Pediatrics
The Johns Hopkins University and Hospital
Baltimore, Maryland 21205

H. Weiner
Department of Psychiatry
Montefiore Hospital and Medical Center
Bronx, New York 10467

J. Werlwas
Department of Psychiatry and Behavioral
Sciences and Department of Pediatrics
The Johns Hopkins University
School of Medicine and Hospital
Baltimore, Maryland 21205

P. C. Whybrow
Department of Psychiatry
Dartmouth Medical School
Hanover, New Hampshire 03755

C. Wiedeking
Department of Psychiatry and Behavioral
Sciences and Department of Pediatrics

The Johns Hopkins University and Hospital
Baltimore, Maryland 21205

I. C. Wilson
Biological Sciences Research Center
University of North Carolina
School of Medicine
Chapel Hill, North Carolina 27514

R. J. Wurtman
Laboratory of Neuroendocrine Regulation
Department of Nutrition and Food Science
Massachusetts Institute of Technology
Cambridge, Massachusetts 02139

Hormones, Behavior, and Psychopathology, edited by
Edward J. Sachar. Raven Press, New York © 1976.

Hormonal Influences on Motivational, Learning, and Memory Processes

D. de Wied, B. Bohus, W. H. Gispen, I. Urban, and Tj. B. van Wimersma Greidanus

Rudolf Magnus Institute for Pharmacology, Medical Faculty, University of Utrecht, Vondellaan 6, Utrecht, The Netherlands

The hypothalamic-pituitary unit manufactures "neuropeptides" which are involved in the formation and maintenance of new behavior. ACTH analogues exert a short-term effect opposite to that of adrenocortical steroids and their derivatives. These compounds may function as modulators of neuronal activity in midbrain limbic structures thereby altering the arousal state and consequently the motivational value of environmental cues. Hypothalamic vasopressin and analogues have a long-term effect on learned behavior, facilitating the consolidation and possibly also retrieval of information. These humoral factors facilitate the generation of adaptive behavioral responses.

The pituitary-adrenal system plays an essential role in homeostatic functions. Numerous aspects of stress-induced pituitary activation in relation to peripheral mechanism of adaptation have been studied, but little attention has been paid to the brain as a possible target organ for these hormones. Clinical observers frequently commented on psychological changes in addition to electroencephalographic (EEG) alterations in hyper- and hypocorticism (1). Many experiments, performed mainly in rodents, during the last decade disclosed the implication of various pituitary and hypothalamic hormones, and of steroids in motivational, learning, and memory processes.

Removal of the adenohypophysis or the neural lobe of the pituitary markedly interferes with, respectively, acquisition and maintenance of avoidance behavior in the shuttle box (2). Extirpation of the whole pituitary similarly impairs avoidance acquisition. It thus seems that lack of pituitary hormones is associated with a behavioral deficiency. Adrenalectomy does not impair avoidance acquisition, but rats with hereditary hypothalamic diabetes insipidus lacking the ability to synthesize vasopressin are inferior in acquiring and maintaining active and passive avoidance behavior (3). These observations point to an intrinsic effect of pituitary hormones on the central nervous system (CNS). Impaired behavior can be readily amended by treatment with ACTH, melanocyte-stimulating hormone (MSH), or vasopressin and with fragments of these hormones, which in themselves are practically devoid of target effects e.g., stimulation of the adrenal cortex in the case of ACTH fragments, or antidiuretic, vasopressor, and other endocrine effects in the case of vasopressin analogues. On the basis of these findings, it was postulated that the pituitary manufactures peptides — neuropeptides — which are involved in the formation and maintenance of new behavior patterns (4).

ACTH ANALOGUES AND BEHAVIOR

The seriously disturbed learning deficit of the hypophysectomized rat can be reversed by either ACTH and its analogues (α-MSH, ACTH 1–10, ACTH 4–10) or by vasopressin and its analogues (5). There is, however, an essential difference between the behavioral effect of ACTH analogues and vasopressin analogues. ACTH-like peptides exert a short-term effect, whereas the effect of vasopressin analogues is of a long term nature. A similar difference in behavioral effect of these two structurally different classes of peptides can be demonstrated in intact rats. A single injection of ACTH analogues (e.g., ACTH 1–10 or ACTH 4–10) delays extinction of a pole-jumping avoidance response for several hours. Arginine vasopressin (AVP), lysine vasopressin (LVP), desglycinamide LVP (DG-LVP), or pressinamide (PA) increase resistance to extinction of a similar pole-jumping avoidance response (6,7), which may last days to weeks. Similar differential effects can be found on extinction of shuttle box avoidance behavior or on retention of a simple one-trial passive avoidance response (8,9).

Effects of pituitary peptides have also been observed in other behavioral situations. Effects of ACTH analogues are a delay of extinction of approach behavior — food running response (10,11), facilitation of reversal learning, increased resistance to a complex brightness discrimination task (12), and facilitation of reversal of CO_2-induced and other forms of retrograde amnesia (13). Recently Bohus (14) showed that ACTH 4–10 delays extinction of a sexually motivated approach response of male rats in a straight runway. Animals treated with ACTH 4–10 ran faster than controls even in the absence of a receptive female in the goal box. Copulation reward during acquisition, however, appeared to be essential for the effect of ACTH 4–10 as observed during the extinction phase of the experiment. The running activity of males that were prevented from copulating was not affected by peptide treatment.

STRUCTURE ACTIVITY STUDIES WITH ACTH ANALOGUES

Detailed structure activity studies with ACTH analogues revealed that the tetrapeptide ACTH 4–7 contained the essential elements required for the behavioral effect of ACTH (Table 1). Since tryptophan is essential for MSH activity, this finding demonstrated a difference in structural require-

TABLE 1. *Effect of progressive shortening of the peptide chain ACTH 1–10 on behavioral potency*

ACTH	1	2	3	4	5	6	7	8	9	10	Approximated potency[a]
ACTH 1–10	H–Ser–Tyr–Ser–Met–Glu–His–Phe–Arg–Trp–Gly–OH										1
ACTH 2–10	H–Tyr–Ser–Met–Glu–His–Phe–Arg–Trp–Gly–OH										1
ACTH 3–10	H–Ser–Met–Glu–His–Phe–Arg–Trp–Gly–OH										1
ACTH 4–10	H–Met–Glu–His–Phe–Arg–Trp–Gly–OH										1
ACTH 5–10	H–Glu–His–Phe–Arg–Trp–Gly–OH										0.5
ACTH 4–9	H–Met–Glu–His–Phe–Arg–Trp–OH										1
ACTH 4–8	H–Met–Glu–His–Phe–Arg–OH										1
ACTH 4–7	H–Met–Glu–His–Phe–OH										1
ACTH 4–6	H–Met–Glu–His–OH										0.3

[a] Determined in the pole-jumping avoidance test.

ments for behavioral activity (ACTH 4–7) and MSH activity (ACTH 6–9) (15).

If the amino acid residue in position 7 was replaced by its D-isomer in either ACTH 1–10, ACTH 4–10, or ACTH 4–7, reversal of the effect in active avoidance behavior was found, i.e., facilitation of extinction (16,17). The reversal of action was found only for analogues with the 7-phenylalanine residue in the D-configuration. Successive replacement of each of the other amino acid residues in the hexapeptide 8-Lys-ACTH 4–9 by D-isomers failed to facilitate extinction of the avoidance response. Thus the reversal is an exception and a privilege of 7-D-Phe-ACTH analogues. All other D-isomer substitutions delayed extinction as found with "all-L" ACTH analogues. Such substitutions often caused potentiation, which was strongest when lysine in position 8 was replaced by its D-isomer. These results again indicate a dissociation between the requirements for behavioral and MSH activity. Although D-isomer substitution of phenylalanine increases MSH activity, a similar substitution of the basic histidyl or arginyl residues decreased this activity. Substitution of arginine by lysine in position 8 is accompanied by loss of steroidogenic activity in ACTH 1–24 (18) and MSH activity in ACTH 1–17 (19). The same occurs when methionine is oxidized to the sulfoxide level (20,21), but it increases the behavioral potency of ACTH 4–10. When tryptophan in position 9 is replaced by phenylalanine, a marked decrease in steroidogenic potency is found (22), but in the presence of D-lysine in position 8 the behavioral activity rises 100-fold. The combination of a number of these changes in 8-Lys-ACTH 4–9 (i.e., an oxidized residue in position 4, a lysine residue in the D-configuration in position 8, and the substitution of tryptophan by phenylalanine in position 9) yielded H–Met(O)–Glu–His–Phe–D–Lys–Phe–OH, which is 1,000 times more potent than ACTH 4–10 (23). This peptide delays extinction of a pole-jumping avoidance response in nanogram quantities following subcutaneous injection. It further appeared to be orally active. A partial explanation for

TABLE 2. *Behavioral effect of a number of oligopeptides related to ACTH and to releasing hormones*

ACTH or hormone	Oligopeptide	Approximated potency[a]
ACTH 4–7	H–Met–Glu–His–Phe–OH	1
	p.Glu–His–Phe–NH$_2$	0.3
TRH	p.Glu–His–Pro–NH$_2$	0.3
	p.Glu–His–Trp–NH$_2$	0.5
LHRH	p.Glu–His–Trp–Ser–Tyr–Gly–Leu–Arg–Pro–Gly–NH$_2$	1

[a] Determined in the pole-jumping avoidance test.

the effectiveness of these substitutions may be found in their protection against enzymatic degradation. The *in vitro* half-life of the various substituted analogues of ACTH 4–9 correlates with the behavioral potency (24).

Thyrotropin-releasing hormone (TRH) also delays extinction of the pole-jumping avoidance response but is less potent. Minor changes in the TRH molecule dramatically alter the intrinsic TSH-releasing activity (25). In contrast, the behavioral effect is not affected by such measures as substitution of proline amide by phenylalanine or tryptophan (Table 2). It is possible, therefore, that some of the recently reported central effects of TRH (26) and those of ACTH analogues are related, possibly owing to the fact that the structure of TRH is related to that of ACTH 4–7. This would also explain the finding that luteinizing hormone-releasing hormone (LH-RH), which contains the structurally closely related N-terminal tripeptide *p*Glu-His-Trp-, is as potent as ACTH 4–7 in delaying extinction of a pole-jumping avoidance response. It is conceivable that a great variety of related sequences is produced by the pituitary gland or originate in the hypothalamus or other brain structures — releasing hormones, other oligopeptides in the brain (27). However, the impaired avoidance acquisition, as found in adenohypophysectomized rats (2), indicates that pituitary peptides are more essential in this respect.

ACTH ANALOGUES AFFECT MOTIVATION IN ANIMAL AND MAN

Data obtained from electrophysiological studies indicate that ACTH and analogues have a central excitatory action in dogs (28). In rats ACTH 4–10 induces a frequency shift in θ activity from 7.0 to 7.5 Hz in hippocampus and thalamus evoked by stimulation of the reticular formation (29). Since similar shifts can be obtained after increasing the stimulus intensity, it follows that ACTH 4–10 facilitates transmission in midbrain limbic structures, suggesting that ACTH analogues increase the state of arousal in these structures. This may determine the motivational influence of environmental stimuli and may result in an increase in the probability of generating stimulus-specific responses. Interestingly, clinical observations are in keeping with the above-mentioned animal studies. Kastin et al. (30) found a significant increase in the averaged somatosensory evoked responses after administration of α-MSH. This increase was greater when subjects paid attention to the sensory stimulation and disappeared when the subjects were distracted. In a recent study (31) ACTH 4–10 exhibited similar effects. In addition, ACTH 4–10 produced an increase in the persistence of a previously habituated EEG response arousal pattern, a decrease in anxiety, an improved intradimensional discrimination, and an improvement in the Benton Visual Retention Test, but not in verbal retention, in healthy volunteers. These authors suggest that ACTH analogues in man affect the arousal state and thereby (visual?) attention. Gaillard and Sanders (32) reported a greater improvement in reaction time during a self-paced reaction test following ACTH 4–10 in healthy volunteers, which is indicative for an effect on motivation.

Thus human data so far available indicate, as in animal studies, that neuropeptides related to ACTH affect the arousal state and thereby (visual) attention. These effects may provide a basis for changes in motivation that have consequences for the performance.

STEROIDS AND BEHAVIOR

The disturbed avoidance acquisition of the hypophysectomized rat cannot be amended by treatment with dexamethasone. Daily administration of corticosterone or dexamethasone, in contrast to aldosterone, facilitates extinction of a shuttle box avoidance response in intact rats (33). This effect is opposite to that of ACTH and its analogues and thus could have been the result of the negative feedback action of glucocorticosteroids on pituitary ACTH release. However, the same steroids are also behaviorally active in hypophysectomized rats. In addition, implantation of cortisol in the median eminence of the hypothalamus, which blocks ACTH release, has a modest facilitatory effect on extinction of a shuttle box avoidance response. As more ACTH release is suppressed, the effect becomes stronger. Conversely, implantation of cortisol in the mesencephalic reticular formation barely reduces ACTH release but markedly facilitates extinction (34). Corticosteroids therefore may have a dual effect on extinction of conditioned avoidance behavior: one through inhibition of ACTH release, and one (probably more important) through a direct action on the CNS. Structure activity studies in the pole-jumping test showed that the influence of steroids on extinction does not exclusively reside in the glucocorticosteroid moiety of the molecule (35,36). Progesterone, 19-norprogesterone, and pregnenolone were as potent as corticosterone. Cholesterol, hydroxydion, testosterone, and estradiol were without effect. Common features of the steroids for facilitation of extinction are a double bond in ring A or B and a keto or hydroxyl group at C_3. A keto group at C_{20} is important for the potency only.

Corticosterone also facilitates extinction of approach behavior—food running response (11). A greater stability in interresponse time has been observed after dexamethasone administration (37). Bohus et al. (38) reported that corticosteroids administered before the learning trial suppresses passive avoidance behavior in a conflict situation of thirst versus fear. Using a similar situation, Endröczi reported facilitation of conditioned reflex activity by dexamethasone. In a less aversive situation, dexamethasone increased the latency to drink in rats kept on a 23-hour water deprivation schedule (39). This suggests a general suppressive effect on behavior, which in fact could be demonstrated on exploratory behavior in an open field. However, Pappas and Gray (40) failed to observe an effect of corticosterone or dexamethasone on the latency to resume licking after a single grid shock in rats trained to lick for water.

Although pituitary-adrenal system hormones affect the formation and maintenance of conditioned behavior, ACTH shares its influence with α- and β-MSH and much smaller analogues, e.g., ACTH 4–10 and ACTH 4–7. Similarly, the behavioral effect of corticosteroids, which is opposite to that of ACTH, is not restricted to those steroids whose production is ACTH-dependent. This makes it rather difficult to assign a specific role to the pituitary-adrenal system in conditioned behavior. However, this system is highly active in many behavioral situations and one may therefore expect that alterations in the level of these hormones affect ongoing behavior associated with stress. The pituitary-adrenal system plays an essential role in adaptation (41). Numerous studies have implicated the role of corticosteroids on peripheral mechanisms of adaptation. The observed behavioral effects of pituitary-adrenal system hormones suggest that these hormones modulate adaptive responses at the level of the CNS as well. Since corticosteroids have been shown to suppress, and ACTH to facilitate, single-unit activity in dorsal hippocampal areas (42), and since ACTH analogues facilitate transmission in midbrain limbic structures, pituitary-adrenal system hormones may function as modulators of these structures, resulting in the expression of the most adequate behavior in a given situation.

VASOPRESSIN ANALOGUES AND BEHAVIOR

Vasopressin and its analogues have a long-term effect on active and passive behavior (6,8,9), affecting the consolidation of learned behavior. These neuropeptides are physiologically involved in memory processes, as can be derived from studies in rats with hereditary diabetes insipidus (DI), which cannot synthesize vasopressin. Acquisition of a shuttle box avoidance response in these animals is significantly retarded, and extinction of the response is markedly facilitated (3). Severe memory impairment becomes manifest when DI rats are subjected to a simple one-trial passive avoidance test (43) wherein rats are trained to avoid a dark box in which they previously experienced an aversive stimulus (electric shocks of varying intensity and duration) (44). Homozygous DI rats exposed to a 1-mA shock lasting 3 or 10 sec fail to exhibit avoidance behavior when tested 1, 2, and 3 days after shock exposure. In contrast, their heterozygous littermates show maximal avoidance under the same conditions. Treatment of homozygous DI rats immediately after the shock trial with AVP or DG-LVP restores passive avoidance behavior in these animals toward that found in untreated heterozygous rats. Homozygous rats express avoidance without vasopressin treatment if tested immediately, and partially when tested 3 hours after the shock trial, indicating that "memory" rather than "learning" processes are disrupted in the absence of vasopressin. As in the studies with ACTH analogues, the behavioral effect of vasopressin is not limited to the whole molecule. DG-LVP, which was originally isolated from hog pituitary material (45), is almost as effective as LVP or AVP in increasing resistance to extinction. This peptide almost completely lacks the classic endocrine effects of vasopressin (46). Recent studies revealed that pressinamide has a similar behavioral effect, although it is much less potent than the parent molecule (Table 3). AVP and pressinamide are very effective in increasing resistance to extinction of a pole-jumping avoidance response following administration into one of the lateral ventricles. Using this route, less than 200 times the amount of AVP and less than 1,000 times that of pressinamide, which is needed systemically to affect avoidance behavior, elicit a response of the same magnitude. This suggests that vasopressin or precursors (e.g., degradation products of this octapeptide) may exert their behavioral effect through the cerebrospinal fluid (CSF). These principles may be released from their production sites probably through the juxtaependymal processes of neurosecretory neurons into the third ventricle. Morphological observations are in keeping with this notion (47). Heller et al. (48) and Vorherr et al. (49)

TABLE 3. Long-term behavioral effects of a number of vasopressin analogues

Vasopressin analogue	Structure	Approximated potency[a]
Arginine-8-vasopressin (AVP)	H–Cys–Tyr–Phe–Gln–Asn–Cys–Pro–Arg–Gly–NH$_2$	1.00
Lysine-8-vasopressin (LVP)	H–Cys–Tyr–Phe–Gln–Asn–Cys–Pro–Lys–Gly–NH$_2$	0.75
Desglycinamide lysine (vasopressin; DG-LVP)	H–Cys–Tyr–Phe–Gln–Asn–Cys–Pro–Lys–OH	0.50
Pressinamide (PA)	H–Cys–Tyr–Phe–Gln–Asn–Cys–NH$_2$	0.10

[a] Determined, following subcutaneous administration, in the pole-jumping avoidance test.

showed a parallel rise in antidiuretic hormone (ADH) activity in blood and CSF in response to severe hemorrhage, dehydration, ether anesthesia, etc. ADH activity in CSF was not detected after intravenous infusion of AVP in dogs (49), but Heller et al. (48) found a slight increase in ADH activity after systemic administration in rabbits.

INTRAVENTRICULAR AVP ANTIBODIES BLOCK MEMORY

If the behavioral effect of vasopressin analogues is exerted through the CSF, serum containing antibodies against vasopressin administered into the CSF might induce a behavioral deficit identical to that found in hereditary hypothalamic DI rats. At the same time, systemic administration of specific AVP antibodies in amounts that would remove the peptide from the peripheral circulation should have no effect on behavior. Passive avoidance behavior was markedly reduced at 4 hours and almost absent at the 24- and 48-hour retention sessions in rats injected 30 min before or immediately after the learning trial with anti-AVP serum administered into one of the lateral ventricles. Intraventricular antioxytocin serum or normal rabbit serum failed to interfere with the behavioral response. In contrast, maximal passive avoidance behavior was observed when rats treated with anti-AVP serum were tested 2 min and 1 hr after the learning trial. Thus learning is normal but memory expression is severely disturbed following removal of AVP from the CSF. Avoidance latencies of rats injected intravenously with 100 times as much anti-AVP serum, 30 min before or immediately after the learning trial, were unaffected in this respect. The removal of systemic AVP thus has no influence on avoidance behavior (50). Whereas intraventricular anti-AVP serum did not affect urine production, water consumption, or urinary vasopressin levels, intravenous administration significantly increased urine production and led to extremely low vasopressin levels in urine as measured by radioimmunoassay (50,51). It is therefore conceivable that the behavioral effects of vasopressin under physiological conditions result from an increased release of this or related peptides into the CSF in response to emotional stress that accompanies various behaviors, in the same way as its release into the bloodstream is augmented in a similar behavioral situation (52).

These peptides do not seem to affect extinction of a food running response (11). However, vasopressin is active in approach behavior. When male rats in a T-maze trained to run for a receptive female are treated with DG-LVP after each acquisition session, they choose the correct arm of the T-maze in a significantly higher percentage. This effect was even stronger during the retention sessions after discontinuation of the treatment (3). Here again the copulation reward appeared to be essential for the behavioral effect of the peptide. Vasopressin and DG-LVP reverse the retrograde amnesia in rats (13). Since Lande et al. (53) found that DG-LVP protects against puromycin-induced memory blockade in mice, vasopressin analogues may affect protein synthesis in the brain. Interestingly, DG-LVP facilitates the development of resistance to the analgesic action of morphine in mice (54). Again, this seems to depend on intrinsic physiological mechanisms since homozygous DI rats, in contrast to heterozygous littermates, are slower in developing resistance to the analgesic action of morphine as measured on the hotplate. The reduction in developing resistance to this action of morphine in DI rats can be amended by chronic treatment with either AVP or DG-LVP. Development of resistance to morphine analgesia, which may be regarded as a form of learning (55), can be prevented by drugs that inhibit protein synthesis (56). Thus mechanisms similar to those in learning and memory processes may be involved. The influence of vasopressin analogues therefore may be of a more general nature, serving the storage of exogenous and endogenous information in the cell.

VASOPRESSIN AND PARADOXICAL SLEEP

Recent electrophysiological studies by us show that rhythmic slow activity (RSA) of homozygous DI rats contains substantially lower hippocampal θ frequencies during paradoxical sleep (PS) than that of heterozygous animals. There were differences in all spectral parameters, and the averaged peak frequency of homozygous DI rats was approximately 1 Hz lower than that of heterozygous DI rats. Thus PS fails to produce normal frequency of RSA in the absence of vasopressin, indicating a qualitatively different PS. Administration of DG-AVP enhanced the generation of higher frequencies in homozygous DI rats, sometimes to the level found in normal controls. Changes in the amount of postlearning RSA after amnesic treatment (57) have been reported. Deprivation of paradoxical sleep has been shown to interfere with the consolidation of learned responses (58–60). In addition, Longo and Loizzo (61) reported that drugs which facilitate memory functions increase θ frequency during the postlearning period. These electrophysiological investigations in DI rats further support the hypothesis that vasopressin is involved in the consolidation of memory processes.

SITE OF ACTION OF BEHAVIORALLY ACTIVE PEPTIDES AND STEROIDS IN THE BRAIN

Indications on the locus of action of behaviorally active neuropeptides and steroids were obtained from lesions and implantation studies. These suggest that the thalamic area and the limbic forebrain are structures sensitive to the behavioral effect of these hormones. Corticosterone facilitates extinction of a pole-jumping avoidance response when implanted in or near the nucleus parafascicularis (62). Corticosterone implantation in the anterior hypothalamus, septum, amygdala, or dorsal hippocampus also facilitates extinction of a shuttle box avoidance response (63). Local applications of ACTH analogues into the mesencephalic-diencephalic area at the posterior thalamic level result in inhibition of extinction (64). In particular, the nucleus parafascicularis is sensitive to these peptides. Vasopressin analogues, however, are also effective when applied to the same structure (65). Thus neuropeptides and steroids seem to act in the same brain area.

Indeed, previous studies with α-MSH had shown that lesions in the nuclei parafascicularis prevent the inhibitory effect of this peptide on extinction of a shuttle box avoidance response (66). Similar lesions also inhibit the effect of ACTH 4–10 on extinction of a pole-jumping avoidance response (67). In contrast, LVP still exerts increased resistance to extinction in rats bearing lesions in the parafascicular nuclei, although the sensitivity of the lesioned rats to the peptide is markedly decreased. Bilateral lesions in the rostral septal area and in the dorsal hippocampus abolish the behavioral effect of vasopressin (67).

The parafascicular nuclei seems to be a preferential locus of action of pituitary-adrenal system hormones. The nonspecific thalamic reticular area has an important integrative function (68,69). It is therefore possible that pituitary-adrenal system hormones modulate transmission in this nodal point of sensory integration. Reciprocal alterations in sensory detection can be demonstrated during alterations in pituitary-adrenal activity (70). When the levels of glucocorticosteroids in plasma are low, sensory detection is high, and vice versa. ACTH analogues and steroids may therefore contribute to the integration of information from the *milieu extérieure* and *intérieure,* which is probably essential for recognizing stimulus-specific cues. The rostral septum and dorsal hippocampus seem to be the site of action of the effect of vasopressin analogues on memory processes. Evidence is indeed accumulating that these structures are concerned with learning and memory mechanisms (71–74).

CONCLUDING REMARKS

The pituitary produces neuropeptides related either to ACTH, exerting a short-term effect, or to vasopressin, exerting a long-term effect on learned behavior. Through these neuropeptides, the pituitary plays an important role in motivational, learning, and memory processes, which enable the organism to cope adequately with environmental changes. Other pituitary hormones may be involved as well: (a) growth hormone, which was shown to stimulate brain development of the fetus when administered to pregnant female rats (75); (b) prolactin, which has been implicated in nursing and nest-building activities (76); and perhaps many others. In addition, hypothalamic and other brain oligopeptides may participate in the formation of various behaviors. LH-RH has been shown to induce sexual behavior by a direct action on the brain (77,78), and TRH has distinct effects in the CNS (26,79,80).

Brain oligopeptides may be released into the CSF, which is a very effective transport medium to midbrain limbic structures; the latter are probably the sites of action of these entities. Pituitary peptides may be transported in a similar fashion since there is some evidence that pituitary principles may enter the brain by retrograde transport along the pituitary stalk or via the basilar cisterns, which seem to connect the hormone-producing cells of the pituitary gland with the liquor (81). Numerous peptides with specific information might exert effects on brain functions, which might facilitate specific behavioral patterns. Others (e.g., ACTH analogues and vasopressin analogues) are involved in the registration, consolidation, and retrieval of information that the organism employs to select the most appropriate response to environmental changes.

REFERENCES

1. Cleghorn, R. A. (1957): Steroid hormones in relation to neuropsychiatric disorders. In: *Hormones, Brain Function and Behavior,* edited by H. Hoogland, pp. 3–25. Academic Press, New York.
2. de Wied, D. (1969): The anterior pituitary and conditioned avoidance behavior. In: *Progress in Endocrinology,* pp. 310–316. Excerpta Medica International Congress Series No. 184, Excerpta Medica, Amsterdam.
3. Bohus, B., van Wimersma Greidanus, Tj. B., and de Wied, D. (1975): Behavioral and endocrine responses of rats with hereditary diabetes insipidus (Brattleboro strain). *Physiol. Behav.,* 14:609–615.
4. de Wied, D. (1969): Effects of peptide hormones on behavior. In: *Frontiers in Neuroendocrinology,* edited by W. F. Ganong and L. Martini, pp. 97–140. Oxford University Press, London.
5. Bohus, B., Gispen, W. H., and de Wied, D. (1973): Effect of lysine vasopressin and ACTH 4–10 on conditioned avoidance behavior of hypophysectomized rats. *Neuroendocrinology,* 11:137–143.
6. de Wied, D. (1971): Long term effect of vasopressin on the maintenance of a conditioned avoidance response in rats. *Nature (Lond.),* 232:58–60.
7. de Wied, D., Bohus, B., and van Wimersma Greidanus, Tj. B. (1974): The hypothalamo-hypophyseal system and the preservation of conditioned avoidance behavior in rats. *Prog. Brain Res.,* 41:417–428.
8. de Wied, D., and Bohus, B. (1966): Long term and short term effect on retention of a conditioned avoidance response in rats by treatment respectively with long acting pitressin or α-MSH. *Nature (Lond.),* 212:1484–1486.
9. Ader, R., and de Wied, D. (1972): Effects of lysine vasopressin on passive avoidance learning. *Psychon. Sci.,* 29:46–48.
10. Kastin, A. J., Miller, L. H., Nockton, R., Sandman, C., Schally, A. V., and Stratton, L.O. (1973): Behavioral aspects of melanocyte stimulating hormone (MSH). *Prog. Brain Res.,* 39:461–470.
11. Garrud, P., Gray, J. A., and de Wied, D. (1974): Pituitary-adrenal hormones and extinction of rewarded behavior in the rat. *Physiol. Behav.,* 12:109–119.
12. Sandman, C. A., Alexander, W. D., and Kastin, A. J. (1973): Neuroendocrine influences on visual discrimination and reversal learning in the albino and hooded rat. *Physiol. Behav.,* 11:613–617.
13. Rigter, H., van Riezen, H. and de Wied, D. (1974): The effects of ACTH- and vasopressin-analogues on CO_2-induced retrograde amnesia in rats. *Physiol. Behav.,* 13:381–388.

14. Bohus, B. (1974): ACTH, ACTH-like peptides and the sexual behavior of the male rat. *Riv. Farmacol. Ter.,* 5:6.
15. Otsuka, H., and Inouye, K. (1964): Synthesis of peptides related to the N-terminal structure of corticotropin. II. The synthesis of L-histidyl-L-phenylalanyl-L-arginyl-L-tryptophan, the smallest peptide exhibiting the melanocyte stimulating and lipolytic activities. *Bull. Chem. Soc. Jap.,* 37:1465–1471.
16. Bohus, B., and de Wied, D. (1966): Inhibitory and facilitatory effect of two related peptides on extinction of avoidance behavior. *Science,* 153:318–320.
17. de Wied, D., Witter, A., and Greven, H. M. (1975): Behaviorally active ACTH analogues. *Biochem. Pharmacol.,* 24:1463–1468.
18. Tesser, G. I., Maier, R., Schenkel-Hulliger, L., Barthe, P. L., Kamber, B., and Rittel, W. (1973): Biological activity of corticotrophin peptides with homo-arginine, lysine or orinithine substitutes for arginine in position 8. *Acta Endocrinol.,* 74:56–66.
19. Chung, D., and Li, C. H. (1967): Adrenocorticotropins. XXXVII. The synthesis of 8-lysine-ACTH 1–17NH$_2$ and its biological properties. *J. Am. Chem. Soc.,* 89:4208–4213.
20. Dedman, M. L., Farmer, T. H., and Morris, C. J. O. R. (1955): Oxidation-reduction properties of adrenocorticotrophic hormone. *Biochem. J.,* 59:xxii.
21. Lo, T.-B., Dixon, J. S., and Li, C. H. (1961): Isolation of methionine sulfoxide analogue of α-melanocyte-stimulating hormone from bovine pituitary glands. *Biochim. Biophys. Acta,* 53:584–586.
22. Hofmann, K., Andreatta, R., Bohn, H., and Moroder, L. (1970): Studies on polypeptides. XLV. Structure-function studies in the β-corticotropin series. *J. Med. Chem.,* 13:339–345.
23. Greven, H. M., and de Wied, D. (1973): The influence of peptides derived from corticotrophin (ACTH) on performance: Structure activity studies. *Prog. Brain Res.,* 39:pp. 429–442.
24. Witter, A., Greven, H. M., and de Wied, D. (1975): Correlation between structure, behavioral activity and rate of biotransformation of some ACTH 4–9 analogues. *J. Pharmacol. Exp. Ther.,* 193:853–860.
25. Burgus, R., Dunn, T. F., Desiderio, D. M., Ward, D. N., Vale, W., Guillemin, R., Felix, A. M., Gillessen, D., and Studer, R. O. (1970): Biological activity of synthetic derivatives related to the structure of hypothalamic TRF. *Endocrinology,* 86:573–582.
26. Prange, A. J., Breese, C. R., Cott, J. M., Martin, B. R., Cooper, B. R., Wilson, I. C., and Plotnikoff, N. P. (1974): Thyrotropin releasing hormone: Antagonism of pentobarbital in rodents. *Life Sci.,* 14:447–455.
27. Reichelt, K. L., and Kvamme, E. (1967): Acetylated and peptide bound glutamate and aspartate in brain. *J. Neurochem.,* 14:987–996.
28. Urban, I., Lopes da Silva, F. H., Storm van Leeuwen, W., and de Wied, D. (1974): A frequency shift in the hippocampal theta activity: An electrical correlate of central action of ACTH analogues in the dog. *Brain Res.,* 70:377–380.
29. Urban, I., and de Wied, D. (1975): The influence of ACTH 4–10 on hippocampal and thalamic synchronized activity in rats. *Brain Res.,* 85:195–196.
30. Kastin, A. J., Miller, L. H., Gonzalez-Barcena, D., Hawley, W. D., Dyster-Aas, K., Schally, A. V., Velasco de Parra, M. L., and Velasco, M. (1971): Psychophysiologic correlates of MSH activity in man. *Physiol. Behav.,* 7:893–896.
31. Miller, L. H., Kastin, A. J., Sandman, C. A., Fink, M., and Van Veen, W. J. (1974): Polypeptide influences on attention, memory and anxiety in man. *Pharmacol., Biochem. Behav.,* 2:663–668.
32. Gaillard, A. W. K., and Sanders, A. F. (1975): Some effects of ACTH 4–10 on performance during a continuous reaction task. *Prog. Brain Res.,* 42:209–210.
33. de Wied, D. (1967): Opposite effects of ACTH and glucocorticosteroids on extinction of conditioned avoidance behavior. In: *Proceedings, 2nd International Congress Hormonal Steroids,* pp. 945–951. Excerpta Medica International Congress Series 132. Excerpta Medica, Amsterdam.
34. Bohus, B. (1968): Pituitary ACTH release and avoidance behavior of rats with cortisol implants in mesencephalic reticular formation and median eminence. *Neuroendocrinology,* 3:355–365.
35. van Wimersma Greidanus, Tj. B. (1970): Effects of steroids on extinction of an avoidance response in rats. A structure-activity relationship study. *Prog. Brain Res.,* 32:185–191.

36. van Wimersma Greidanus, Tj. B., Wijnen, H., Deurloo, J., and de Weid, D. (1973): Analysis of the effect of progesterone on avoidance behavior. *Horm. Behav.*, 4:19–30.
37. Wertheim, G. A., Conner, R. L., and Levine, S. (1967): Adrenocortical influences on free-operant avoidance behavior. *J. Exp. Anal. Behav.*, 10:555–563.
38. Bohus, B., Grubits, J., Kovács, G., and Lissák, K. (1970): Effect of corticosteroids on passive avoidance behavior of rats. *Acta Physiol. Acad. Sci. Hung.*, 38:381–391.
39. Endröczi, E. (1972): *Limbic System and Pituitary-Adrenal Function*. Académiai Kiadó, Budapest.
40. Pappas, B. A., and Gray, P. (1971): Cue value of dexamethasone for fear-motivated behavior. *Physiol. Behav.*, 6:127–130.
41. Selye, H. (1950): *The Physiology and Pathology of Exposure to Stress*, p. 6. Acta Inc., Montreal, Canada.
42. Pfaff, D. W., Silva, M. T. A., and Weiss, J. M. (1971): Telemetered recording of hormone effects on hippocampal neurons. *Science*, 172:394–395.
43. de Wied, D., Bohus, B., and van Wimersma Greidanus, Tj. B. (1975): Memory deficit in rats with hereditary diabetes insipidus. *Brain Res.*, 85:152–156.
44. Ader, R., Weijnen, J. A. W. M., and Moleman, P. (1972): Retention of a passive avoidance response as a function of the intensity and duration of electric shock. *Psychon. Sci.*, 26:125–128.
45. Lande, S., Witter, A., and de Wied, D. (1971): Pituitary peptides: An octapeptide that stimulates conditioned avoidance acquisition in hypophysectomized rats. *J. Biol. Chem.*, 246:2058–2062.
46. de Wied, D., Greven, H. M., Lande, S., and Witter, A. (1972): Dissociation of the behavioral and endocrine effects of lysine vasopressin by tryptic digestion. *Br. J. Pharmacol.*, 45:118–122.
47. Wittkowski, W. (1968): Elektronenmikroskopische Studien zur intraventrikulären Neurosekretion in den Recessus infundibularis der Maus. *Z. Zellforsch.*, 92:207–216.
48. Heller, H., Hasan, S. H., and Saifi, A. Q. (1968): Antidiuretic activity in the cerebrospinal fluid. *J. Endocrinol.*, 41:273–280.
49. Vorherr, H., Bradbury, M. W. B., Hoghoughi, M., and Kleeman, C. R. (1968): Antidiuretic hormone in cerebrospinal fluid during endogenous and exogenous changes in its blood level. *Endocrinology*, 83:246–250.
50. van Wimersma Greidanus, Tj. B., Dogterom, J., and de Wied, D. (1975): Intraventricular administration of anti-vasopressin serum inhibits memory in rats. *Life Sci.*, 16:637–644.
51. van Wimersma Greidanus, Tj. B., Buys, R. M., Hollemans, H. J. G., and de Jong, W. (1974): A radioimmunoassay of vasopressin: A note on pituitary vasopressin content in Brattleboro rats. *Experientia*, 30:1217–1218.
52. Thompson, E. A., and de Wied, D. (1973): The relationship between the antidiuretic activity of rat eye plexus blood and passive avoidance behaviour. *Physiol. Behav.*, 11:377–380.
53. Lande, S., Flexner, J. B., and Flexner, L. B. (1972): Effect of corticotropin and desglycinamide[9]-lysine vasopressin on suppression of memory by puromycin. *Proc. Natl. Acad. Sci. U.S.A.*, 69:558–560.
54. Krivoy, W. A., Zimmermann, E., and Lande, S. (1974): Facilitation of development of resistance to morphine analgesia by desglycinamide[9]-lysine-vasopressin. *Proc. Natl. Acad. Sci. U.S.A.*, 71:1852–1856.
55. Cohen, M., Keats, A. S., Krivoy, W. A., and Ungar, G. (1965): Effect of actinomycin on morphine tolerance. *Proc. Soc. Exp. Biol. Med.*, 119:381–384.
56. Cox, B. M., and Osman, O. H. (1970): Inhibition of the development of tolerance to morphine in rats by drugs which inhibit ribonucleic acid or protein synthesis. *Br. J. Pharmacol.*, 38:157–170.
57. Landfield, P. W., McGaugh, J. L., and Tusa, R. J. (1972): Theta rhythm: A temporal correlate of memory storage processes in the rat. *Science*, 175:87–89.
58. Leconte, P., and Bloch, V. (1970): Déficit de la retention d'un conditionnement après privation de Sommeil Paradoxal chez le rat. *C. R. Acad. Sci. (Paris) [D]*, 271:226–229.
59. Stern, W. C. (1971): Acquisition impairments following rapid eye movement sleep deprivation in rats. *Physiol. Behav.*, 7:345–352.
60. Fishbein, W. (1971): Disruptive effects of rapid eye movement sleep deprivation on long-term memory. *Physiol. Behav.*, 6:279–282.

61. Longo, V. G., and Loizzo, A. (1973): Effects of drugs on the hippocampal theta-rhythm: possible relationships to learning and memory processes. In: *Pharmacology and the Future of Man,* Vol. 4, edited by F. E. Bloom and J. H. Acheson, pp. 46–54. Karger, Basel.

62. van Wimersma Greidanus, Tj. B., and de Wied, D. (1969): Effect of intracerebral implantation of corticosteroids on extinction of an avoidance response in rats. *Physiol. Behav.,* 4:365–370.

63. Bohus, B. (1970): Central nervous structure and the effect of ACTH and corticosteroids on avoidance behaviour: A study with intracerebral implantation of corticosteroids in the rat. *Prog. Brain Res.,* 32:171–184.

64. van Wimersma Greidanus, Tj. B., and de Wied, D. (1971): Effects of systemic and intracerebral administration of two opposite acting ACTH-related peptides on extinction of conditioned avoidance behavior. *Neuroendocrinology,* 7:291–301.

65. van Wimersma Greidanus, Tj. B., Bohus, B. and de Wied, D. (1973): Effects of peptide hormones on behavior. In: *Progress in Endocrinology,* pp. 197–201. Excerpta Medica International Congress Series No. 273. Excerpta Medica, Amsterdam.

66. Bohus, B., and de Wied, D. (1967): Failure of α-MSH to delay extinction of conditioned avoidance behavior in rats with lesions in the parafascicular nuclei of the thalamus. *Physiol. Behav.,* 2:221–223.

67. van Wimersma Greidanus, Tj. B., Bohus, B., and de Wied, D. (1975): CNS sites of action of ACTH, MSH and vasopressin, related to avoidance behavior. In: *Anatomical Neuroendocrinology,* edited by W. E. Stumpf and L. D. Grand. Karger, Basel (*in press*).

68. Cardo, B. (1965): Rôle de certains noyaux thalamiques dans l'élaboration et la conservation de divers conditionnements. *Psychol. Franc.,* 10:334–351.

69. Cardo, B. (1967): Effets de la stimulation du noyau parafasciculaire thalamique sur l'acquisition d'un conditionnement d'évitement chez le rat. *Physiol. Behav.,* 2:245–248.

70. Henkin, R. I. (1970): The effects of corticosteroids and ACTH on sensory systems. *Prog. Brain Res.,* 32:270–294.

71. Fried, P. A. (1972): The septum and behaviour: A review. *Psychol. Bull.,* 78:292–310.

72. Isaacson, R. L., and Kimble, D. P. (1972): Lesions of the limbic system: Their effects upon hypothesis and frustration. *Behav. Biol.,* 7:767–793.

73. Olds, J., Disterhoft, J. F., Segal, M., Kornblith, C. L., and Hirsh, R. (1972): Learning centers of rat brain mapped by measuring latencies of conditioned unit responses. *J. Neurophysiol.,* 35:202–219.

74. Hirsh, R. (1974): The hippocampus and contextual retrieval of information from memory: A theory. *Behav. Biol.,* 12:421–444.

75. Sara, V. R., and Lazarus, L. (1974): Prenatal action of growth hormone on brain and behaviour. *Nature (Lond.),* 250:257–258.

76. Zarrow, M. X., Gandelman, R., and Denenberg, V. H. (1971): Prolactin: Is it an essential hormone for maternal behavior in the mammal? *Horm. Behav.,* 2:343–354.

77. Moss, R. L., and McCann, S. M. (1973): Induction of mating behavior in rats by luteinizing hormone-releasing hormone, *Science,* 181:177–179.

78. Pfaff, D. W. (1973): Luteinizing hormone-releasing factor potentiates lordosis behavior in hypophysectomized ovariectomized female rats. *Science,* 182:1148–1149.

79. Dyer, R. G., and Dyball, R. E. J. (1974): Evidence for a direct effect of LRF and TRF on single unit activity in the rostral hypothalamus. *Nature (Lond.),* 252:486–488.

80. Plotnikoff, N. P. (1975): Prolyl-leucyl-glycine amide (PLG) and thyrotrophin releasing hormone (TRH): Dopa potentiation and biogenic amine studies. *Prog. Brain Res.,* 42:11–23.

81. Allen, J. P., Kendall, J. W., McGilvra, R., and Vancura, C. (1974): Immunoreactive ACTH in cerebrospinal fluid. *J. Clin. Endocrinol.,* 38:586–593.

Hormones, Behavior, and Psychopathology, edited by
Edward J. Sachar. Raven Press, New York © 1976.

Behavioral Effects of Hypothalamic Polypeptide Hormones in Animals and Man

M. A. Lipton, G. R. Breese, A. J. Prange, Jr., I. C. Wilson, and B. R. Cooper

Biological Sciences Research Center, University of North Carolina, School of Medicine, Chapel Hill, North Carolina 27514

We reviewed the behavioral effects of peripheral hormones in man in an earlier publication (1). Whybrow and Hurwitz refined and extended this work, and reported it elsewhere in this volume. In another chapter de Wied considered the behavioral effects of pituitary hormones and their fragments in animals. This leaves us the task of reviewing the behavioral effects of a newly discovered series of hormones, the so-called hypothalamic polypeptide-releasing factors. Most of what we know of this subject is derived from animal studies. Therefore we emphasize this body of work, referring to findings in man when appropriate. By way of orientation, let us repeat a hypothesis we stated previously (2–4): Hypothalamic releasing hormones may exert brain effects apart from their actions on the anterior pituitary gland. These brain effects may have neurological or behavioral consequences, and in clinical disorders these consequences may manifest as either benefits or aggravations.

Before presenting the evidence to support this hypothesis, we add that our knowledge of the extrapituitary functions of the hypothalamic hormones is limited for several reasons. Although these substances have been under investigation for a long time, it is only within the last few years that they have been isolated, identified, and synthesized (5–8). Since they are present in the hypothalamus in tiny quantities (roughly a few thousandths of a microgram per gram of tissue) and since they were measured until very recently by bioassay, their isolation represents a major triumph in biochemistry. Chemically they are moderately small polypeptides, a property that has made their synthesis relatively easy. These compounds are now available in adequate quantities for investigation in animals and man. Unfortunately there are still some constraints that limit both basic and clinical experimentation with these polypeptides. At present most of what is known about their actions is derived from the strategy of administering these hormones in normal and pathological states. The recently developed radioimmune assays are sufficiently sensitive to measure some of these hormones in areas of the hypothalamus and other portions of animal brains.

Unfortunately the assays still lack the sensitivity and specificity required for measuring these hormones in peripheral tissues, blood, or urine. Even though it is likely that methodology shall soon be improved sufficiently to permit assay in other tissues and urine, we are now required to measure hypothalamic function indirectly by measuring pituitary or peripheral hormones which reflect the action of the polypeptide hormones. When such endocrine disorders are found, it is often difficult to determine whether they are primary or secondary to hypothalamic dysfunction.

Our own interest in the behavioral effects of hypothalamic hormones derives from a longstanding interest in the relationships between endocrine state and affective disorders (1). Our demonstration several years ago that L-triiodothyronine (9,10) accelerates the antidepressant action of imipramine in women, along with other considerations, led us to the concept that a diminution of thyroid function may be among the many factors contributing to the etiology of depressive disorders. This diminution may be small, so that depressed patients often fall within the normal range (11). These studies were followed by a demonstration that thyroid-stimulating hormone (TSH) was also effective in accelerating the antidepressant action of imipramine (12). These findings motivated us to investigate higher segments of the hypothalamic pituitary axis, and when thyrotropin-releasing hormone (TRH) became available for human investigation we examined its possible effects in depressed women. The results we obtained were unexpected insofar as a single intravenous injection of 600 μg TRH led to a rapid, although partial and brief, improvement in the depressive symptomatology (13,14). This improvement, although clinically incomplete, occurred without antidepressant medication and was sufficiently interesting to stimulate work on mechanisms. We therefore expanded our efforts to study the effects of TRH in other clinical syndromes and began parallel work with animals and with other hypothalamic hormones. The results led us to formulate the generic hypothesis stated above. In this hypothesis the action of the hypothalamic releasing hormones on the pituitary is not questioned. What is suggested is that they have additional effects. In this chapter we review the findings supporting this hypothesis.

ACTIONS OF RELEASING HORMONES

TRH in Animals

TRH (L-pyroglutamyl-L-histidyl-D-proline amide) is a potent releaser of pituitary TSH (15) and prolactin (16). The physiological significance of its release of prolactin is uncertain (17). TRH was the first of the hypothalamic factors to be identified and synthesized (5,18).

The first evidence that TRH exerts actions on the brain not mediated by the pituitary came from pharmacological experiments. A common screening test for potential antidepressants uses animals given the monoamine oxidase inhibitor pargyline and L-dihydroxyphenylalanine (L-DOPA) (19). The increased activity of animals so treated is further enhanced by antidepressant drugs. TRH was active in this test, even when mice were hypophysectomized (20,21) and rats thyroidectomized (21,22). TRH also potentiated the behavioral activation in hypophysectomized mice caused by the administration of pargyline and 5-hydroxytryptophan (5-HTP) (23). Another screening test for antidepressants employs the yohimbine-treated dog. In this test Hine et al. (24) found that TRH behaved more like amphetamine than like either a tricyclic drug or a monoamine oxidase inhibitor. These actions also are consistent with the concept of a direct brain action of TRH. More recently TRH has been shown to antagonize the sleep and hypothermia caused by pentobarbital (25,26) and ethanol (27), even in hypophysectomized animals. Although many of the effects of TRH in such experiments resemble those of amphetamine, the antagonism of ethanol shows a difference because amphetamine does not antagonize ethanol sedation; rather, it prolongs it. TRH also opposes the action of other centrally acting drugs including chloral hydrate, reserpine, diazepam, and several barbiturates other than pentobarbital (28).

The antagonism of pentobarbital by TRH is not reduced by hypophysectomy and is not imitated by TSH or T_3 (28). The constituent amino acids of TRH and proline amide are without effect. The antagonism does not appear to depend on an alteration of barbiturate metabolism (28). Recent work (28,29) has shown that certain synthetic derivatives of TRH, which have virtually no effect on TSH release, are still potent antagonists of pentobarbital. This suggests that brain and pituitary receptors for TRH have different properties and therefore are different entities. In still other studies, pentobarbital-treated mice were given norepinephrine (NE) intracisternally. Administered in this fashion NE antagonized pentobarbital-induced hypothermia but had little effect on pentobarbital-induced sleep (28). This suggests that the effects of TRH on noradrenergic fibers are not crucially involved in the mechanisms by which TRH reverses pentobarbital anesthesia in mice (28).

The evidence described thus far is based on the effect of TRH in animals that have been first treated with drugs. There is also anatomical, behavioral, biochemical, and electrophysiological data to support the hypothesis of extrapituitary actions of TRH. Some of these were obtained with untreated animals. Thus TRH has been found to be widely distributed throughout the brain (30,31). Its concentration is highest in the hypothalamus, but the mass of the remainder of the brain is so much greater than this small specific region that approximately two-thirds of the total TRH is found outside the hypothalamus (31). Its origin and specific role in these brain regions is unknown. However, high-affinity binding sites for TRH have been found in the brain as well as the pituitary (32).

Keller et al. (33) found that when rats are given TRH intraperitoneally there is an increased rate of incorporation of radioactive tyrosine into brain NE and a rise in brain 3-methoxy-4-hydroxyphenylglycol (MHPG), the primary metabolite of NE in brain. These findings were confirmed (28) and extended by Horst and Spirt (34), who also performed *in vitro* studies using synaptosome preparations in which pituitary, thyroid, or other possible peripheral effects of TRH could not play a part.

Segal and Mandell (35) infused microgram quantities of TRH into the lateral ventricle of free-roving rats and found that animals so treated showed increased activity. Renaud and Martin (36) reported a potent depressant action of TRH applied iontophoretically on the activity of a certain population of neurons in several areas of the rat brain. This depressant activity resembled that obtained with γ-aminobutyric acid and glycine.

None of these bits of evidence are conclusive, but they tend to converge in support of our hypothesis. Clearly much more must be done to test the utility and limits of the hypothesis. TRH must be examined as a possible transmitter and as a receptor modulator. Methods must be developed to alter the TRH content of the brain so that the biological consequences can be measured. Above all, new behavioral studies in animals are required. It is curious that large quantities of TRH can be administered to animals parenterally without gross biological or behavioral effects. Yet it is likely that subtle effects will be found, probably in the realm of adaptive behavior. A search for such effects is under way in our laboratory and others.

Behavioral Effects of TRH in Man

The availability of synthetic TRH has added a new tool to the diagnostic resources of the "thyroidologist" (37,38). With it, some cases of hypothalamic hypothyroidism have been found. These patients characteristically show low baseline TSH levels but a rise after TRH administration (39). It would be of uncommon interest to the psychiatrist to determine if the behavioral characteristics of such patients differs significantly from those of other hypothyroid patients. Unfortunately such patients are rare (39) and have not yet been examined by psychiatrists. We are therefore unable to describe the consequences of a hypothalamic TRH deficiency.

In our studies of depression, we found that rapid intravenous administration of TRH (0.6 mg) to otherwise untreated unipolar depressed women caused a rapid, although brief and partial, antidepressant response (13,14). This first demonstration of a behavioral effect of a hypothalamic polypeptide in humans has been confirmed by several workers (40–42). On the other hand, others have been unable to replicate our findings (43–45). In particular, a number of trials involving repeated oral administration of 300 mg TRH to small numbers of patients have failed to demonstrate clinically useful effects (44–46). The controversial role of TRH in relation to depression therefore warrants some discussion.

Early reports of the action of TRH in depressed patients demonstrated transient beneficial effects but did not claim definitive clinical utility. Comparison of the results of a single dose of 600 μg TRH with a full course of electroconvulsive or imipramine therapy is not warranted for several reasons. The half-life of exogenously administered TRH in plasma is very brief (47) and its ability to penetrate the blood-brain barrier unknown. Given the short half-life of administered TRH and its susceptibility to destruction by deamidation and by peptidase activity, it is not at all surprising that oral administration produced no noticeable results. Whereas TRH is clearly not an effective antidepressant in the sense that it treats the illness as well as a full course of tricyclic treatment antidepressant medication, many investigators have noted a transient effect on the mood of both depressed and normal women (48). Granted that similar transient effects can follow the administration of amphetamine, it is nonetheless of unusual interest that a natural compound made in the hypothalamus should have these mood-altering properties. The mechanism by which this is achieved warrants further examination. From the perspective of clinical therapeutics, it is possible that synthetic congeners of TRH with more prolonged biological activity or greater specificity for action in the central nervous system (CNS) may be found useful. The antagonism of CNS depressants by TRH, studied thus far only in animals, may offer such an area.

In the course of our studies of the antidepressant response produced by TRH, we encountered another interesting finding that may be an even more valuable contribution to the progress in understanding and treating depressive disorders. In our depressed patients pituitary secretion of TSH after TRH stimulation is deficient when compared to the responses of normal subjects (14,41–43). A few normals have been reported to show no measurable response (49,50), and a genetic and metabolic study of this fault with special reference to affective disorder would be of great interest.

Consideration of the diminished pituitary response to TRH of depressed patients suggests that it cannot be attributed to increased negative feedback on the pituitary gland from elevated thyroid hormones. In our patients chemical measures of thyroid function are at the low end of the usual normal limits, and in the depressed patients of other investigators there is no reason to suspect hyperthyroidism. A more likely interpretation implies that in depression hypothalamic TRH activity is diminished, leading to a paucity of releasable TSH (14). A direct test of this will require measuring TRH in cerebrospinal fluid, blood, or urine. Whereas this notion requires exploration, it is not the simplest idea that will cover the most neuroendocrine information pertaining to depression. In depression there is also a deficiency of growth hormone response when evoked by insulin hypoglycemia (51,52). Somatotropin release-inhibiting factor (SRIF, somatostatin) inhibits the release of growth hormone (53,54), and recently it has been shown also to inhibit TRH-induced release of TSH (55). It is reasonable to suggest, therefore, that in depressions there is excessive activity of SRIF (56). This notion fits well with the inhibitory effects of intracisternally administered SRIF on the behavior of rats (35).

In a double-blind, placebo-controlled crossover study of 10 normal women, we found that intravenous TRH promptly caused a sense of relaxation and mild euphoria (48). In this study, as well as in a study of Chase et al. (57) which showed large doses of TRH by slow intravenous infusion to patients with parkinsonism to be devoid of neurological effects, we noted an increased tendency on the part of the subjects to socialize. Similarly, in our early work with TRH and depression, we noted that withdrawal and emotional flatness, when present, were often improved along with the other symptoms. This suggested a trial with the hormone in schizophrenia and we were further inclined toward this venture by the observation of Tiwary et al. (58) that a dose of TRH given for diagnostic purposes to a mentally deficient boy with cerebral gigantism greatly improved his disorganized behavior for many hours. We reported our preliminary single-blind experience with TRH in four schizophrenic women (59), and our double-blind experience with eight more patients was similar. The patients employed in these studies all had several hospital admissions for schizophrenia and had shown good response to phenothiazines; when they relapsed it was usually because they interrupted medication against advice. All patients had been withdrawn from drugs for several weeks before being given intravenous TRH, and none were in an excited state. Of the 12 patients studied, only one failed to show a favorable response. In general, improvement became apparent within 6 hours, achieved a maximum in 30 hours, and lasted approximately 1 week before relapse was complete. In three of the 12 patients, improvement was sufficient to permit discharge from the hospital and management with supportive psychotherapy alone. In the other patients improvement was far from complete but could be noted in the sphere of thought disorder as well as affect.

In our clinic TRH has also shown promise in a preliminary single-blind study in producing a relaxed state with moderate relief of depression in men with alcohol withdrawal symptoms; Huey (60) reported similar findings in a small series.

Itil (40) showed that the intravenous injection of TRH causes a characteristic change in the electroencephalogram of conscious human subjects. The changes caused by TRH resemble those caused by amphetamine, methylphenidate, isocarboxyzide, and protriptyline, a stimulating tricyclic antidepressant. His findings are elaborated in the discussion of this chapter. It is important to keep in mind that, even if TRH mimics CNS stimulants in pharmacological tests in animals or electroencephalographically in humans, the fact that it is an endogenous substance produced by the brain (which nonetheless mimics some properties of synthetic drugs that are not endogenous) is of intrinsic interest.

It appears from our work and that of others, that TRH has activity in a variety of psychiatric conditions which violate diagnostic boundaries. At first this appears absurd and led Hollister (61) to refer to it as resembling psychoanalysis in its therapeutic claims. In fact no such claims were made. Instead, observations were noted in which TRH exerted biological activity in a number of psychopathological states. This activity was manifest as transient and partial improvement. It must be emphasized that TRH is a naturally occurring constituent of the brain. Norepinephrine, a more thoroughly studied natural constituent, has been implicated not only in depression and schizophrenia, but there is evidence, summarized by Kety (62), that it influences arousal, aggression, certain types of appetitive behavior, and perhaps certain types of adaptive behavior expressed in learning and memory. It is therefore not inconceivable that TRH should also exert manifold effects. Furthermore, the behavioral manifestations of endocrinopathies or of vitamin deficiencies also violate psychiatric diagnosis boundaries. The therapeutic effects of some drugs are also not limited to specific psychiatric syndromes. Imipramine, for example, has use in depression (63), enuresis (64), and hyperkinesis in children (65). Even the symptoms by which we recognize mental disorders overlap in different syndromes and are not diagnosis-specific. Since TRH is endogenous and is widely distributed in the brain, it might be expected to exert a medley of effects. Clearly we are not yet able to describe the actions of TRH in molecular or biochemical terms. In psychological terms, however, this polypeptide hormone appears to increase coping behavior. Thus it is not astonishing that TRH seems to exert beneficial effects—even if partial, transient, and variable in degree—in depression, schizophrenia, hyperkinesis, and alcohol withdrawal. It is imperative to know the full range of physiological actions of TRH and other releasing factors from the perspective of biology and psychology, rather than from the point of view of psychopathology alone.

SRIF

SRIF, a tetradecapeptide (6) inhibits the release of growth hormone in the rat and man (53,54), as well as the release of pituitary TSH (55). Behavioral studies of the effects of this hormone in man have not been conducted. The effects described here are therefore limited to animal studies. Segal and Mandell (35) infused SRIF into the lateral ventricle of the rat and found that animals thus treated showed significant reduction in spontaneous activity without appearing sedated. Siler and his colleagues (54) gave large intravenous doses of SRIF to monkeys and noted anecdotally that they seemed tranquilized.

SRIF is the only hypothalamic peptide we have found, apart from TRH, that is active in altering the response of mice to pentobarbital. SRIF has the slight but statistically significant effect of prolonging sedation and exaggerating hypothermia (29,66). SRIF has not been tested in hypophysectomized animals, but the experiments performed to date suggest that it may have direct actions on the brain. The prolongation of sedation suggests that SRIF and TRH may be set in opposition at the central level, just as they appear to be at the pituitary level.

L-Prolyl-L-leucyl-L-glycine Amide

L-Prolyl-L-leucyl-L-glycine amide, a tripeptide, is often designated as melanocyte-stimulating hormone release-inhibiting factor (MIF) because it opposes the release of melanocyte-stimulating hormone (MSH) in certain assay systems (7,67); in others it is inactive (68). In animal tests MIF is active in the pargyline-L-DOPA test in hypophysectomized mice (69), but it is not active in the pargyline-5-HTP test (23). MIF, unlike TRH, does not alter pentobarbital sleeping time or hypothermia in rodents (23), but it antagonizes the tremor caused by oxotremorine (70).

In man MIF has been reported to possess a certain degree of antiparkinsonism activity and to oppose L-DOPA-induced dyskinesia (71). Ehrensing and Kastin (72) presented preliminary evidence to suggest that MIF may have antidepressant activity. In one of their patients, the symptoms of tardive dyskinesia yielded to administration of this tripeptide.

Luteinizing Hormone-Releasing Hormone

Luteinizing hormone-releasing hormone (LHRH) is a decapeptide that causes the release of follicle-stimulating hormone (FSH) and luteinizing hormone (LH) from the pituitary gland (8). Its behavioral effects have not been studied in man. Investigation of this compound in animals has offered the most direct evidence of a central action of a hypothalamic polypeptide. Moss and McCann (73) gave ovariectomized rats estrogen in doses too small to reinstate mating behavior. When LHRH was then given, these behaviors occurred with normal frequency. Pfaff (74) independently repeated and extended these experiments using hypophysectomized, ovariectomized rats and also found LHRH to reinstate mating behavior in animals primed with small doses of estrogen. TRH had no effect in this system (73); LHRH has no effect on pentobarbital responses in rodents (28,29).

Sexual behavior involves internal preparation of the animal for mating behavior, sensory inputs, integration of these, the elaboration of cues to attract the male, and finally a motor response manifested as receptive behavior toward the male. The neuroanatomical tracts and the number and type of neurotransmitters and receptors involved must be vast. It is therefore impressive to learn that a hypothalamic polypeptide, in the absence of the pituitary, can initiate so complex a pattern of behavior. One can only speculate that the neuronal and synaptic patterns already exist and that they are activated by the polypeptide. The mechanisms by which this takes place are totally unknown.

DISCUSSION

It is difficult to assimilate and organize the rapidly accumulating information concerning brain-behavioral effects of hypothalamic hormones. The difficulties are compounded by the absence of methods crucial for further investigation, especially in man. Nonetheless, the possibilities are exciting, and we think that this area of investigation is destined to contribute substantially to basic neurobiological knowledge. Table 1 lists the salient findings which support the generic hypothesis that hypothalamic releasing fac-

TABLE 1. *Evidence for brain-behavioral effects of hypothalamic releasing hormones*

I. Plausibility of direct brain effects
   A. TRH
      1. Occurrence in brain of species in which it lacks pituitary-thyroid effects
      2. Distribution in brain outside hypothalamus
      3. Maximum (as regards TSH release) acute (i.v.) and chronic (oral) doses have minimal thyroid effects
      4. Some normal humans show no TSH response
II. Evidence for brain effects
   A. Pharmacological
      1. TRH
         a. Pargyline–L-DOPA potentiation
         b. Pargyline–5-HTP potentiation
         c. Antagonism of depressants
         d. Congeners: dissociation of TSH release and pentobarbital antagonism
      2. SRIF
         a. Pentobarbital potentiation
      3. MIF
         a. Pargyline–L-DOPA potentiation
         b. Oxotremorine antagonism
         c. Partial relief of Parkinson's disease
   B. Biochemical
      1. TRH
         a. Increased catecholamine turnover in rat brain
   C. Physiological
      1. TRH
         a. By iontophoresis, dose-related inhibition of neuronal firing
         b. EEG changes like those of stimulants
III. Evidence for behavioral effects
   A. TRH
      1. Increased activity after intraventricular infusion in rats
      2. Partial relief of certain clinical disorders
      3. Relaxation and mild euphoria in normal women
   B. SRIF
      1. Decreased activity after intraventricular infusion in rats
   C. MIF
      1. Partial relief of depression
   D. LHRH
      1. Reinstatement of sexual behavior in ovariectomized, hypophysectomized rats

tors can exert direct brain effects, which may in turn have behavioral consequences. Table 1 includes data from other laboratories (grouped under the heading "Plausibility of Brain Effects." These findings require a brief elaboration.

We already mentioned the wide distribution of TRH in rat brain, with most of the TRH being found outside of the hypothalamus; even though it is most highly concentrated within the hypothalamus, TRH is found in brain throughout the phylum chordata (30). In some species (e.g., the axolotl) it occurs in the brain but seems incapable of stimulating the pituitary thyroid axis (75). This suggests the possibility of this hormone having other functions (30,75). There are two additional reasons for thinking that TRH may have other functions not involving its effects on the pituitary. First, in humans the administration of an intravenous bolus of TRH usually causes a sharp rise in plasma TSH, but the thyroid consequences of this rise are generally slight (76,77). Second, some normal subjects show no TSH response to a dose of TRH that is generally supramaximal for TSH release. The other bits of evidence noted in the table have already been discussed.

The relevance of hypothalamic hormones to psychopathological states offers exciting possibilities, but the clinical conditions under which their actions are investigated will be better understood only as the brain effects of these hormones are further defined. The role of the hypothalamus in instinctive behavior (e.g., eating, drinking, aggression, and sexuality) has been made clear from physiological experiments, but it is not yet known whether these effects depend on alterations in the hypothalamic polypeptides or are a product of neurons located in the hypothalamus which employ the usual transmitters. Both types of neurons are found in the hypothalamus. The synthesis and release of several hypothalamic hormones is known to be influenced by the biogenic amines (78,79). The established neurotransmitters excite or inhibit transmission across many synapses. What determines which of many potential neuronal circuits is activated or inhibited is not yet understood. Conceivably the hypothalamic polypeptides might offer such specificity and in this sense might be more directly related to the regulation of drives and affects than are the amines themselves. Alternatively, the hypothalamic polypeptides may somehow modulate the functions of the biogenic amines. For example, if TRH is additionally shown to effect noradrenergic fibers in the brain, it may have relevance to the observation that TRH exerts some benefit in only certain depressed patients.

Most of our biochemical notions about mental disorders involve the biogenic amines. In large measure they have been derived from the noted effects of drugs on these conditions. Current theories have been first the products and second the progenators of therapeutic advances. The monoamine theories are readily subject to two species of criticism: They are insufficient to account for all relevant observations (80), and they are not adequately specific. Relevant to the latter point, it may be noted that brain catecholamine function is said to be aberrant in depression (81–83), schizophrenia (84), and hyperkinetic syndrome (85). Although there is a tendency to describe the affective disorders in terms of quantitative changes in the availability of biogenic amines and the psychotic disorders in terms of qualitative changes in their chemical structure, it seems that discriminating factors have not yet been fully identified. Whether hypothalamic releasing factors in their aggregate can supply discrimination is uncertain but possible. Whether they can add to the sufficiency of our ideas about causality of mental disorders is likewise uncertain, but probable.

TRH appears to have arisen early in phylogenetic history, and this may be true of other releasing hormones and other polypeptides in the brain as well. One suspects that these simple, primitive molecules may have retained generalized functions, even as they have acquired discrete ones. The search for such generalized functions appears to be worthy of the efforts of both basic scientist and clinician, for neither can be satisfied with his present understanding of the physiology and functional pathology of the nervous system. Impediments to progress fall into two related categories: The first is the need for methods to measure accurately the minute quantities of these hormones in available tissues; the other is the development of appropriate experimental models or of pathological examples found in nature. Progress in both areas may be predicted and will undoubtedly lead to further understanding.

ACKNOWLEDGMENT

This work was funded in part by grant HD-03110 from the National Institute of Child Health and Human Development.

REFERENCES

1. Prange, A. J., Jr., and Lipton, M. A. (1972): Hormones and behavior: some principles and findings. In: *Psychiatric Complications of Medical Drugs*, edited by R. I. Shader, pp. 213–249. Raven Press, New York.
2. Prange, A. J., Jr., and Wilson, I. C. (1975): Behavioral effects of thyrotropin releasing hormone in animals and man: A generic hypothesis. *Psychopharmacology*, 11:22–24.
3. Prange, A. J., Jr., Breese, G. R., Wilson, I. C., and Lipton, M. A. (1975): Brain behavioral effects of hypothalamic releasing hormone: a generic hypothesis. In: *Anatomical Neuroendocrinology*. Karger, Basel (*in press*).
4. Prange, A. J., Jr., Breese, G. R., Wilson, I. C., and Lipton, M. A. (1975): Pituitary and suprapituitary hormones: Brain-behavioral effects. *Semin. Psychiatry* (*in press*).
5. Burgus, R., Dunn, T. F., Desiderio, D., and Guillemin, R. (1969): Molecular structure of the hypothalamic thyrotropin-releasing factor (TRF) of ovine origin; demonstration of the pyroglutamylhistidylprolinamide sequence by mass spectrometry. *C.R. Acad. Sci. (Paris)* [*Ser. D*], 269:1870–1873.
6. Brazeau, P., Vale, W., Burgus, R., Ling, N., Butcher, M., Rivier, J., and Guillemin, R. (1973): Hypothalamic polypeptide that inhibits the secretion of immunoreactive pituitary growth hormone. *Science*, 179:77–79.

7. Celis, M. E., Taleisnik, S., and Walter, R. (1971): Regulation of formation and proposed structure of the factor inhibiting the release of melanocyte-stimulating hormone. *Proc. Natl. Acad. Sci. U.S.A.*, 68:1428–1433.
8. Matsuo, H., Arimura, A., Nair, R. M. G., and Schally, A. V. (1971): Synthesis of the porcine LH- and FSH-releasing hormone by the solid-phase method. *Biochem. Biophys. Res. Commun.*, 45:822–827.
9. Prange, A. J., Jr., Wilson, I. C., Rabon, A. M., and Lipton, M. A. (1969): Enhancement of imipramine antidepressant activity by thyroid hormone. *Am. J. Psychiatry*, 126:457–469.
10. Wilson, I. C., Prange, A. J., Jr., McClane, T. K., Rabon, A. M., and Lipton, M. A. (1970): Thyroid-hormone enhancement of imipramine in nonretarded depressions. *N. Engl. J. Med.*, 282:1063–1067.
11. Prange, A. J., Jr., Wilson, I. C., Knox, A. E., McClane, T. K., Breese, G. R., Martin, B. R., Alltop, L. B., and Lipton, M. A. (1972): Thyroid-imipramine clinical and chemical interaction: Evidence for a receptor deficit in depression. *J. Psychiatr. Res.*, 9:187–205.
12. Prange, A. J., Jr., Wilson, I. C., Knox, A., McClane, T. K., and Lipton, M. A. (1970): Enhancement of imipramine by thyroid stimulating hormone: Clinical and theoretical implications. *Am. J. Psychiatry*, 127:191–199.
13. Prange, A. J., Jr., and Wilson, I. C. (1972): Thyrotropin releasing hormone (TRH) for the immediate relief of depressions: A preliminary report. *Psychopharmacologia*, 26 (suppl.): 82.
14. Prange, A. J., Jr., Wilson, I. C., Lara, P. P., Alltop, L. B., and Breese, G. R. (1972): Effects of thyrotropin-releasing hormone in depression. *Lancet*, 2:999–1002.
15. Fleischer, N., Burgus, R., Vale, W., Dunn, T., and Guillemin, R. (1970): Preliminary observations on the effect of synthetic thyrotropin releasing factor on plasma thyrotropin levels in man. *J. Clin. Endocrinol. Metab.*, 31:109–112.
16. Bower, C. Y., Friesen, G. H., Hwang, P., Guyda, H. J., and Folkers, K. (1971): Prolactin and thyrotropin release in man by synthetic pyroglutamyl-histadyl-prolineamide. *Biochem. Biophys. Res. Commun.*, 45:1033–1041.
17. Toft, A. D., Boyns, A. R., Cole, E. N., and Irvine, W. J. (1974): Prolactin response to thyrotropin releasing hormone in thyrotoxicosis. *J. Endocrinol.*, 61:515–516.
18. Bowers, C. Y., Schally, A. V., Enzmann, F., Boler, J., and Folkers, K. (1970): Porcine thyrotropin releasing hormone is (pyro) glu-his-pro(NH₂). *Endocrinology*, 86:1143–1153.
19. Everett, G. M. (1966): The dopa response potentiation test and its use in screening for antidepressant drugs. In: *Proceedings of the First International Symposium on Antidepressant Drugs.* Excerpta Medica International Congress Series No. 122. Excerpta Medica, Amsterdam.
20. Plotnikoff, N. P., Prange, A. J., Jr., Breese, G. R., Anderson, M. S., and Wilson, I. C. (1972): Thyrotropin releasing hormone: Enhancement of DOPA activity by a hypothalamic hormone. *Science*, 178:417–418.
21. Plotnikoff, N. P., Prange, A. J., Jr., Breese, G. R., Anderson, M. S., and Wilson, I. C. (1974): The effects of thyrotropin-releasing hormone on DOPA response in normal, hypophysectomized, and thyroidectomized animals. In: *The Thyroid Axis, Drugs and Behavior*, edited by A. J. Prange, Jr., pp. 103–113. Raven Press, New York.
22. Plotnikoff, N. P., Prange, A. J., Jr., Breese, G. R., and Wilson, I. C. (1974): Thyrotropin releasing hormone: Enhancement of DOPA activity in thyroidectomized rats. *Life Sci.*, 14:1271–1278.
23. Plotnikoff, N. P. (1973): *Personal communication.*
24. Hine, B., Sanghvi, I., and Gershon, S. (1973): Evaluation of Thyrotropin-releasing hormone as a potential antidepressant agent in the conscious dog. *Life Sci.*, 13:1789–1797.
25. Breese, G. R., Cooper, B. R., Prange, A. J., Jr., Cott, J. M., and Lipton, M. A. (1974): Interactions of thyrotropin-releasing hormone with centrally acting drugs. In: *The Thyroid Axis, Drugs, and Behavior*, edited by A. J. Prange, Jr., pp. 115–127. Raven Press, New York.
26. Prange, A. J., Jr., Breese, G. R., Cott, J. M., Martin, B. R., Cooper, B. R., Wilson, I. C., and Plotnikoff, N. P. (1974): Thyrotropin releasing hormone: Antagonism of pentobarbital in rodents. *Life Sci.*, 14:447–455.
27. Breese, G. R., Cott, J. M., Cooper, B. R., Prange, A. J., Jr., and Lipton, M. A. (1974): Antagonism of ethanol narcosis by thyrotropin releasing hormone. *Life Sci.*, 14:1053–1063.

28. Breese, G. R., Cott, J. M., Cooper, B. R., Prange, A. J., Jr., Wilson, I. C., Lipton, M. A., and Plotnikoff, N. P. (1975): Effects of thyrotropin releasing hormone (TRH) on the actions of pentobarbital and other centrally acting drugs. *J. Pharmacol. Exp. Ther.* (*in press*).

29. Prange, A. J., Jr., Breese, G. R., Jahnke, G. D., Martin, B. R., Cooper, B. R., Cott, J. M., Wilson, I. C., Alltop, L. B., and Lipton, M. A. (1975): Modification of pentobarbital sedation by natural and synthetic polypeptides: Dissociation of central and pituitary effects. *Life Sci.* (*in press*).

30. Jackson, I. M. D., and Reichlin, S. (1974): Thyrotropin releasing hormone (TRH) distribution in hypothalamus and extra hypothalamic brain tissues of mammalian and submammalian chordates. *Endocrinology*, 95:854–862.

31. Winokur, A., and Utiger, R. D. (1974): Thyrotropin releasing hormone: Regional distribution in rat brain. *Science*, 185–265.

32. Burt, R. D., and Snyder, S. H. (1974): A second site for binding of thyrotropin releasing hormone to rat brain. In: *Proceedings, Society of Neuroscience, Fourth Annual Meeting*, Abstract 101.

33. Keller, H. H., Bartholini, G., and Pletscher, A. (1974): Enhancement of cerebral noradrenaline turnover by thyrotropin-releasing hormone. *Nature (Lond.)*, 248:528–529.

34. Horst, W. D., and Spirt, N. (1974): A possible mechanism for the antidepressant activity of thyrotropin releasing hormone. *Life Sci.*, 15:1073–1082.

35. Segal, D. S., and Mandell, A. J. (1974): Differential behavioral effects of hypothalamic polypeptides. In: *The Thyroid Axis, Drugs, and Behavior*, edited by A. J. Prange, Jr., pp. 129–133. Raven Press, New York.

36. Renaud, L. P., and Martin, S. B. (1975): Thyrotropin releasing hormone (TRH): Depressant action on central neuronal activity. *Brain Res.*, 86:150–154.

37. Hershman, J. M. (1974): Clinical application of thyrotropin-releasing hormone. *N. Engl. J. Med.*, 290:886–890.

38. Snyder, P. J., Jacobs, L. S., Rabello, M. M., Sterling, F. H., Shore, R. N., Utiger, R. D., and Daughaday, W. H. (1974): Diagnostic value of thyrotropin-releasing hormone in pituitary and hypothalamic diseases. *Ann. Intern. Med.*, 81:751–757.

39. Pittman, J. A., Jr., (1974): Thyrotropin-releasing hormone. *Adv. Intern. Med.*, 19:303–325.

40. Itil, T. M., Patterson, C. D., Polvan, N., Mehta, D., and Bergey, B. (1975): Clinical and CNS effects of oral and i.v. thyrotropin releasing hormone (TRH). *Psychopharmacol. Bull.*, 11:29.

41. Kastin, A. J., Ehrensing, R. H., Schalch, D. S., and Anderson, M. S. (1972): Improvement in mental depression with decreased thyrotropin response after administration of thyrotropin-releasing hormone. *Lancet*, 2:740–742.

42. van der Vis-Melsen, M. J. E., and Wiener, J. D. (1972): Improvement in mental depression with decreased thyrotropin response after administration of thyrotropin-releasing hormone. *Lancet*, 2:1415.

43. Takahashi, S., Kondo, H., Yoshimura, M., and Ochi, Y. (1973): Antidepressant effect of thyrotropin-releasing hormone (TRH) and plasma thyrotropin levels in depression. *Folia Psychiatr. Neurol. Jap.*, 27:305–314.

44. Dimitrikoudi, M., Hanson-Norty, E., and Jenner, F. A. (1974): TRH in psychoses. *Lancet*, 1:456.

45. Drayson, A. M. (1974): TRH in cyclical psychoses. *Lancet*, 1:312.

46. Sorensen, R., Svendson, K., and Schou, M. (1974): TRH in depression. *Lancet*, 1:865–866.

47. May, P., and Donabedian, R. K. (1973): Factors in blood influencing the determination of thyrotropin releasing hormone. *Clin. Chim. Acta*, 46:377–382.

48. Wilson, I. C., Prange, A. J., Jr., Lara, P. P., Alltop, L. B., Stikeleather, R. A., and Lipton, M. A. (1973): TRH (lopremone): Psychobiological responses of normal women. I. Subjective experiences. *Arch. Gen. Psychiatry*, 29:15–21.

49. Anderson, M. S., Bowers, C. Y., Kastin, A. J., Schalch, D. S., Schally, A. V., Snyder, P. J., Utiger, R. D., Wilber, J. F., and Wise, A. J. (1971): Synthetic thyrotropin-releasing hormone: A potent stimulator of thyrotropin secretion in man. *N. Engl. J. Med.*, 285:1279–1283.

50. Donabedian, R. (1974): *Personal communication*.

51. Mueller, P. S., Heninger, G. R., and McDonald, R. K. (1972): Studies on glucose utilization and insulin sensitivity in affective disorders. In: *Recent Advances in the Psychobiology of the Depressive Illnesses,* edited by Williams, Katz, and Shield, pp. 235–248. N.I.M.H., Rockville, Md.
52. Sachar, E. J., Finkelstein, J., and Hellman, L. (1971): Growth hormone responses in depressive illness. I. Response to insulin tolerance test. *Arch. Gen. Psychiatry,* 25:263–269.
53. Brazeau, P., Rivier, J., Vale, W., and Guillemin, R. (1974): Inhibition of growth hormone secretion in the rat by synthetic somatostatin. *Endocrinology,* 94:184–187.
54. Siler, T. M., VandenBerg, G., and Yen, S. S. C. (1973): Inhibition of growth hormone release in humans by somatostatin. *J. Clin. Endocrinol. Metab.,* 37:632–634.
55. Vale, W., Brazeau, P., Rivier, C., Rivier, J., Grant, G., Burgus, R., and Guillemin, R. (1973): Inhibitory hypophysiotropic activities of hypothalamic somatostatin. *Fed. Proc.,* 32:211.
56. Reichlin, S., and Vale, W. (1974): *Personal communication.*
57. Chase, T. N., Woods, A. C., Lipton, M. A., and Morris, C. E. (1974): Hypothalamic releasing factors and Parkinson disease. *Arch. Neurol.,* 31:55–56.
58. Tiwary, C. M., Frias, J. L., and Rosenbloom, A. L. (1972): Response to thyrotropin in depressed patients. *Lancet,* 2:1086.
59. Wilson, I. C., Lara, P. P., and Prange, A. J., Jr. (1973): Thyrotropin-releasing hormone in schizophrenia. *Lancet,* 2:43–44.
60. Huey, L. Y., Janowsky, D. S., Mandell, A. J., Judd, L. L., and Pendery, M. (1975): Preliminary studies on the use of thyrotropin releasing hormone in manic states, depression, and dysphoria of alcohol withdrawal. *Psychopharmacol. Bull.,* 11:29.
61. Hollister, L. E. (1975): New developments in psychotherapeutic drugs. *Psychiatr. Digest* March:11.
62. Kety, S. S. (1972): Norepinephrine in the central nervous system and its correlations with behavior. In: *Brain and Behavior,* edited by A. E. Karczmar and J. C. Eccles, pp. 115–129. Springer-Verlag, New York.
63. Kuhn, R. (1957): Uber die Behandlung depressiver Zustande mit einem Iminodibenzylderivat (G 22355). *Schwiez. Med. Wochenschr.,* 14:1135–1140.
64. Poussiant, A., and Ditman, K. S. (1965): A controlled derivative study of imipramine (Tofranil) in the treatment of childhood enuresis. *J. Pediatr.,* 67:283–290.
65. Rapoport, J. L., Zuinn, P. O., Bradbard, G., Riddle, D., and Brooks, E. (1974): Imipramine and methylphenidate treatments of hyperactive boys. *Arch. Gen. Psychiatry,* 30:789–793.
66. Prange, A. J., Jr., Breese, G. R., Jahnke, G. D., Cooper, B. R., Cott, J. M., Wilson, I. C., Lipton, M. A., and Plotnikoff, N. P. (1975): Parameters of alteration of pentobarbital response by hypothalamic polypeptides. *Neuropsychobiology (in press).*
67. Kastin, A. J., Schally, A. V., and Viosca, S. (1971): Inhibition of MSH release in frogs by direct application of L-prolyl-L-leucylglycinamide to the pituitary. *Proc. Soc. Exp. Biol. Med.,* 137:1437–1439.
68. Vale, W., Grant, G., and Guillemin, R. (1973): Chemistry of the hypothalamic releasing factors—studies on structure-function relationships. In: *Frontiers in Neuroendocrinology,* edited by W. F. Ganong and L. Martini, pp. 375–413. Oxford University Press, New York.
69. Plotnikoff, N. P., Kastin, A. J., Anderson, M. S., and Schally, A. V. (1971): DOPA potentiation by a hypothalamic factor, MSH release-inhibiting hormone (MIF). *Life Sci.,* 10:1270–1283.
70. Plotnikoff, N. P., Kastin, A. J., Anderson, M. S., and Schally, A. V. (1972): Oxotremorine antagonism by a hypothalamic hormone, melanocyte-stimulating hormone release-inhibiting factor (MIF) (36558). *Proc. Soc. Exp. Biol. Med.,* 140:811–814.
71. Kastin, A. J., and Barbeau, A. (1972): Preliminary clinical studies with L-prolyl-L-leucylglycine amide in Parkinson's disease. *Can. Med. Assoc. J.,* 107:1079–1081.
72. Ehrensing, R. H., and Kastin, A. J. (1974): Melanocyte-stimulating hormone-release inhibiting hormone as an antidepressant: A pilot study. *Arch. Gen. Psychiatry,* 30:63–65.
73. Moss, R. L., and McCann, S. M. (1973): Induction of mating behavior in rats by luteinizing hormone-releasing factor. *Science,* 181:177–179.
74. Pfaff, D. W. (1973): Luteinizing hormone-releasing factor potentiates lordosis behavior in hypophysectomized ovariectomized female rats. *Science,* 182:1148–1149.

75. Taurog, A., Oliver, C., and Porter, J. C. (1973): Metamorphosis in mexican axoltl treated with TSH, TRH, or LATS. Presented at the 49th Meeting of the American Thyroid Association, Seattle.
76. Hollander, C. S., Mitsuma, T., Shenkman, L., Woolf, P., and Gershengorn, M. C. (1972): Thyrotropin-releasing hormone: Evidence for thyroid response to intravenous injection in man. *Science,* 175:209–210.
77. Prange, A. J., Jr., Wilson, I. C., Lara, P. P., Wilber, J. F., Breese, G. R., Alltop, L. B., and Lipton, M. A. (1973): TRH (lopremone): Psychobiological responses of normal women. II. Pituitary-thyroid responses. *Arch. Gen. Psychiatry,* 29:28–32.
78. Reichlin, S., and Mitnick, M. (1973): Biosynthesis of hypothalamic hypophysiotropic factors. In: *Frontiers in Neuroendocrinology,* edited by W. F. Ganong and L. Martini, pp. 61–88. Oxford University Press, New York.
79. Schally, A. V., Arimura, A., and Kastin, A. J. (1973): Hypothalamic regulatory hormones. *Science,* 179:341–350.
80. Prange, A. J., Jr. (1972): Discussion in: Schildkraut, J. J.: Neuropharmacological studies of mood disorders. In: *Disorders of Mood,* edited by Zubin and Freyhan. Johns Hopkins Press, Baltimore.
81. Prange, A. J., Jr. (1964): The pharmacology and biochemistry of depression. *Dis. Nerv. Syst.,* 25:217–221.
82. Bunney, W. E., Jr., and Davis, J. M. (1965): Norepinephrine in depressive reactions: A review. *Arch. Gen. Psychiatry,* 13:483–494.
83. Schildkraut, J. J. (1965): The catecholamine hypothesis of affective disorders: A review of supporting evidence. *Am. J. Psychiatry,* 122:509–522.
84. Stein, L., and Wise, C. D. (1971): Possible etiology of schizophrenia: Progressive damage to the noradrenergic reward system by 6-hydroxydopamine. *Science,* 171:1032–1039.
85. Wender, P. H. (1971): *Minimal Brain Dysfunction in Children.* Wiley-Interscience, New York.

Hormones, Behavior, and Psychopathology, edited by
Edward J. Sachar. Raven Press, New York © 1976.

Neurophysiological Effects of Hormones in Humans: Computer EEG Profiles of Sex and Hypothalamic Hormones

Turan M. Itil

*Division of Biological Psychiatry, Department of Psychiatry, New York Medical College,
New York, New York 10029*

Whereas obvious central nervous system (CNS) effects of sex hormones and hypothalamic regulatory hormones have been demonstrated in animal studies, their effects on human brain function, particularly after oral single-dose administration, were not clearly identified. Based on the experience that drugs effective on human behavior produce systematic effects on human brain function, and that these effects can be demonstrated by computer analysis of the scalp-recorded electroencephalogram (EEG) in humans, we carried out a series of investigations with clinically interesting hormones. Although our immediate aims were to determine immediate CNS actions of the studied compounds, the overall goal of our program has been to develop natural substances with psychotropic potentials.

MATERIAL AND METHODS

The most interesting substances investigated to date are mesterolone (MST), a male hormone; cyproterone acetate (CYP), an antimale hormone; thyrotropic releasing hormone (TRH); and melanocyte-stimulating hormone inhibiting factor (MIF). The studies were done in both normal volunteers and patients using the methods of quantitative pharmaco-EEG (1,2). This method was developed to establish the CNS effects of drugs and to predict their possible psychiatric use based on the computer-analyzed EEG (CEEG) and a series of statistical procedures. In this method, in addition to the procedures of clinical pharmacology, EEG recordings are taken before and after oral or parenteral administration of drugs. The type and number of subjects, time of postdrug recordings, use of control drugs, and number of recording leads depend on the type of drug studied. Either continuous recordings before and up to 4 hours after drug are done, or 5- to 20-min recordings are made before, and 1, 2, 3, etc. hours after drug (or placebo) administration. Half of the recording is done during "resting time" (RR) and the other half during "reaction time" (RT) measurements when the subjects must respond to random acoustic stimuli by pushing a button. In addition to the paper chart on which the EEG is recorded according to the international 10–20 system, two to four EEG leads (two leads are always right occipital to right ear and right occipital to anterior vortex) are also recorded on an analogue tape recorder. The EEG record is analyzed on- or off-line by an additional computer (IBM System 7) using period analysis or power spectrum programs (3). These programs provide continuous analysis of 22 CEEG measurements during every 16-sec epoch. The measurements include average frequency and frequency variability and eight different frequency bands for the primary-wave and first-derivative analyses, and average absolute amplitude and amplitude variability. Since only four of 22 CEEG measurements of the computer analysis are considered for clinical electroencephalography in neurological diagnosis, and the other 18 CEEG measurements are not used or cannot be determined by conventional visual (eyeball) evaluation of the records, to prevent any misunderstanding we prefer to call the activities analyzed "computerized cerebral biopotentials" or simply "computerized EEG" (CEEG).

In quantitative pharmaco-EEG, the mean and sigmas of these 22 measurements are used for various statistical applications to demonstrate the effects of drugs. The pre- minus postdrug CEEG measurements in terms of t-values provide the "computer EEG profiles" of drugs. The "real" CEEG profile is determined by subtracting the placebo-induced changes from the drug-induced changes.

RESULTS

Investigations with MST

MST $(17\beta$-hydroxy-1α-methyl-5α-androstane-3,1) is an orally effective androgen with some dissociation between effects on pituitary hormone secretions and peripheral androgenic effects. Although it is active as an androgenic compound, it does not block the luteinizing hormone (LH) and follicle-stimulating hormone (FSH) secretion of the anterior pituitary up to 150 mg daily. MST is considered to be very effective in the treatment of disorders attributable to androgen deficiency, such as disturbances of sexual libido and potency, decline of physical activity and mental alertness in the middle-aged or elderly man, infertility, and hypogonadism. When the patients with androgen deficiency were treated with MST, certain behavioral alterations were observed, e.g., increased mental alertness, mood elevation, improvement of memory and concentration, and enhancement of psychomotor performance. An increase of energy, drive, and interest have also been reported during MST treatment.

Using the methods of quantitative pharmaco-EEG, it was established that MST, given in single oral dosages of 1 to 1,600 mg, produces systematic changes in human brain function. CEEG profiles of MST in low

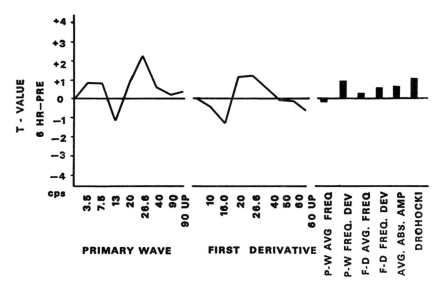

FIG. 1. CEEG profile of MST. In the abscissa are 22 CEEG measurements, and in the ordinate are the changes from before to 6 hours after drug administration in terms of t-values ($p = 0.05$ by the t-test). After the mean of the dosages from 100 to 1,600 mg, in the primary-wave measurement in 12 patients there was an increase of slow waves (1.3 to 3.5 and 3.5 to 7.5 cps) and fast waves (over 13 cps waves), and a decrease of 7.5- to 13-cps activity; there was also a decrease of slow waves (less than 16 cps) and an increase of activities in the frequency range 16 to 40 cps in the first-derivative measurements. The temporal shape of the CEEG profile of MST was similar to those seen after administration of tricyclic antidepressants.

dosages (1 and 10 mg) are similar to profiles of psychostimulant drugs, and with high dosages (100 to 1,600 mg) the profiles resemble those observed after tricyclic antidepressants particularly those with less sedative properties (Figs. 1 and 2).

Based on these findings, it was predicted that MST may be effective in patients with depressive symptomatology (4). The preliminary uncontrolled open clinical trials supported this prediction. MST was found to be effective not only in elderly depressive patients but also in young patients who had no symptoms of hormonal deficiency. Double-blind control trials with MST are under way.

Investigations with CYP

CYP (6-chloro-17-acetoxy-1α, 2α-methylene-pregna-4,6-dien-3,20-dione acetate), an antiandrogen, is currently used in pubertas precox and in patients suffering from pathological sexual hyperactivity. In addition to inhibition of all sexual parameters (sexual desire, drive, and erection), a

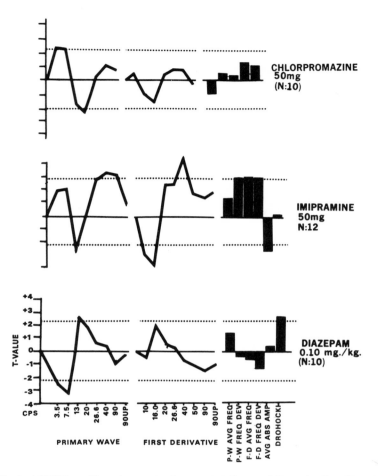

FIG. 2. Typical CEEG profiles of neuroleptics, anxiolytics, and thymoleptics. The abscissa and ordinate are the same as in Fig. 1. Chlorpromazine (a well-known neuroleptic) increased slow waves and decreased faster activities in the range 7.5 to 20 cps in the primary-wave measurements. Imipramine (a representative tricyclic antidepressant) produces both slow and fast activities, and a decrease in α-waves (7.5 to 13 cps) in the primary-wave measurements; as well as a decrease of slow waves and increase of fast activities in the first-derivative measurements. Diazepam (a typical benzodiazepine anxiolytic) shows increased activities in the frequency range 20 to 26.6 cps in both primary-wave and first-derivative measurements. After diazepam, a decrease of slow waves is seen in the primary-wave measurements. Dotted line, $p = 0.05$ by the t-test.

decrease of aggressive drive, the occurrence of apathy, a decrease of energy and drive, as well as depression and restlessness were noted as undesirable side effects during treatment with CYP.

Based on the studies with quantitative pharmaco-EEG, it was established that CYP has systematic effects on human brain function in single oral dosages of 1 to 800 mg, administered in normal volunteers. After high

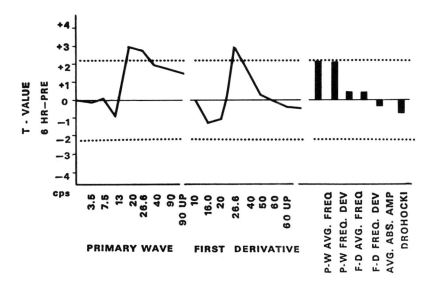

FIG. 3. CEEG profile of CYP. The abscissa and ordinate are the same as in Fig. 1. In dosages of the mean of 50 to 800 mg, CYP increased the 20- to 40-cps waves in both primary-wave and first-derivative measurements in 12 patients. The temporal shapes of the CEEG profile of CYP were similar to those of the anxiolytic compounds. Dotted line, $p = 0.05$ by the t-test.

dosages (50 to 800 mg), CYP produces an increase of 20- to 40-cps β-waves and a decrease of slower waves in both primary-wave and first-derivative measurements. Accordingly, CEEG profiles of CYP in high dosages were similar to those seen with benzodiazepine anxiolytics (Fig. 3). In lower dosages (1 to 10 mg) CYP produces an increase of slow waves and very fast activity and a decrease of β- and α-waves in the primary-wave measurements and a decrease of slow and increase of fast activity in the first-derivative measurements. The CEEG profiles of CYP in low dosages resemble sedative tricyclic antidepressants.

Based on CEEG profiles, antianxiety and antidepressant properties of CYP were predicted (4). The subsequent uncontrolled open pilot trials supported the EEG prediction. In young males with anxiety syndromes and female patients with premenstrual tension-depression, CYP showed therapeutic effects. Even though double-blind control trials would have proved that CYP indeed has psychotropic properties, its practical use in patients is questionable, since CYP has psychotropic activity in dosages that have significant hormonal action.

Investigations with TRH

TRH is a tripeptide that controls the secretion of thyrotropin in interaction with thyroid hormones. Animal pharmacological and neurophysiological studies have indicated that it is active in the CNS (5,6). In some studies TRH was found effective in depressive patients (7,8), although in others there was no demonstrable therapeutic effect in such patients (9,10). We studied CNS effects of oral and parenteral TRH in depressed patients as well as in normal volunteers.

CEEG Investigations in Depressed Patients

In three patients 500 and 1,000 μg TRH and saline were administered parenterally during the drug-free period. It was established that the 1,000-μg dose particularly produced systematic alterations (3 hours after administration) in CEEG measurements (11). The changes are characterized by a decrease of very slow and very fast activities, an increase of potential in the frequency range 13 to 20 cps in the primary-wave measurement, and an increase of very slow and fast activities in the first-derivative measurements (Fig. 4).

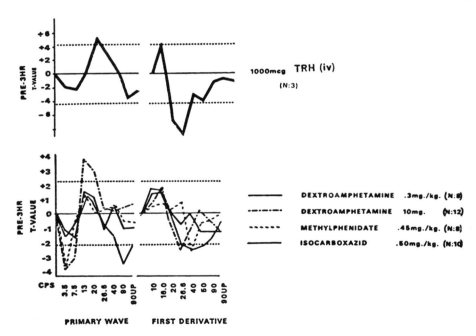

FIG. 4. CEEG profiles of intravenous and oral TRH compared to those of the psychostimulants. The abscissa and ordinate are the same as in Fig. 1 (the last six measurements are not included). **Top:** CEEG changes 3 hours after 1,000 μg TRH is administered intravenously. There is a decrease of slow and very fast activity and an increase of 7.5- to 20-cps waves in the primary-wave measurements, as well as an increase of very slow and a decrease of fast activities in the first-derivative measurements. **Bottom:** CEEG profiles of dextroamphetamine (two studies) in which methylphenidate and isocarboxazid have been superimposed. The shape of the CEEG profiles of oral and parenteral TRH show striking similarities to those of psychostimulant compounds.

The effects of oral TRH were studied in three other depressed patients. In a double-blind study patients received 100 mg TRH for 3 to 5 days and placebo for 5 to 7 days during a 2-week treatment period. EEG was recorded every day before and 3 hours after oral TRH or placebo administration. The changes in EEG measurements from the prestudy level to that 3 hours after each drug (or placebo) administration in terms of *t*-values were plotted (CEEG profile). When these were superimposed, it was observed that the CEEG profiles after oral and parenteral treatment were very similar. These profiles, which were characterized by an increase of 7.5- to 20-cps waves and a decrease of very slow and very fast activities in the primary-wave measurements, together with an increase of slow and a decrease of fast activities in the first-derivative measurements, resemble those CEEG profiles we observed after giving psychostimulant compounds which we called "psychostimulant" CEEG profiles (11).

CNS Effects of TRH in Normal Volunteers

In a recent study the effects of different dosages of oral (20, 50, and 300 mg) and parenteral (600 and 1,200 μg i.v.) formulations of TRH were studied in a group of 10 normal volunteers in a double-blind crossover research design with placebo (oral) and saline (intravenous), as well as dextroamphetamine (10 mg) controls using the methods of quantitative pharmaco-EEG. It was established that:

1. CEEG profiles after 10 mg dextroamphetamine administered orally in each subject before the study were very similar to the profiles seen in previous studies. Accordingly, the data collection and analysis procedures were functioning properly and the subjects responded adequately to this compound.

2. Qualitatively, according to the CEEG profiles, both TRH formulations (oral and intravenous) produced effects on human brain function similar to those seen after psychostimulant compounds. Particularly, 600 μg TRH (intravenous) and 20 and 50 mg TRH (oral) showed very similar CEEG profiles, which were obtained after oral and parenteral administration of TRH in depressive patients.

Investigations with MIF

MIF (L-prolyl-L-leucylglycinamide), a tripeptide that regulates the release of melanocyte-stimulating hormone, is released from the hypothalamus of many species; its physiological role in man in unknown. Based on animal pharmacology in which MIF potentiates L-DOPA in pargyline pretreated animals and antagonizes the tremors produced by oxotremorine, the antiparkinson effects of MIF have been predicted. In low dosages no significant changes in background EEG were observed, but in higher dosages (30 mg/kg) the neurocortical background electrical activity changed from a sleep-like pattern to a pattern characteristic of continuous arousal. Based on these observations, we were interested in investigating CNS effects of MIF in healthy volunteers using the methods of quantitative pharmaco-EEG. It was established that MIF produces different effects in subjects with α-background EEGs than in patients with β-background EEGs. MIF, in all dosages, produces statistically significant effects in computer EEG measurements of subjects with predominant α-record in comparison to placebo. The CEEG profiles of low and moderate dosages of MIF (50 and 500 mg) are characterized by an increase of very slow waves and a decrease in 7.5- to 13-cps activity. The shape of the CEEG profiles in these dosages were similar to those seen after administration of CNS depressants such as tricyclic antidepressants. In higher dosages (1,000 and 1,500 mg), MIF decreases very slow waves and increases activities in α- and slow β-frequency range. After high dosages of MIF the CEEG profiles resemble those seen with psychostimulant compounds. Subjects with fast β-background EEGs did not show significant alterations in computer EEG measurements.

DISCUSSION

The important implication of the results of our studies is the fact that we are able to demonstrate the immediate CNS effects of hormonal substances using CEEG measurement. This was particularly significant for the MIF study since there was no evidence to determine if this substance penetrates the blood-brain barrier in humans. Our investigations have shown that the dose-related alterations in computer EEG provides information on the pharmacokinetics and bioavailability of hormonal substances.

A more important finding is the predictive value of the CEEG. Because of the tricyclic antidepressant-like CEEG profile of MST in high dosage, similar "psychotropic" properties of this compound were predicted. Preliminary clinical trials suggest that MST may indeed be effective in some depressive patients.

Without having scientific evidence from well-controlled double-blind studies, it has frequently been suggested that testosterone has therapeutic effects on male depression. In 1954 Bleuler (12) reported that androgens have mood-elevating effects and that testosterone in high dosages produces therapeutic effects in depressive patients. In a recent study Klaiber et al. (13) found that depressed men have elevated levels of plasma monoamine oxidase activity that can be returned to normal by administering testosterone. A decrease in depressive symptomatology was observed during testosterone therapy. Sachar et al. (14), however, reported that the mean plasma testosterone level of depressed male patients was not significantly changed after recovery from illness. According to our studies MST in high dosage is dramatically effective in some depressive patients with elevating depressive symptomatology within 1 to 2 weeks. In other subjects, however, no effects were observed. The reasons for these discrepancies as to whether hormonal levels play a role in the effectiveness of MST and/or other androgens in male depression or if some other factors are involved are not known.

Anxiolytic-type clinical effects were predicted for CYP, an antiandrogen, based on the shape of CEEG profiles. Pilot clinical trials seem to confirm this prediction. Unfortunately, CYP cannot be truly studied in patients, since in the dosages predicted to be anxiolytic it produces significant hormonal changes. It is interesting that the effects of CYP in EEG and in clinical observations were not associated with any noticeable sedative properties. Obviously an anxiolytic compound without sedative properties will be very significant for clinical practice.

Even without knowledge on pharmacological and/or neurophysiological properties of TRH we could, based on CEEG profiles alone, predict that this compound will have "psychostimulant"-type effects on human behavior. In our clinical studies we were very impressed with the systematic effects of TRH on "instinctive behavior." The majority of the patients showed an increase of interest, desire, and drive for food, sex, work, and hobbies after TRH administration. The effect on depressed mood was less impressive, although some patients showed definite improvement in "depressive symptomatology." We speculated that the "antidepressive" effects of TRH may be seen only in patients in whom depression may be the result of an inhibition of "instinctive" functions (11).

The CNS effects of MIF are much more complicated than those of the other hormonal substances. Because of the bimodal effects (in low dosages CNS depression, and in high dosages CNS stimulation) and because of the biological variability in responses (subjects with α-background activity showed marked response to MIF whereas those with β-pattern showed no significant response), we are cautious in making any definite prediction concerning the eventual psychotropic properties of MIF.

SUMMARY

The immediate CNS effects of two steroid sex hormones (MST and CYP) and two hypothalamic regulating hormones (TRH and MIF) were established in humans based on the scalp-recorded CEEG. Furthermore, it was found that MST and MIF also produce dose-related alterations in human brain function. The most striking but still speculative finding was the prediction of the antidepressant properties of MST (an androgen) in high dosage; the anxiolytic effects of CYP, an antiandrogen; and psychostimulant properties of TRH, a hypothalamic regulating hormone. Although these predictions were supported by pilot clinical trials, well-controlled double-blind investigations are required to substantiate the value of CEEG for the development of psychotropic hormones.

REFERENCES

1. Itil, T. M., Guven, F., Cora, R., Hsu, W., Polvan, N., Ucok, A., Sanseigne, A., and Ulett, G. A. (1971): Quantitative pharmaco-electroencephalography using frequency analyzer and digital computer methods in early drug evaluations. In: *Drugs, Development, and Brain Functions,* edited by W. L. Smith, pp. 145–166. Charles C Thomas, Springfield, Ill.
2. Itil, T. M. (1974): Quantitative pharmaco-electroencephalography. In: *Modern Problems of Pharmacopsychiatry, Vol. 8: Psychotropic Drugs and the Human EEG,* edited by T. M. Itil, pp. 43–75. Karger, New York.
3. Shapiro, D. M., Itil, T. M., Noach, M., Bagley, L., Spence, J., Drossman, M. (1975): HZI Systems I and II, an EEG, EP and CNV analysis package for the IBM System 7 (*in press*).
4. Itil, T. M., Cora, R., Akpinar, S., Herrmann, W. M., and Patterson, C. (1974): "Psychotropic" action of sex hormones: Computerized EEG in establishing the immediate CNS effects of steroid hormones. *Curr. Ther. Res.,* 16:1147–1170.
5. Plotnikoff, N. P., Prange, A. J., Jr., Breese, G. R., Anderson, M. S., and Wilson, I. C. (1972): Thyrotropin-releasing hormone: Enhancement of DOPA activity by a hypothalamic hormone. *Science,* 178:417.
6. Inoué, M., Saji, Y., Mikoda, T., and Nagawa, Y. (1974): Central nervous actions of thyrotropin-releasing hormone (TRH). Presented at the 45th Kinki Regional Meeting of the Japanese Pharmacological Society, Tokushima, Japan, June 8.
7. Prange, A. J., Jr., Wilson, I. C., Lara, P. P., Alltop, L. B., and Breese, G. R. (1972): Effects of thyrotropin-releasing hormone in depression. *Lancet,* 2:999–1002.
8. Kastin, A. B., Ehrensing, R. H., Schalch, D. S., and Anderson, M. S. (1972): Improvement in mental depression with decreased thyrotropin response after administration of thyrotropin-releasing hormone. *Lancet,* 2:740–742.
9. Hollister, L. E., Berger, P., Ogle, F. L., Arnold, R. C., and Johnson, A. (1975). *Psychopharmacol. Bull.,* 11:27–28.
10. Takahashi, S., Condo, H., Yoshimura, M., and Ochi, Y. (1973): Antidepressant effect of thyrotropin-releasing hormone (TRH) and the plasma thyrotropin levels in depression. *Folia Psychiatr. Neurol. Jap.,* 27:305–314.
11. Itil, T. M., Patterson, C. D., Polvan, N., Mehta, S., and Bergey, B. (1975): Clinical and computerized EEG findings after oral and i.v. thyrotropin-releasing hormone in depressed subjects. *Dis. Nerv. Syst.,* 36:529–536.
12. Bleuler, M. (1954): *Endokrinologische Psychiatrie,* pp. 193–204. G. Thieme Verlag, Stuttgart.
13. Klaiber, E. L., Broverman, D. M., Vogel, W., and Kobayashi, Y. (1975): The use of steroid hormones in depression. In: *The Proceedings of the Meeting on Psychotropic Hormones* (*in press*).
14. Sachar, E. J., Halpern, F., Rosenfeld, R. S., Gallagher, T. F., and Hellman, L. (1973): Plasma and urinary testosterone levels in depressed men. *Arch. Gen. Psychiatry,* 28:15–18.

Hormones, Behavior, and Psychopathology, edited by
Edward J. Sachar. Raven Press, New York © 1976.

Hormonal Alteration of Imipramine Response: A Review

A. J. Prange, Jr., *I. C. Wilson, G. R. Breese, and M. A. Lipton

*Departments of Psychiatry and Pharmacology, Biological Sciences Research Center of The Child Development Institute, University of North Carolina School of Medicine, Chapel Hill, North Carolina 27514; and *Division of Research, North Carolina Department of Mental Health, Raleigh, North Carolina 27611*

The purpose of this chapter is to review our experience in using peripheral hormones (i.e., hormones of nonbrain origin) to accelerate the therapeutic action of imipramine. The data referred to, as well as the details of experimental design, have been presented elsewhere. This allows, perhaps requires, a modification of style. We discuss the various studies seriatim, compromising between the chronology of their performance and the logic that developed as results unfolded.

The theme to be pursued began with the observation of accidental toxicity (1). A woman factitiously hyperthyroid presented for treatment of depression, and when given usual doses of imipramine she developed paroxysmal auricular tachycardia. This sign disappeared when her desiccated thyroid medication, given as replacement therapy, was reduced to suitable amounts.

ENHANCEMENT OF IMIPRAMINE RESPONSE BY THYROID HORMONES

Animal Studies

We performed a series of animal studies to examine systematically if thyroid hormones and imipramine interact to produce increased toxicity. It did indeed appear that imipramine is more lethal in hyperthyroid mice (2). Although the drug appeared less lethal in hypothyroid mice (3), an attempt to replicate this finding was unsuccessful (4). Desipramine was also more lethal in hyperthyroid mice, whereas its lethality in hypothyroid mice was unchanged (4). This work was substantially confirmed and extended by Avni (5). It is important to note, however, that the enhancement of lethality by hyperthyroidism is not limited to antidepressants; it applies to a wide variety of drugs (6–8).

More detailed animal experimentation (8–10) was performed later, but it is convenient to indicate some of the main findings in the present context. We found that L-triiodothyronine (T3) enhanced the activity of imipramine in the pargyline-DOPA mouse activation test. Indeed, the hormone was active when given alone. Brain dopamine accumulated more rapidly in test animals treated with T3 than in controls. It should be recalled that the pargyline-DOPA test has been used to screen drugs for probable anti-depressant activity (11). However, some substances that are not anti-depressants are active in the test (12).

We also found that T3 and imipramine show marked interaction in producing hyperthermia. Studies of the actions of T3 on catecholamine parameters were less conclusive. The hormone had no effect on temperature loss in the cold after acute release of catecholamines by 6-hydroxydopamine injection intracisternally, whereas imipramine enhanced hypothermia. The hormone also failed to affect amphetamine-induced running behavior in rats. Neither the hormone nor the drug influenced activity of tyrosine hydroxylase in brain.

Park and his colleagues (13,14) studied the interactions of thyroid state and phenothiazines. Both toxicity and certain behavioral effects of these drugs were potentiated by thyroid hormone administration.

Imipramine-T3 in Depressed Women and Men

When we began our systematic clinical work, we were equipped only with the observations that in a patient the combination of thyroid hormones and imipramine had apparently produced paroxysmal auricular tachycardia and that in mice the drug is more lethal when animals are grossly hyperthyroid. Our clinical endeavors were based on the postulation that if interaction between drug and hormone at large doses manifests as enhanced toxicity, then interaction at small controlled doses might manifest as enhanced therapeutic action.

After several informal trials, we selected T3 as the preferred thyroid hormone. It acts more rapidly than thyroxine, and when stopped it disappears more rapidly (15). Thus toxicity, if it occurred, could be foreshortened. Twenty-five micrograms orally per day seemed adequate. This dose is only one-third the amount used in a thyroid-suppression diagnostic test (16).

Women

Our first study was performed in a group of hospitalized depressed women with retarded depression (17). The selection of this subtype was based on our notion that any biological treatment has its best chance of success in patients of this description. All, or nearly all, the patients were unipolar although we did not pay systematic attention to the unipolar-bipolar dichotomy at the time. In a design influenced by caution we started imipramine (50 mg p.o. t.i.d. on day 1) and on the morning of day 4 added T3 or matched placebo. All substances were then continued throughout the 28 days of the trial, and "blind" ratings were performed thrice weekly.

Study 1 (Table 1) shows the responses of our patients as assessed by the Hamilton Rating Scale for Depression (HRS) (18). Women receiving imipramine plus T3 reached remission (scores of 7 or less) within approximately half the time required for women receiving imipramine plus placebo. In this study, as in all others to be described, we also employed the Self Rating Depression Scale (SDS) (19). In this study, as in all others, it yielded approximately the same response curves as the HRS, although with somewhat more variance. We examined each item of the HRS for between-group differences and found that all favored T3, the advantage being greatest on "psychic anxiety" and least on "hypochondriasis." The important point is that all aspects of the depression syndrome were benefited by T3. This is of special interest as T3 toxicity, had it occurred, could have become manifest on any of several subscales, e.g., insomnia, somatic anxiety, and agitation.

Next we performed an identical study in nonretarded depressed women (20). This group consisted of all depressed women available for study who failed to meet the previous criteria for retardation. A somewhat more heterogeneous group, they were nevertheless all afflicted by primary depression (21), as were the patients in the first study, and were mainly unipolar. A few were frankly agitated. Study 2 (Table 1) shows the main clinical results. Again, the efficacy of T3 (compared to placebo) in accelerating the action of imipramine was clear and of approximately the same magnitude as in retarded patients. HRS findings were supported by self-rating scores; T3 advantage applied to all aspects of the syndrome; T3 did not increase toxicity.

Men

In parallel with our studies in women we pursued two similar studies in men (22). Study 3 (Table 1) shows the main clinical results of comparing T3 to placebo in imipramine-treated, retarded, depressed men; study 4 (Table 1) shows the results in nonretarded, depressed men. It can readily be seen that T3 added nothing to the imipramine treatment of men in either diagnostic category.

Since the T3 advantage in women is equal in retarded and nonretarded patients, results from the two studies can be combined. Since the T3 advantage in men is also equal in the two diagnostic categories (i.e., nil), these studies can also be combined. These computations are presented in Table 1 (all women, T3 versus placebo; all men, T3 versus placebo). They not only sharpen the difference in T3 response between the sexes but also show that imipramine alone exerts its antidepressant action substantially faster in men than in women. Thus a sex-related determinant of drug action emerges.

TABLE 1. *Clinical responses of depressed patients given imipramine plus placebo or T3*

Study	Sex	Diagnosis	Treatment[a]	No. of subjects	Clinical response on days 0 to 28[b]												
					0	2	5	7	9	12	14	16	19	21	23	26	28
1	F	Retarded	IMP + P	10	25	25	22	19	21	21	18	16	14	14	13	12	11
			IMP + T3	10	25	23	15	12	9	9	8	7	7	6	4	5	5
2	F	Not retarded	IMP + P	10	25	25	22	19	16	14	11	12	9	9	6	7	6
			IMP + T3	10	25	23	13	7	7	7	6	4	5	5	4	3	3
3	M	Retarded	IMP + P	7	22	19	13	14	10	10	9	8	7	6	7	6	5
			IMP + T3	6	24	19	11	7	7	7	6	5	5	6	5	6	3
4	M	Not retarded	IMP + P	3	25	19	12	7	8	9	6	10	8	13	9	10	9
			IMP + T3	4	25	18	12	10	10	8	8	8	9	7	7	11	9
Total	F	All women	IMP + P	20	25	25	22	19	18	17	15	14	12	11	10	10	9
			IMP + T3	20	25	23	14	10	8	8	7	6	6	5	4	4	4
Total	M	All men	IMP + P	10	25	19	11	12	9	9	8	9	7	7	7	7	7
			IMP + T3	10	23	21	14	8	8	8	7	6	7	7	7	9	7

[a] IMP, imipramine. P, placebo. T3, triiodothyronine.
[b] Hamilton Rating Scale scores obtained during hormone administration are underlined. Standard errors, omitted here, are given in the primary publications along with statistical treatment.

It is possible to conclude that T3, when given to imipramine-treated women, only confers the rapid response that characterizes men given the drug alone. Moreover, men may fail to profit from T3 because they have less to gain.

Imipramine-TSH and T3 Alone in Depressed Women

Imipramine-TSH

Before the above comparisons were available we had begun a study directed partly to the question of the specificity of action of T3 (in imipramine-treated patients) and partly to the thyroid state of depressed patients.

Twenty consecutive women with primary unipolar depression, retarded or nonretarded, were assigned randomly to one of two treatment groups (23). Each patient in the experimental group was given an intramuscular injection of 10 international units (IU) of thyroid-stimulating hormone (TSH) on day 1 and again on day 8. Patients in the control group received saline injections. Patients in both groups received imipramine (50 mg p.o. t.i.d.) beginning on day 1 and continuing throughout.

Table 2 shows the HRS results. TSH injection greatly accelerated the antidepressant action of imipramine, the second dose of TSH appearing superfluous. We had hoped to answer two questions by performing this study. The first was this: Would thyroxine (T4) accelerate imipramine as readily as T3 does? Since TSH evokes secretion of more T4 than T3 (15), the answer is probably affirmative. However, TSH does evoke some T3, perhaps all that is needed for potentiation of imipramine; moreover, a substantial peripheral conversion of T4 to T3 is now known to occur (24). Thus the affirmative answer is only tentative, and the question would best be tested by administration of T4, although even in such a design peripheral conversion would confound interpretation. The question is not entirely pedantic, however, for a consensus has developed that T3 and T4 differ in more than potency and onset of action (25). The second question was this: What is the thyroid reserve of depressed patients? We found that 24 hours

TABLE 2. *Clinical responses of depressed women, retarded or nonretarded, given imipramine plus diluent or TSH*

Treatment[a]	No. of subjects	Clinical response on days 0 to 28[b]												
		0	2	4	7	9	11	14	16	18	21	23	25	28
IMP + Diluent	10	28	24	21	18	15	14	14	13	11	10	10	8	9
IMP + TSH	10	28	18	10	8	6	5	7	6	5	5	4	3	4

[a] TSH or diluent was given on days 1 and 8. IMP, imipramine. TSH, thyroid-stimulating hormone.
[b] Hamilton Rate Scale scores.

after the first injection of TSH, bound T4 doubled. This is a normal response (26).

T3 Alone

In a fourth study of women (22), 27 patients with primary unipolar depression, retarded or nonretarded, were randomly assigned to one of three treatment groups: imipramine plus placebo; placebo plus T3; imipramine plus T3 (Table 3). Imipramine, 50 mg t.i.d. (or placebo), was started on day 1 and given throughout; T3 (or placebo) was started on day 3. This was increased from 25 to 50 μg on day 7. It was stopped on day 11. Table 3 shows the main results. The two groups of patients receiving T3 (with or without imipramine) improved more rapidly than the group receiving imipramine (plus placebo) until day 9, when both T3 groups began to worsen, as noted on day 12. By the time administered hormone had probably largely disappeared (day 14), the patients in the two groups who had received it began to improve at about the same rate as the imipramine-placebo patients. Late in this study, although not in the others, some of the most improved patients in the various treatment groups discontinued their participation, thus rendering less meaningful comparisons between groups.

The findings of main interest can be summarized as follows: again, T3 accelerated the action of imipramine; T3 alone temporarily exerted an antipressant action not inferior to that of imipramine; T3 became toxic regardless of whether it was accompanied by imipramine administration shortly after a daily dose of 62.5 μg had been achieved. The latter finding was substantiated by examination of side effect scales and clinical notes.

Related Studies

Feighner et al. (27) were the first group to attempt to replicate our clinical work. They gave depressed patients imipramine (200 mg/day p.o.) and T3

TABLE 3. Clinical responses of depressed women, retarded or nonretarded, given imipramine and T3

Treatment[a]	No. of subjects	Clinical response on days 0 to 28[b]											
		0	2	5	7	9	12	14	16	19	21	23	26	28
IMP + P	10	25	21	23	15	14	12	11	11	9	6	5	7	4
P + T3	8	25	17	15	11	10	13	9	7	7	8	3	4	6
IMP + T3	9	25	20	13	9	7	9	8	9	9	10	6	7	7

[a] IMP, imipramine. P, placebo. T3, triiodothyronine. Note that T3 was given briefly and in increasing dosage (see text).
[b] Hamilton Rating Scale scores. Those obtained during hormone administration are underlined.

(25 μg/day p.o. for 10 days). They found only a nonsignificant trend for T3-treated patients to improve more rapidly than control patients. We have discussed an interpretation of this study elsewhere (28).

Wheatley (29) conducted a multioffice double-blind study of 52 depressed outpatients. During the 3-week course of the study, patients received one of three treatments: amitriptyline (100 mg) plus T3 (40 μg) daily; amitriptyline (100 mg) plus T3 (20 μg) daily; amitriptyline (100 mg) plus placebo daily. Patients who received the hormone improved faster and attained lower severity scores at the conclusion of the study than patients who received amitriptyline alone. On several measures the larger dose of hormone was superior to the small dose. Women profited from the hormone more than men, as in our studies. Within the treatment cell of amitriptyline plus T3 (20 μg), patients with low thyroid indices responded better than patients with high indices. In both T3 treatment groups the hormone caused suppression of endogenous T4 secretion, as expected, suggesting that in depression the thyroid gland is normally suppressible. Side effects were equally distributed between groups, and Wheatley attributed them solely to amitriptyline.

Coppen et al. (30) performed a double-blind study of 30 patients with severe unipolar depression. Treatments consisted of L-tryptophan alone, imipramine alone, and each of these agents combined with T3 (25 μg/day p.o.) given for the first 14 days of the study. The main focus of interest in the present context is T3. The hormone did not alter response to the amino acid. However, it clearly enhanced the response to imipramine, and this effect was limited to female patients, all of whom reached the rather unusual HRS score of zero by the 28th day of the study. Side effects appeared to be diminished by T3. This can be regarded as a reflection of the improved therapeutic responses of T3-treated patients, if one appreciates the overlap of items on side effect and depression measures.

T3 is not the only hormone that has been used in an attempt to accelerate the response of depressed patients to antidepressant drugs. McClure and Cleghorn (31) pretreated four patients with dexamethasone, a synthetic adrenal glucocorticoid substance, and then administered imipramine. They noted an enhanced therapeutic response. Feighner et al. (27), however, were unsuccessful in their attempt to potentiate either imipramine or electroconvulsive therapy (ECT) response with dexamethasone.

The use of T3 to treat patients resistant to tricyclics is a separate issue and is discussed below.

Comment on Clinical Utility

The six original studies outlined above, along with the complementary work of others, require consideration from the standpoint of clinical utility. T3 accelerates the onset of antidepressant actions of imipramine and amitriptyline, and the phenomenon probably holds for other tricyclics as well. It may also hold for monoamine oxidase inhibitors (6,32), since an action of T3 on tricyclic metabolism appears to be an insufficient explanation for what we have observed (*vide infra*), but this possibility has not been tested. T3, at the dosage level of 25 μg/day, does not increase the side effects of tricyclic treatment. It appears to *reduce* side effects, but this is probably in itself a "side effect" of antidepressant efficacy. T3 should probably be used only in euthyroid patients, but this is hardly a limitation, for the vast majority of depressed patients are euthyroid. In hypothyroid patients the endocrine fault should be corrected before tricyclic treatment is started (5,33). T3 (25 μg/day) never produces hyperthyroidism but only a slight shift in that direction well within the range of euthyroidism. Larger daily doses of T3 may be more effective, but only marginally; 50 μg/day seems well tolerated, at least briefly, but 62.5 μg/day causes side effects even when imipramine is not given.

For safety we delayed starting T3 until imipramine treatment had been established, but this seems unnecessary. Informal clinical experience shows that it can be started the same day as the tricyclic drug. The duration of administration of T3 as a tricyclic adjunct appears to be flexible. Ten days seems too brief, but 14 days is usually sufficient. It can be given until remission is achieved and then stopped altogether a few days later.

We have not performed systematic studies of the use of T3 in tricyclic-resistant patients. However, our informal observations as well as anecdotal reports by others suggest that T3 corrects refractoriness in approximately half such patients. In an early single-blind trial Earle (34) found an even better result, and Hatotani et al. (33) reported similar experience. The most extensive study of this problem was performed by Ogura et al. (35). They identified 44 depressed patients, all of whom had responded poorly to tricyclics. The group included men and women, young and old, unipolar and bipolar patients. Sixty-six percent of these patients showed a good to excellent response when T3 was added, usually within 4 days. Recently Goodwin (36) also found T3 to be effective. Imipramine-resistant men seem about as responsive to T3 as imipramine-resistant women. Refractoriness to tricyclics is a problem in only a few patients, whereas latency of response is a problem in most. We developed the adjunctive use of T3 as a remedy for the latter, and it would be wry if it were reserved for the former.

T3 alone appears to have antidepressant effects. This finding is consistent with several earlier reports that thyroid hormones have therapeutic effects in a variety of mental illnesses (37–41). The use of T3 alone, however, appears to be limited by toxicity, and our observation of therapeutic effect has greater theoretical than practical clinical value. It may be a general principle that the use of peripheral hormones alone to produce central effects (other than the relief of symptoms of endocrine deficiency) is severely limited by peripheral toxicity. The work of Campbell et al. (42,43) may offer an exception to this generalization. They found that by adjusting individual dosage they could use T3 alone successfully to treat psychotic children without incurring undue toxicity. In a preliminary clinical study Park (44) suggested that T3 may enhance the antipsychotic effects of chlorpromazine.

THEORETICAL CONSIDERATIONS

Thyroid State of Depressed Patients

Findings

Thus far we have focused on clinical utility, but we have maintained an active interest in the endocrinological implications of our work as well. What is the thyroid state of depressed patients? Although depression appears to be the rule in hypothyroidism (45,46), and we usually find hypothyroidism in perhaps 5% of depressed women, we also find hypothyroidism in "normal" volunteers (47). In the studies described above we usually found our female patients oriented toward the low side of normal in regard to the chemical measures of thyroid state, and they had correspondingly slow normal ankle reflex time (ART). The latter generally accelerated slightly but significantly with the use of a small dose of T3. This acceleration has generally been sustained as long as T3 has been given, suggesting that T3 has acted more as a replacement than as a superfluous agent that would suppress endogenous hormone secretion to the extent of its own potency. These findings can be taken as "soft" signs that depressed patients are not truly euthyroid in all respects. On the other hand, thyroid reserve seems normal; i.e., in depressed patients the thyroid gland responds appropriately to the stimulus of injected TSH. Moreover, as Wheatley (29) showed, depressed patients given T3 tend to show chemical, as opposed to physiological, evidence of suppression of endogenous hormone secretion. One is left with the conclusion that the vast majority of depressed patients are within the range of euthyroidism as assessed by usual standards. This generalization, however, may obscure certain subtle but important points. We considered these in detail and discussed the criteria for euthyroidism elsewhere (48).

Recently we showed that in depression the pituitary TSH response to injected thyrotropin-releasing hormone is often deficient (49), and this has been broadly confirmed (50–52). Sachar et al. (53) earlier showed that in postmenopausal depressed women the pituitary growth hormone response to insulin injection is deficient. These findings suggest that in depression there is a hypothalamic disturbance. Since growth hormone is exquisitely responsive to stress (54), it appears that the hypothalamic fault, with its endocrine components, may reduce the coping capacity of the organism.

A Formulation

It is probably of limited value to attempt to describe the thyroid state of depressed patients without reference to the parameter of time. Depression is rarely a steady state; it is generally episodic, and in a few patients it may be regularly cyclic. The disorder has a natural progression and a natural resolution. It seems reasonable to believe that any biological function, certainly including thyroid state, influenced as it is by the hypothalamus, may vary with the overall affective tone of the organism, even if only as a consequence of depression rather than as a contributing cause. However, there are better grounds than mere plausibility to think that thyroid state may change during the course of depression; and there is also reason to believe that this change may have more than secondary significance. In the formulation which follows, we try to impose order on these considerations and to defend them with citations where possible. The notions to be described are not original. They have been expressed in nuclear form by Whybrow and co-workers (46,55). They are consistent with the concept of Hamburg and Lunde (56) that a limited capacity to respond to stress with increased secretion of thyroid hormones would amount to an adaptive disability.

Reflection on our work and consultation with colleagues led us to believe that the patients whose responses are reported above were studied and treated relatively early in the course of depression. These patients have generally had thyroid indices that were somewhat lower than the population means. The diagnosis of hypothyroidism is made in an occasional patient, and we have not yet seen a depressed patient in our hospital who has shown even borderline hyperthyroidism. Whybrow et al. (45) showed that depression is an almost invariable accompaniment of hypothyroidism. We suggest that diminished thyroid state constitutes a predisposition to depression.

In a study of this problem, Dewhurst (57) found that thyroid indices of depressed patients are often elevated. There is no contradiction between his findings and ours if one assumes that his patients were studied later in the course of depression. This assumption, of course, cannot be tested. However, in a study by Coppen et al. (30), in which one of the present authors participated, patients were in fact seen later than were the patients discussed above, and Coppen's patients showed higher thyroid indices. Some were borderline hyperthyroid by usual chemical criteria, although severity of depression was approximately the same as in our patients.

Still later in the course of depression, as improvement begins, thyroid function seems to take one of two courses. We suggest that it usually subsides toward population mean values, although exceptionally it may continue to increase. Lidz (58) and Kleinschmidt and Waxenberg (59) reported that hyperthyroidism sometimes follows severe depression.

In the absence of longitudinal studies using modern chemical techniques in untreated depressed patients, and borrowing from study to study as it does, the above formulation must be regarded as tentative. There is, however, a variety of findings to lend its plausibility. Mason (60) found the thyroid system to be a slow responder to stress in monkeys. Responses are clear but delayed. In an early endocrinological study of depression Board et al. (61) refer to the thyroid system as a third-order sequential responder, after the adrenal medullary and the adrenocortical systems.

We submit that, just as diminished thyroid function early in depression may represent a persisting contribution to cause, so may enhanced thyroid function later in depression represent a restorative effort of the organism. This formulation is frankly teleological. However, there is evidence to support it. Whybrow et al. (55) found that the greater the thyroid activation of depressed patients as assessed by ART or serum cholesterol, the more prompt was the response to imipramine. It is possible that by giving T3 to depressed patients one accomplishes promptly what the organism would later usually accomplish unaided, i.e., a tendency toward remission and a readiness to respond to treatment. It is uncertain how this generalization may relate to the sex differences we have described. Nevertheless, it is true that men, both normal and depressed, tend to maintain faster ART than women (17,55,62). If the above formulation is accurate, T3 should enhance the response to treatments other than drugs. With this in mind we are currently testing its capacity to potentiate the therapeutic effects of ECT. A bonus of this effort is the possibility that T3 will offset the memory fault often caused by ECT. Thyroid deficiency causes memory defect (45); perhaps a slight thyroid excess will counteract memory loss when caused by a nonthyroidal influence.

Possible Effects of Thyroid Hormones on Imipramine Metabolism

If one were to assume that depressed patients are at least slightly hypothyroid, then an explanation for T3 acceleration of imipramine response would be readily available: T3 acts as replacement therapy, tends to restore euthyroidism, and thus promotes recovery from any other illness present, in this case depression. However, the most we can assume about thyroid state once depression has been established is: (a) There may be a hypothalamic dysfunction which could have thyroid and other consequences. (b) An enhanced thyroid state, endogenous or imitated by treatment, might promote recovery. In any case, we are left to search for the mechanism(s) by which enhanced thyroid state promotes the antidepressant action of imipramine. None of the possibilities to be mentioned excludes any other. Indeed it seems likely that when a drug and a hormone each exert a broad range of effects these effects impinge on each other in several ways.

Clinical Inferences

A variety of indirect evidence suggests that an effect on drug distribution or metabolism by T3 cannot be a sufficient explanation for our observations. First, if T3 increased imipramine levels or free available imipramine, not only enhanced therapeutic action would be expected but enhanced toxicity as well. In the doses we have used in our clinical studies, T3 has not increased imipramine toxicity but has instead reduced it. This was a focal point in the study by Coppen et al. (30), and the result was clear. Second, T3 exerts an antidepressant effect by itself, however brief and clinically impracticable.

Animal Study

We examined drug distribution in rats that were thyroidectomized, hypophysectomized, or thyroid-treated (8,10). Animals were given radioactive imipramine acutely following chronic endocrine manipulation and chronic imipramine treatment. Radioactivity was then examined in homogenates of various organs and in an extract of the homogenates containing imipramine (and its product desipramine). In both the crude homogenates and their extracts, minor changes were found in the various peripheral tissues from animals in which endocrine state had been altered. However, in no case was there significant change in radioactivity in preparations of brain. None of our findings precludes the possibility of a T3 effect on imipramine metabolism. However, the findings taken together do suggest that other factors are probably operative.

Possible Modes of Interaction of Imipramine and Thyroid Hormones

A leading biochemical hypothesis suggests that central catecholaminergic mechanism are deficient in depression (63–65). While this concept has guided much productive research and remains a cornerstone of theory, it appears insufficient by itself to govern all the information that has accrued (66,67). At least indoleaminergic mechanisms must also be taken into account (68–71).

Catecholamine Mechanisms

In order to understand the interdigitation of two substances, systems are sought that are influenced by both. Catecholaminergic systems offer such a meeting ground for imipramine (72,73) and thyroid hormones (15,74,75). Imipramine promotes the activity of norepinephrine (NE) by blocking the reuptake inactivation of released neurohormone (76,77). Thyroid hormones appear to promote catecholamine activity as well, but the mechanism of this action is less certain.

Extensive reviews of thyroid hormone-catecholamine interactions are available, and another cannot be undertaken here. Suffice it to say that unanimity is lacking for all generalizations; yet certain ones have found substantial support and form the basis for continuing investigation.

Hypothyroid animals appear hypoadrenergic; hyperthyroid animals look hyperadrenergic. Yet the former generally show increased turnover of catecholamines (78–81) and the latter show decreased turnover (81,82). Such findings have led to the concept that catecholamine changes are compensatory for induced thyroid changes. The question then arises as to where in the catecholamine system thyroid hormones may exert their effects. An attractive supposition is that hypothyroidism diminishes and hyperthyroidism increases the sensitivity of catecholamine receptors. This of course is an inference, and unfortunately the study of receptors must usually be indirect.

We have already cited our animal work, but it is necessary to reconsider some of it in the present context. We found that T3 did not further increase hypothermia in the cold caused by 6-hydroxydopamine administration (8,10). Since increased temperature loss after 6-hydroxydopamine is due to catecholamine release (83), this finding does not support the notion that T3 sensitizes catecholamine receptors. In a similar way T3 did not enhance amphetamine-induced hyperactivity in rats. In these experiments T3 was given acutely, and it is possible that chronic treatment would produce quite different results. Engstrom et al. (84) recently advanced pharmacological evidence to suggest that T4 does increase catecholamine receptor sensitivity in brain. In contradistinction to the bulk of previous reports, these workers found that thyroid treatment *increased* brain catecholamine turnover.

The fact that the enzyme adenyl cyclase is an integral part of receptor systems (85) offers a biochemical approach to the study of receptors. Findings are not yet conclusive, but the weight of evidence suggests that T3 has no effect on the adenyl cyclase system in brain (86), nor for that matter in heart (87) or liver (88). On the other hand, T4 increases the stimulation of adenyl cyclase by epinephrine in fat (89). Adenyl cyclase appears to be a family of enzymes that are differentially responsive (90). Therefore additional studies may be required to exhaust the possibility of thyroid effects.

Murphy and his colleagues (91) approached the problem of receptor sensitivity in depressed patients by examining their platelets, guided by the finding that NE affects platelets and α-adrenergic receptors in other tissues similarly (92–94). They measured the inhibition by NE of prostaglandin (PG) E_1-induced stimulation of ^{3}H-cyclic adenosine monophosphate (AMP) formation. They found no difference in this effect between platelets from depressed patients and from controls. Assuming that changes in brain would be reflected in platelets, a point discussed by Murphy et al., one must conclude that the finding is a criticism of the idea we advanced that sensitivity of brain monoamine receptors may be diminished in depression (95). As Murphy et al., point out, their finding does not traduce the validity of the T3-imipramine interaction. It only suggests that if the interaction involves receptors it is not reparative of a deficit but compensatory for some other fault.

Motivated in part by our clinical reports, Frazer et al. (96) undertook a thorough study of imipramine-T3 actions and interactions on receptor systems in rat brain. T3 did not alter the basal net synthesis of cyclic AMP in brain nor did it alter the stimulation produced by NE. Chronic imipramine treatment reduced cyclic AMP formation caused by NE stimulation. When T3 was added, the imipramine effect was partly offset. In an addendum the authors suggested that chronic imipramine treatment may reduce the effects of NE by producing receptor "subsensitivity" from chronic overexposure of receptors to this agonist. We suggest that if this interpretation is correct they have in fact shown an important T3 effect, i.e., that it offsets an antiadrenergic activity of chronic imipramine. One must suppose, provisionally at least, that imipramine is antidepressant in spite of possible receptor-inhibiting properties.

The above comments introduce rather than exhaust the range of thyroid-adrenergic interactions and therefore this category of thyroid-imipramine interactions. Recently Dratman (75) published an elegant review of the modes of action of thyroid hormones. She advanced the fundamental concept that thyroid hormones, like catecholamines, are elaborations of the parent amino acid tyrosine.

Indoleamine Mechanisms

Central indoleamine function should probably be taken into account in formulating the biochemical substrate of depression. We proposed a scheme (71) in which catecholamine and indoleamine findings could be integrated, a "permissive hypothesis."

Indolamine-thyroid interactions have been less thoroughly studied than catecholamine-thyroid interactions. Nevertheless, a number of points of convergence have been identified. Parratt and West (97) showed that T4 treatment raises the 5-hydroxytryptamine (5-HT) level in some tissues of the rat. Skillen et al. (98), however, found no change in male or female rats in brain levels of 5-HT, 5-hydroxytryptophan decarboxylase, or monoamine oxidase after thyroid or propylthiouracil feeding. On the other hand, Toth and Csaba (99), working with rabbits, found that thyroid feeding caused a marked increase in brainstem 5-HT whereas thyroidectomy caused a small decrease.

Allergic phenomena, in which 5-HT may play a role, have also been investigated in reference to modification by thyroid state. Parratt and West (97) found that T4 increased the sensitivity of rats to 5-HT and to allergenic substances that release 5-HT and histamine. Spencer and West (100) confirmed this work and extended it by showing that hydrocortisone antagonizes these effects of T4. The possibility of subtle interactions in brain between slight changes in thyroid state and indoleamine activity have not been investigated to our knowledge.

Cholinergic Mechanisms

Recent findings suggest the importance of cholinergic mechanisms in affective disorders. Since physostigmine, a cholinesterase inhibitor and therefore a "procholinergic" substance, appears to produce improvement in mania (101); and since tricyclic antidepressants exert anticholinergic properties (72,73), it is reasonable to postulate diminished cholinergic activity as a substrate for mania and increased cholinergic activity as a substrate for depression. Indeed an early theory of the antidepressant effect of imipramine depended on its anticholinergic activity (102). This idea was set aside because anticholinergic manifestations of the drug's action (e.g., dry mouth) appear early whereas antidepressant effects are often delayed, and because a more appealing explanation of the drug's antidepressant action became available (63–65). The recent findings, however, suggest that the anticholinergic properties of tricyclics are not without antidepressant significance. Thus the actions of imipramine and of T3 may converge within the cholinergic system. We are ignorant, however, of studies of possible actions of thyroid hormones on this system.

Alteration of Imipramine Response by Sex Hormones

Imipramine-Ethinyl Estradiol in Women

After completing the six thyroid-related clinical studies, considering the results of animal studies, and contemplating the myriad actions of thyroid hormones, we sought a theme that would be simple and coherent enough to guide additional clinical investigation. We chose sex. Depression is more common in women than in men (103). It appears that men respond more promptly than women to imipramine. Thus, crudely put, being male protects against depression, and when depression occurs being male favors a better response to an antidepressant drug. Moreover, our findings with T3 are consistent with this formulation. Hellman et al. (104) gave T3 to normal men and women. In both sexes this produced a change in the metabolic pathways of adrenal steroids toward products with enhanced androgenic activity. We lack evidence that this occurred in our depressed patient, but if it did our female patient would have shifted toward a sex steroid balance characteristic of men, who customarily demonstrate a more rapid response to imipramine.

The maneuver that would emerge from this chain of reasoning is the use of androgens in both men and women to enhance imipramine response. We considered this procedure for female patients. However, consultation with senior gynecologists revealed that the administration of androgens for dysmenorrhea was once a frequent procedure but that masculinization sometimes occurred and did not always resolve completely with cessation of treatment. We therefore abandoned the idea and considered use of a hormone of opposite sign, an estrogen. The use of such a substance of course would run counter to the rationale we had built. However, there are other findings that suggest, at least in a general way, the use of an estrogenic substance as a possible accelerator of imipramine. Some oral contraceptives contain estrogenic substances, and all oral contraceptives appear capable of producing mood changes of either sign in some women (105–109). We considered it possible, therefore, that ethinyl estradiol, an estrogen incorporated in some oral contraceptives, might enhance the antidepressant activity of imipramine.

We decided to test this proposition in a population of depressed patients who were inmates at the North Carolina Correctional Center for Women. Thirty women were studied (22). All showed primary depression in the sense that no other diagnosable mental disorder was antecedent or concomitant. All but three showed unipolar depression, with approximately equal distribution between first and subsequent attacks. In somewhat more than half the patients depression appeared to be a response to incarceration. In the others depression either was clearly evident by history before the criminal act or was recurrent during the term of imprisonment. Patients were not housed in the infirmary but reported there for medication and for weekly assessment by use of the HRS and SDS. During working hours they remained at their assigned tasks of light labor. The patients who became toxic were of course admitted to the infirmary. Patients were in excellent general health and were taking no medication whatever. Special measures were employed to exclude pelvic or breast disorders. Patients were assigned to one of three treatment groups: placebo alone; imipramine (50 mg p.o. t.i.d.) plus placebo; imipramine plus ethinyl estradiol (50 μg/day p.o. for the first 14 days of the study).

Table 4 shows the main results. Patients who received only placebo

TABLE 4. *Clinical responses of depressed women, retarded or nonretarded, given placebo, imipramine, or imipramine plus ethinyl estradiol*

Treatment[a]	No. of subjects	Clinical response on days −7 to +42[b]							
		−7	0	7	14	21	28	35	42
P-P	10	23	25	22	21	24	24	22	22
IMP-P	10	23	26	16	8	6	6	6	5
IMP-EE25	5	27	28	10	6	6	11	4	4
IMP-EE50	5	24	25	12	12	14	7	9	8

[a] P, placebo, IMP, imipramine. EE, ethinyl estradiol. Ethinyl estradiol was given in either of two daily doses (25 or 50 μg) from day 1 through day 14. The hormone has a long biological half-life.

[b] Hamilton Rating Scale scores obtained during hormone administration are underlined.

showed no change whatever during 6 weeks of observation. This testifies to the authenticity of the depression syndrome under study. It is also interesting to compare this utterly null effect of inactive placebo in an impoverished therapeutic setting to its generally better performance in an enriched setting. For example, in an early clinical trial performed in a university teaching hospital, we found imipramine to be only slightly and statistically insignificantly better than placebo (110). The narrow difference was not due so much to lack of imipramine effect as to the efficacy of placebo.

The present focus of interest, however, is not placebo but ethinyl estradiol. Table 4 shows that imipramine patients were substantially improved by day 14. Prior to that time patients who had received 50 μg ethinyl estradiol as an adjunct improved as much or slightly more, but by day 14 they had worsened. This apparent depressive worsening was in fact fully accounted for by toxicity. Three of the five patients who received ethinyl estradiol showed hypotension and marked drowsiness. One of the three developed urinary retention, and another showed a coarse tremor in her hands. All five had rather more than the usual complaints of dry mouth and metallic taste. These women looked and complained as if they had received imipramine overdosage. A review of our procedure revealed that this had not occurred. Symptoms of toxicity persisted for approximately a week (day 21) after ethinyl estradiol was stopped, as per the protocol, on day 14.

At this point we incorporated a simple change in design. Women who were to have received 50 μg ethinyl estradiol were given 25 μg instead. The study technique was otherwise unchanged. Table 4 shows that the patients who received the lesser dose of hormone improved slightly faster than patients who received only imipramine. The differences, however, were statistically insignificant. These patients showed no enhanced toxicity. The only statistically significant HRS finding in this study was that pertaining to the massive advantage for imipramine (with or without hormone) compared to placebo.

Khurana (111) reported severe toxicity in a patient taking imipramine and an estrogenic substance chronically. Two anecdotal reports have reached us in which severe toxicity attended imipramine-estrogen treatment started simultaneously.

The mechanism of imipramine-estrogen interaction is unknown, although it seems likely that this phenomenon, unlike the imipramine-T3 interaction, is related to an effect of hormone on drug kinetics. We think the most likely explanation is that estrogen competes for imipramine binding sites in blood and that after estrogen a given amount of administered imipramine produces higher levels in plasma and tissues. Anton-Tay and Wurtman (112) showed in rats that ovariectomy increases brain NE turnover. If this change is regarded as compensatory, one might expect estrogen administration to increase overall noradrenergic activity, although perhaps reducing NE turnover, without necessarily enhancing toxicity. Since we found enhanced toxicity without clear enhancement of imipramine's antidepressant effect, the phenomenon described by Anton-Tay and Wurtman probably did not contribute prominently to what we observed. Estrogen, at least in large doses in animals, inhibits MAO (113). However, we employed average doses of estrogen, and the toxicity we observed did not resemble the pattern that has been observed for tricyclic-MAO inhibitor toxicity (114) but quite definitely the toxicity that attends the excessive use of tricyclic drug alone. Such toxicity seems to be mainly the result of anticholinergic activity (115). For this reason the possibility that estrogenic substances may exert anticholinergic effects appears deserving of study.

Oral contraceptives, including estrogen-containing preparations, can produce a variety of mood effects, including depression. Recently Adams et al. (116) showed that in some women who become depressed when taking oral contraceptives there is chemical evidence of pyridoxine deficiency. In these women the depressive symptoms are pyridoxine-responsive. These observations, while of great clinical significance, probably do not help explain the phenomenon we have observed.

There is at least one more mechanism by which estrogen might produce dysphoria, i.e., its action to increase the plasma proteins that bind thyroid hormones (117). Administered estrogen, by increasing T4 binding space, may produce secondary hypothyroidism when thyroid reserve is inadequate. Again, this series of events probably has little relevance for the phenomenon we observed, for that phenomenon suggested neither aggravated depression nor hypothyroidism but rather exaggerated imipramine toxicity.

We have no desire to overstate the clinical significance of our observation. We noted markedly enhanced imipramine-like toxicity when imipramine and ethinyl estradiol were started simultaneously. We have not given either substance to a patient whose treatment with the other was established. This presumably is often done in clinical practice and presumably with impunity, although Khurana's report (111) sounds a note of caution.

From a superficial point of view our findings and formulations seem to contradict the work of Klaiber and his colleagues (118). They identified a group of chronically depressed women who were resistant to all usual forms of treatment including ECT. They then showed that these women were resistant to estrogen; i.e., they showed little inhibition of plasma MAO after doses of estrogen that would inhibit this system in other women. The authors claimed that these patients usually showed improvement after they had been given large doses of estrogen alone. The study by Klaiber's group and the one by us are so different as to offer no contradiction at all. They studied a carefully defined subgroup, whereas we dealt with unselected patients; they used a large dose of estrogen, and we used a usual dose; they used estrogen alone, and we employed it as an adjunct to imipramine; their patients showed estrogen resistance, and in ours there was no reason to suspect this fault.

Imipramine-Methyl Testosterone in Men

Our experience with ethinyl estradiol in depressed women reinforced our rationale for a trial of testosterone in depressed men. We reasoned that if being "more female" worsens the patient's response to imipramine (if only in the realm of toxicity), and if being male improves it as well as reducing the risk for depression, then being "more male" might further improve the response to imipramine. In preparing for a double-blind study we decided to perform an informal preliminary trial in five men with unipolar depression (119). We selected methyltestosterone as the androgenic agent because it largely avoids destruction by liver and therefore can be given orally. We selected a daily dose of 15 mg. We were assured by endocrinological consultants that this was a very small dose, but we proposed to err on the side of underdosage and safety in our preliminary work.

After a drug washout period that varied according to previous drug intake, patients were started on imipramine (50 mg p.o. t.i.d.) and methyltestosterone (15 mg p.o. in the morning). Four of the five patients given this treatment developed an acute paranoid reaction that began 1 to 4 days after initiation of treatment and resolved 1 to 5 days after methyltestosterone was stopped. In the first patient in whom we saw this phenomenon we suspected the precipitation of delirium tremens. He had a remote history of excessive alcohol intake but had had no apparent somatic consequences. In any case, hallucinations did not occur and sleep was not notably affected. He was entirely preoccupied with delusional ideas concerning bizarre sexual activity on the part of his wife. In the other three patients there was no suggestion of delirium tremens and alcohol intake had been moderate or nil. All developed fearful delusions with ideas of reference or persecution. In none of the five patients was there more than the usual degree of nuisance side effects that usually accompany imipramine treatment.

Table 5 presents some baseline data and the rate of recovery of each patient as judged by "open" weekly HRS ratings. All patients were initially free of paranoid symptoms. Review of diagnostic material revealed that four patients exhibited feelings of guilt as a prominent symptom, and that these were the same four who later developed paranoid symptoms. Antidepressant response in four of the five men was rather rapid, but it is impossible to say without a control group if it was more rapid than it would have been had imipramine been given alone. In patient 2 recovery was gradual and required 5 weeks. The recovery from depression of patient 4, the only man who did not develop paranoid symptoms, was remarkably rapid.

As in our imipramine-ethinyl estradiol experiment, this attempt to enhance the antidepressant action of imipramine was obscured by toxicity. As in the ethinyl estradiol experiment, drug and hormone were started simultaneously, and the question is moot whether equivalent toxicity would have occurred if the body had been given an opportunity to adapt to either before

TABLE 5. *Baseline observations and clinical course of five men with unipolar depression receiving imipramine-methyl testosterone*

Response	Clinical response in five patients				
	1	2	3	4	5
Initial IMPS scores (standardized)					
Hostile belligerence	30	28	12	18	14
Intrapunitiveness	64	50	52	44	56
Excitement	4	18	10	0	4
Retardation	48	6	24	46	30
Paranoid projection	0	0	0	0	0
Guilt as an initial symptom	+	+	+	0	+
Follow-up Hamilton scores					
Day 0	30	31	28	27	25
Day 8	22	20	12	0	13
Day 15	8	19	13	0	9
Day 22	0	20	5	0	7
Became paranoid on testosterone	+	+	+	0	+

the other was administered. The cause of the paranoid syndrome is uncertain. Testosterone, like estrogen, is taken up by brain discretely (120) and appears to enhance the activity of noradrenergic systems (112). This might account for an enhanced antidepressant response, although how it would account for paranoid reaction is uncertain. In our original report we discussed this from the standpoint of biochemical modulation of aggressive drive and its behavioral expression (119).

We suspect, but lack that data to demonstrate, that the paranoid reaction after methyltestosterone-imipramine is a dose-related event. It seems possible that the use of smaller doses of methyltestosterone would avoid paranoid symptoms and enhance antidepressant response. We believe this treatment might be especially applicable to depressed male alcoholic patients, who often show diminished gonadal function (121). Testosterone in such a population might act partly as replacement therapy. On the other hand, the penchant of alcoholic patients to develop paranoid symptoms represents a precaution.

SUMMARY

Sex seems to be related to the predisposition for depression and to the speed of response to tricyclic antidepressants. Women become depressed more often than men, and when they do their response to tricyclics is slower.

Studies by our group and by others demonstrate that the response of depressed patients to imipramine, and probably other tricyclics, can be substantially altered by small induced changes in endocrine state. This in turn suggests the importance of endocrine state for antidepressant response and suggests that the endocrine state, particularly the thyroid state, of depressed patients be reinvestigated using the sophisticated biochemical measures now available; moreover, there should be special reference to longitudinal changes that may run parallel to and be part of the phases of depression. The question of the biological state that corresponds to readiness to respond to antidepressant treatment, and perhaps to remit spontaneously, is of special interest.

Current biochemical theories of depression are centered around the vagaries of brain biogenic amines. Thus endocrine findings, whether derived from naturalistic studies or from experimental intervention, tend to be interpreted in terms of their effects on biogenic amines. This is a valuable exercise and may be entirely correct. However, biogenic amines influence discharge and perhaps synthesis of hypothalamic releasing hormones, which seem to have direct CNS activity in addition to influencing pituitary tropic hormones, which influence target gland hormones, at least some of which enter brain where they influence biogenic amines. In a circular interdependent system such as this, it may be no more profitable to ask which is the most important component than to ask which is the most important component of an automobile giving it motility and direction. To begin to unravel the contributions of some components, it may be useful to study the effects of brain biogenic amine manipulations on the disposition and metabolism of the hormones that are known to affect amine metabolism. For example, monosodium-L-glutamate administration to neonatal rats causes dopamine depletion discretely limited to the tuberoinfundibular system, and this hypothalamic lesion leads to gross endocrine changes (122).

There are only crude observations available on which to make guesses about the complementary activities of biogenic amines and peripheral hormones. However, one can risk the suggestion that amines may regulate the relative activity between neuronal circuits. Thus, for example, imipramine-enhanced biogenic amine activity almost never result in paranoid symptoms unless certain circuits are turned on, as it were, by testosterone treatment. Circuit selection by hormones may be most apparent when energetics are heightened by drugs. Energetics of course may be influenced by more than biogenic amine factors, and circuit selection by more than peripheral hormones. These hormones have molecular representatives at the pituitary and hypothalamic levels, and these substances also appear to have direct brain-behavioral effects. The matter is surely complex. In this chapter we have tried to describe some simple maneuvers that have clear clinical advantage, or disadvantage, and which taken together indicate the need for modern investigations of the ancient problem of hormone-behavioral relationships.

ACKNOWLEDGMENTS

This work was supported in part by United States Public Health Service Career Scientist award MH-22536 to A.J.P.; Career Development Award

HD-24585 to G.R.B.; and by United States Public Health Service grants MH-15631, MH-16522, and HD-03110.

REFERENCES

1. Prange, A. J., Jr. (1963): Paroxysmal auricular tachycardia apparently resulting from combined thyroid-imipramine treatment. *Am. J. Psychiatry,* 119:994–995.
2. Prange, A. J., Jr., and Lipton, M. A. (1962): Enhancement of imipramine mortality in hyperthyroid mice. *Nature (Lond.),* 196:588–589.
3. Prange, A. J., Jr., Lipton, M. A., and Love, G. N. (1963): Diminution of imipramine mortality in hypothyroid mice. *Nature (Lond.),* 197:1212–1213.
4. Prange, A. J., Jr., Lipton, M. A., and Love, G. N. (1964): The effect of altered thyroid status on desmethylimipramine mortality in mice. *Nature (Lond.),* 204:1204–1205.
5. Avni, J., Edelstein, E. L., Khazan, N., and Sulman, F. G. (1967): Comparative study of imipramine, amitriptyline and their desmethyl analogues in the hypothyroid rat. *Psychopharmacologia,* 10:426–430.
6. Carrier, R. N., and Buday, P. V. (1961): Augmentation of toxicity of monoamine oxidase inhibitors by thyroid feeding. *Nature (Lond.),* 196:588.
7. Prange, A. J., Jr., Lipton, M. A., and Love, G. N. (1963): The effects of thyroid status on the toxicity of imipramine and the toxicity of drugs with common actions in the mouse. Presented at the Third International Congress of Chemotherapy, July 22–27, Stuttgart, Germany.
8. Breese, G. R., Prange, A. J., Jr., and Lipton, M. A. (1974): Pharmacological studies of thyroid-imipramine interactions in animals. In: *The Thyroid Axis, Drugs, and Behavior,* edited by A. J. Prange, Jr., pp. 29–48. Raven Press, New York.
9. Prange, A. J., Jr., and Bakewell, W. E. (1966): Effects of imipramine and reserpine on body size and organ size in euthyroid and hyperthyroid growing rats. *J. Pharmacol. Exp. Ther.,* 151:409–412.
10. Breese, G. R., Traylor, T. D., and Prange, A. J., Jr. (1972): The effect of triiodothyronine on the disposition and actions of imipramine. *Psychopharmacologia,* 25:101–111.
11. Everett, G. M. (1966): The dopa response potentiation test and its use in screening for antidepressant drugs. In: *Proceedings of the First International Symposium on Antidepressant Drugs.* Excerpta Medica International Congress Series No. 122. Excerpta Medica, Amsterdam.
12. Sigg, E. B., and Hill, R. T. (1967): *Neuropsychopharmacology,* edited by H. Brell. Excerpta Medica, Amsterdam.
13. Park, S., Happy, J. M., and Prange, A. J., Jr. (1972): Thyroid action on behavioral-physiological effects and disposition of phenothiazines. *Eur. J. Pharmacol.,* 19:357–365.
14. Park, S., Happy, J. M., and Prange, A. J., Jr. (1974): Thyroid action on behavioral-physiological effects and disposition of phenothiazines. In: *The Thyroid Axis, Drugs, and Behavior,* edited by A. J. Prange, Jr., pp. 63–74. Raven Press, New York.
15. Tata, J. R. (1964): Biological action of thyroid hormones at the cellular and molecular levels. In: *Actions of Hormones on Molecular Processes,* edited by E. Litwack and D. Kritchevsky, pp. 58–131. Wiley, New York.
16. Williams, E. S., Ekins, R. P., and Ellis, S. M. (1969): Thyroid suppression test with serum thyroxine concentration as index of suppression. *Br. Med. J.,* 4:338–340.
17. Prange, A. J., Jr., Wilson, I. C., Rabon, A. M., and Lipton, M. A. (1969): Enhancement of imipramine antidepressant activity by thyroid hormone. *Am. J. Psychiatry,* 126:457–469.
18. Hamilton, M. (1960): A rating scale for depression. *J. Neurol. Neurosurg. Psychiatry,* 23:56–62.
19. Zung, W. W. K. (1965): A self-rating depression scale. *Arch. Gen. Psychiatry,* 12:63–70.
20. Wilson, I. C., Prange, A. J., Jr., McClane, T. K., Rabon, A. M., and Lipton, M. A. (1970): Thyroid-hormone enhancement of imipramine in nonretarded depression. *N. Engl. J. Med.,* 282:1063–1067.
21. Kline, N. S. (1961): Depression: Diagnosis and treatment. *Med. Clin. North Am.,* 45:1041–1053.

22. Prange, A. J., Jr. (1971): Therapeutic and theoretical implications of imipramine-hormone interactions in depressive disorders. In: *Proceedings of the V World Congress of Psychiatry.* Excerpta Medica, Amsterdam.
23. Prange, A. J., Jr., Wilson, I. C., Knox, A., McClane, T. K., and Lipton, M. A. (1970): Enhancement of imipramine by thyroid stimulating hormone: Clinical and theoretical implications. *Am. J. Psychiatry,* 127:191–199.
24. Sterling, K., Refetoff, S., and Selenkow, H. A. (1970): T3 thyrotoxicosis: Thyrotoxicosis due to elevated serum triiodothyronine levels. *J.A.M.A.,* 213:571–575.
25. Sterling, K., Brenner, M. A., and Newman, E. S. (1970): Conversion of thyroxine to triiodothyronine in normal human subjects. *Science,* 169:1099–1100.
26. Williams, E. S., Ekins, R. P., and Ellis, S. M. (1969): Thyroid stimulation test with serum thyroxine concentration as index of thyroid response. *Br. Med. J.,* 4:336–338.
27. Feighner, J. P., King, L. J., Schuckit, M. A., Croughan, J., and Briscoe, W. (1972): Hormonal potentiation of imipramine and ECT in primary depression. *Am. J. Psychiatry,* 128:1230–1238.
28. Prange, A. J., Jr. (1972): Discussion of paper by Feighner et al. *Am. J. Psychiatry,* 128:55–58.
29. Wheatley, D. (1972): Potentiation of amitriptyline by thyroid hormone. *Arch. Gen. Psychiatry,* 26:229–233.
30. Coppen, A., Whybrow, P. C., Noguera, R., Maggs, R., and Prange, A. J., Jr. (1972): The comparative antidepressant value of L-tryptophan and imipramine with and without attempted potentiation by liothyronine. *Arch. Gen. Psychiatry,* 26:234–241.
31. McClure, D. J., and Cleghorn, R. A. (1968): Suppression studies in affective disorders. *Can. Psychiatr. Assoc. J.,* 13:477–488.
32. Bailey, S. d'A., Bucci, L., Gosline, E., Kline, N. S., Park, Q. H., Rochline, D., and Saunders, J. C. (1959): Comparison of iproniazid with other amine oxidase inhibitors. *Ann. N.Y. Acad. Sci.,* 80:652–668.
33. Hatotani, N., Tsujimura, R., Nishikubo, M., Yamaguchi, T., Endo, M., and Endo, J. (1974): Endocrinological studies of depressive states, with special reference to hypothalamo-pituitary function. In: *First World Congress of Biological Psychiatry, Sept. 24–28, Buenos Aires, Argentine,* Abstr. 101.
34. Earle, B. V. (1970): Thyroid hormone and tricyclic antidepressants in resistant depressions. *Am. J. Psychiatry,* 126:1667–1669.
35. Ogura, C., Okuma, T., Uchida, Y., Imai, S., Yogi, H., and Sunami, Y. (1974): Combined thyroid (triiodothyronine)-tricyclic antidepressant treatment in depressive states. *Folia Psychiatr. Neurol. Jap.,* 28:179–186.
36. Goodwin, F. K. Personal communication.
37. Reiss, M., and Haigh, C. P. (1954): Various forms of hypothyroidism in mental disorder. *Proc. R. Soc. Med.,* 47:889–893.
38. Fedlmesser-Reiss, E. E. (1958): The application of triiodothyronine in the treatment of mental disorders. *J. Nerv. Ment. Dis.,* 127:540–545.
39. Danziger, L., and Kindwall, J. A. (1953): Thyroid therapy in some mental disorders. *Dis. Nerv. Syst.,* 14:2–12.
40. Flach, F., Celion, C. I., and Rawson, R. W. (1958): Treatment of psychiatric disorders with triiodothyronine. *Am. J. Psychiatry,* 114:841–842.
41. Treadway, C. R., Prange, A. J., Jr., Doehne, E. F., Edens, C. J., and Whybrow, P. C. (1967): Myxedema psychosis: Clinical and biochemical changes during recovery. *J. Psychiatr. Res.,* 5:289–296.
42. Campbell, M., Fish, B., David, R., Shapiro, T., Collins, P., and Koh, C. (1973): Liothyronine treatment in psychotic and nonpsychotic children under six years. *Arch. Gen. Psychiatry,* 29:602–608.
43. Campbell, M., and Fish, B. (1974): Triiodothyronine in schizophrenic children. In: *The Thyroid Axis, Drugs, and Behavior,* edited by A. J. Prange, Jr., pp. 87–102. Raven Press, New York.
44. Park, S. (1974): Enhancement of antipsychotic effect of chlorpromazine with L-triiodothyronine. In: *The Thyroid Axis, Drugs, and Behavior,* edited by A. J. Prange, Jr., pp. 75–86. Raven Press, New York.
45. Whybrow, P. C., Prange, A. J., Jr., and Treadway, C. R. (1969): Mental changes accompanying thyroid gland dysfunction. *Arch. Gen. Psychiatry,* 20:48–63.

46. Whybrow, P., and Ferrell, R. (1974): Thyroid state and human behavior: Contributions from a clinical perspective. In: *The Thyroid Axis, Drugs, and Behavior*, edited by A. J. Prange, Jr., pp. 5–28. Raven Press, New York.
47. Wilson, I. C., Prange, A. J., Jr., Lara, P. P., Alltop, L. B., Stikeleather, R. A., and Lipton, M. A. (1973): TRH (lopremore): Psychobiological responses of normal women. *Arch. Gen. Psychiatry*, 29:15–21.
48. Prange, A. J., Jr., and Lipton, M. A. (1972): Hormones and behavior: Some principles and findings. In: *Psychiatric Complications of Medical Drugs*, edited by R. I. Shader, pp. 213–249. Raven Press, New York.
49. Prange, A. J., Jr., Wilson, I. C., Lara, P. P., Alltop, L. B., and Breese, G. R. (1972): Effects of thyrotropin-releasing hormone in depression. *Lancet*, 2:999–1002.
50. Kastin, A. J., Ehrensing, R. H., Schalch, D. S., and Anderson, M. S. (1972): Improvement in mental depression with decreased thyrotropin response after administration of thyrotropin-releasing hormone. *Lancet*, 2:740–742.
51. Takahashi, S., Kondo, H., Yoshimura, M., and Ochi, Y. (1973): Antidepressant effect of thyrotropin-releasing hormone (TRH) and the plasma thyrotropin levels in depression. *Folia Psychiatr. Neurol. Jap.*, 27:305–314.
52. Coppen, A., Montgomery, S., Peet, M., Bailey, J., Marks, V., and Woods, P. (1974): Thyrotropin-releasing hormone in the treatment of depression. *Lancet*, 2:433–435.
53. Sachar, E. J., Finkelstein, J., and Hellman, L. (1971): Growth hormone responses in depressive illness. I. Response to insulin tolerance test. *Arch. Gen. Psychiatry*, 25:263–269.
54. Greene, W. A., Conron, G., Schalch, D. S., and Schreiner, B. R. (1970): Psychologic correlates of growth hormone and adrenal secretory responses of patients undergoing cardiac catheterization. *Psychosom. Med.*, 32:599–614.
55. Whybrow, P. C., Coppen, A., Prange, A. J., Jr., Noguera, R., and Bailey, J. E. (1972): Thyroid function and the response to liothyronine in depression. *Arch. Gen. Psychiatry*, 26:242–245.
56. Hamburg, D. A., and Lunde, D. T. (1967): Relation of behavioral genetic, and neuroendocrine factors to thyroid function. In: *Genetic Diversity and Human Behavior*, edited by J. N. Spuhler, pp. 135–170. Aldine Publishing Co., Chicago.
57. Dewhurst, D. E., El Kabir, D. J., Harris, G. W., and Mandelbrote, B. M. (1969): Observations on the blood concentration of thyrotrophic hormone (TSH) in schizophrenia and the affective states. *Br. J. Psychiatry*, 115:1003–1011.
58. Lidz, T. (1949): Emotional factors in the etiology of hyperthyroidism. *Psychosom. Med.*, 2:2–8.
59. Kleinschmidt, H. J. and Waxenberg, S. E. (1956): Psychophysiology and psychiatric management of thyrotoxicosis: A two year follow-up study. *J. Mt. Sinai. Hosp.*, 23:131–153.
60. Mason, J. W. (1968): A review of psychoendocrine research on the pituitary-thyroid system. *Psychosom. Med.*, 30:666–681.
61. Board, F., Wadeson, R., and Persky, H. (1957): Depressive affect and endocrine functions. *Arch. Neurol. Psychiatry*, 78:612–620.
62. Ziegler, L. H., and Levine, B. S. (1925): The influence of emotional reactions on basal metabolism. *Am. J. Med. Sci.*, 169:68–76.
63. Prange, A. J., Jr., (1964): The pharmacology and biochemistry of depression. *Dis. Nerv. Syst.* 25:217–221.
64. Schildkraut, J. J. (1965): The catecholamine hypothesis of affective disorders: A review of supporting evidence. *Am. J. Psychiatry*, 122:509–522.
65. Bunney, W. E., and Davis, J. M. (1965): Norepinephrine in depressive reactions. *Arch. Gen. Psychiatry*, 13:483–494.
66. Prange, A. J., Jr. (1972): Discussion of Schildkraut. In: *Disorders of Mood*, edited by J. Zubin and F. A. Freyhan. Johns Hopkins Press, Baltimore.
67. Shopsin, B., Wilk, S., Sathananthan, G., Gershon, S., and Davis, K. (1974): Catecholamines and affective disorders revised: A critical assessment. *J. Nerv. Ment. Dis.*, 158:369–383.
68. Coppen, A. (1967): The biochemistry of affective disorders. *Br. J. Psychiatry*, 113:1237–1264.
69. Lapin, I. P., and Oxenkrug, G. F. (1969): Intensification of the central serotoninergic processes as a possible determinant of the thymoleptic effect. *Lancet*, 1:132–136.

70. Glassman, A. (1969): Indoleamines and affect disorders. *Psychosom. Med.*, 31:107–114.
71. Prange, A. J., Jr., Wilson, I. C., Lynn, C. W., Alltop, L. B., and Stikeleather, R. A. (1974): L-Tryptophan in mania. *Arch. Gen. Psychiatry*, 30:56–62.
72. Klerman, G. I., and Cole, J. O. (1965): Clinical pharmacology of imipramine and related antidepressant compounds. *Pharmacol. Rev.*, 17:101–141.
73. Gyermek, L. (1966): The pharmacology of imipramine and related antidepressants. *Int. Rev. Neurobiol.*, 9:95–143.
74. Waldstein, S. S. (1966): Thyroid-catecholamine interrelations. *Ann. Rev. Med.*, 17:123–132.
75. Dratman, M. B. (1974): On the mechanism of action of thyroxin, an amino acid analog of tyrosine. *J. Theor. Biol.*, 46:255–270.
76. Glowinski, J., and Axelrod, J. (1964): Inhibition of uptake of tritiated noradrenaline in the intact rat brain by imipramine and related compounds. *Nature (Lond.)*, 204:1318–1319.
77. Glowinski, J., Axelrod, J., and Iversen, L. L. (1966): Regional studies of catecholamines in the rat brain. IV. Effects of drugs on the disposition and metabolism of H3-norepinephrine and H3-dopamine. *J. Pharmacol. Exp. Ther.*, 153:30–41.
78. Landsberg, L., and Axelrod, J. (1968): Increased turnover of cardiac norepinephrine (NE) following hypophysectomy. *Fed. Proc.*, 27:602.
79. Landsberg, L., and Axelrod, J. (1968): Reduced accumulation of ^{3}H-norepinephrine in the rat heart following hypophysectomy. *Endocrinology*, 82:175–178.
80. Lipton, M. A., Prange, A. J., Jr., Dairman, W., and Udenfriend, S. (1968): Increased rate of norepinephrine biosynthesis in hypothyroid rats. *Fed. Proc.*, 27:399 (Abstr.).
81. Emlen, W., Segal, D. S., and Mandell, A. J. (1972): Thyroid state: Effects on pre- and postsynaptic central noradrenergic mechanism. *Science*, 175:79–82.
82. Prange, A. J., Jr., Meek, J. L., and Lipton, M. A. (1970): Catecholamines: Diminished rate of synthesis in rat brain and heart after thyroxine pretreatment. *Life Sci.*, 9:901–907.
83. Breese, G. R., Moore, R., and Howard, J. L. (1972): Central actions of 6-hydroxydopamine and other phenylethylamine derivatives on body temperature in the rat. *J. Pharmacol. Exp. Ther.*, 180:591–602.
84. Engstrom, G., Svensson, T. H., and Waldeck, B. (1974): Thyroxine and brain catecholamines: Increased transmitter synthesis and increased receptor sensitivity. *Brain Res.*, 77:471–483.
85. Sutherland, E. W., and Rall, T. W. (1960): The relation of adenosine-3′,5′-phosphate and phosphorylase to the actions of catecholamines and other hormones. *Pharmacol. Rev.*, 12:265–299.
86. Schmidt, M. J., and Robison, G. A. (1972): The effect of neonatal thyroidectomy on the development of the adenosine 3′,5′-monophosphate system in the rat brain. *J. Neurochem.*, 19:937–947.
87. Frazer, A., Hess, M. E., and Shanfeld, J. (1969): The effects of thyroxine on rat heart adenosine 3′,5′-monophosphate, phosphorylase b kinase and phosphorylase a activity. *J. Pharmacol. Exp. Ther.*, 170:10–16.
88. Jones, J. K., Ismail-Beigi, F., and Edelman, I. S. (1972): Rat liver adenyl cyclase activity in various thyroid states. *J. Clin. Invest.*, 51:2498–2501.
89. Krishna, G., Hynie, S., and Brodie, B. B. (1968): Effects of thyroid hormones on adenyl cyclase in adipose tissue and on free fatty acid mobilization. *Proc. Natl. Acad. Sci. U.S.A.*, 59:884–889.
90. Weiss, B. (1970): *Biogenic Amines as Physiologic Regulators*, edited by J. J. Blum, pp. 35–73. Prentice-Hall, Englewood Cliffs, N. J.
91. Murphy, D. L., Donnelly, C., and Moskowitz, J. (1974): Catecholamines receptor function in depressed patients. *Am. J. Psychiatry*, 131:1389–1391.
92. Moskowitz, J., Harwood, J. P., Reid, W. D., and Krisha, G. (1971): The interaction of norepinephrine and prostaglandin E_1 on the adenyl cyclase system of human and rabbit blood platelets. *Biochim. Biophys. Acta*, 230:279–285.
93. Johnson, M., and Ramwell, P. W. (1973): Implications of prostaglandins in hematology. In: *Prostaglandins and Cyclic AMP*, edited by R. H. Kahn and W. E. Lands, pp. 275–304. Academic Press, New York.
94. Siggins, G. R., Hoffer, B. J., and Bloom, F. E. (1969): Cyclic adenosine monophosphate: Possible mediator for norepinephrine effects on cerebellar Purkinje cells. *Science*, 165:1018–1020.

95. Prange, A. J., Jr., Wilson, I. C., Knox, A. E., McClane, T. K., Breese, G. R., Martin, B. R., Alltop, L. B., and Lipton, M. A. (1972): Thyroid-imipramine clinical and chemical interaction: Evidence for a receptor deficit in depression. *J. Psychiatr. Res.,* 9:187–205.
96. Frazer, A., Pandey, G., Mendels, J., Neeley, S., Kane, M., and Hess, M. E. (1974): The effect of triiodothyronine in combination with imipramine on (^{3}H)-cyclic AMP production in slices of rat cerebral cortex. *Neuropharmacology,* 13:1131–1140.
97. Parratt, J. R., and West, G. B. (1960): Hypersensitivity and the thyroid gland. *Int. Arch. Allergy Appl. Immunol.,* 16:288–302.
98. Skillen, R. G., Thienes, C. H., and Strain, L. (1961): Brain 5-hydroxytryptamine, 5-hydroxytryptophan decarboxylase, and monoamine oxidase of normal, thyroid-fed, and propylthiouracil-fed male and female rats. *Endocrinology,* 69:1099–1102.
99. Toth, S., and Csaba, B. (1966): The effect of thyroid gland on 5-hydroxytryptamine (5-HT) level of brain stem and blood in rabbits. *Experientia,* 22:755–756.
100. Spencer, P. S. J., and West, G. B. (1962): Further observations on the relationship between the thyroid gland and the anaphylactoid reaction in rats. *Int. Arch. Allergy Appl. Immunol.,* 20:321–343.
101. Janowsky, D. S., El-Yousef, M. K., Davis, J. M., and Sekerke, H. J. (1972): A cholinergic-adrenergic hypothesis of mania and depression. *Lancet,* 2:632–635.
102. Abood, L. G., Ostfeld, A. M., and Biel, J. H. (1959): Structure-activity relationships of 3-piperidyl benzilates with psychotogenic properties. *Arch. Int. Pharmacodyn. Ther.,* 120:186–200.
103. Silverman, C. (1968): *The Epidemiology of Depression.* Johns Hopkins Press, Baltimore.
104. Hellman, L., Bradlow, H. L., Zumoff, B., and Gallagher, T. F. (1961): The influence of thyroid hormone of hydrocortisone production and metabolism. *J. Clin. Endocrinol. Metab.,* 21:1231–1247.
105. Kaye, B. M. (1963): Oral contraceptives and depression. *J.A.M.A.,* 186:522.
106. Pincus, G. (1966): Control of conception by hormonal steroids. *Science,* 153:493–500.
107. Kane, F. J., Daly, R. J., Ewing, J. A., and Keeler, M. H. (1967): Mood and behavioural changes with progestational agents. *Br. J. Psychiatry,* 113:265.
108. Daly, R. J., Kane, F. J., Jr., and Ewing, J. A. (1967): Psychosis associated with the use of a sequential oral contraceptive. *Lancet,* 2:444–445.
109. Herzberg, B. N., Johnson, A. L., and Brown, S. (1970): Depression symptoms and oral contraceptives. *Br. Med. J.,* 4:142–145.
110. Prange, A. J., Jr., McCurdy, R. L., and Cochrane, C. M. (1967): The systolic blood pressure response of depressed patients to infused norepinephrine. *J. Psychiatr. Res.,* 5:1–13.
111. Khurana, R. C. (1972): Estrogen-imipramine interaction. *J.A.M.A.,* 222:702–703.
112. Anton-Tay, F., and Wurtman, R. J. (1968): Norepinephrine: Turnover in rat brains after gonadectomy. *Science,* 159:1245.
113. Kobayashi, T., Kobayashi, T., Kato, J., and Minaguchi, H. (1966): Cholinergic and adrenergic mechanisms in the female rat hypothalamus with special reference to feedback of ovarian steroid hormone. In: *Steroid Dynamics,* edited by T. Nakao and J. Tait, pp. 305–307. Academic Press, New York.
114. Schuckit, M., Robins, E., and Feighner, J. (1971): Tricyclic antidepressants and monoamine oxidase inhibitors. *Arch. Gen. Psychiatry,* 24:509–514.
115. Granacher, R. P., and Baldessarini, R. J. (1975): Physostigmine: Its use in acute anticholinergic syndrome with antidepressant and antiparkinson drugs. *Arch. Gen. Psychiatry,* 32:375–380.
116. Adams, P. W., Wynn, V., Rose, D. P., Seed, M., Folkard, J., and Strong, R. (1973): Effect of pyridoxine hydrochloride (vitamin B_6) upon depression associated with oral contraception. *Lancet,* 1:897–904.
117. Doe, R. P., Mellinger, G. R., Swain, W. R., and Seal, U. S. (1967): Estrogen dosage effects on serum proteins: A longitudinal study. *J. Clin. Endocrinol. Metab.,* 27:1081–1086.
118. Klaiber, E. L., Broverman, D. M., Vogel, W., Kobayashi, Y., and Moriarty, D. (1972): Effects of estrogen therapy on plasma MAO activity and EEG driving responses of depressed women. *Am. J. Psychiatry,* 128:1492–1498.
119. Wilson, I. C., Prange, A. J., Jr., and Lara, P. P. (1974): Methyltestosterone with imipramine in man: Conversion of depression to paranoid reaction. *Am. J. Psychiatry,* 131:21–24.

120. Stumpf, W. E. (1970): Localization of hormones by autoradiography and other histochemical techniques: A critical review. *J. Histochem. Cytochem.*, 18:21–29.
121. Mendelson, J. H., and Mello, N. K. (1972): Alcohol, aggression and androgens. In: *Aggression*, edited by S. H. Frazier, pp. 225–247. Williams & Wilkins, Baltimore.
122. Nemeroff, C. B., Ervin, G. N., Bissette, G., Harrell, L. E., and Grant, L. D. (1975): Growth and endocrine alterations after neonatal monosodium L-glutamate: Evaluation of the role of arcuate dopamine, neuron damage. Presented at the Society of Toxicology 14th Annual Meeting, Williamsburg, Va.

Hormones, Behavior, and Psychopathology, edited by
Edward J. Sachar. Raven Press, New York © 1976.

Effects of Prenatal Hormone Exposure on Physical and Psychological Development in Humans and Animals: With a Note on the State of the Field

June Machover Reinisch*

Department of Developmental Psychology, Teachers College, Columbia University, New York, New York 10027

There is ample experimental evidence to suggest that the presence of steroid hormones both during pregnancy in long-gestation mammals and shortly after birth in short-gestation mammals affects a variety of substrata in the developing organism. The effects are manifested morphologically, physiologically, histologically, neurologically, and behaviorly by the animal throughout the life span, and seem to be dependent on the critical period during development when the hormones were present. The vast majority of this evidence comes from experimental animal research; however, a small but growing number of studies exist concerning human subjects. This chapter first reviews the experimental animal literature that is the foundation on which human research is predicated. A discussion of the problems obtained in doing research with human subjects is followed by a review of the available human data, including some preliminary results from a study just completed by the author. Finally, the current controversies in the field of prenatal hormone effects are discussed.

ANIMAL DATA

There are several levels on which critical period hormones influence development. Under normal conditions the presence of androgen during different stages of early life affects the development of internal and external genital morphology, the structure and physiological functioning of the central nervous system (CNS), sexual behavior, and sexually dimorphic non-mating behavior. Recently there has been the suggestion that the presence of androgen during a critical period of development also affects the function of other major systems, e.g., the liver. (Estrogen, exogenously introduced, has many effects that parallel those of the androgens; however, their presence is probably not a factor in normal development.)

---

* Present address: Department of Psychology, Rutgers College, Rutgers University, New Brunswick, New Jersey 08903

The basic model for understanding the normal sequence of events influenced by pre- or neonatal steroid hormones begins with the production of
testosterone by the early-differentiated testis. The ovary remains undifferentiated until a much later time in development and has no known effect on its
course during this period (1–8). "The basic pattern of differentiation in the
absence of fetal gonads is feminine; the ovary plays no indispensable role
in the development of male structures. On the contrary, the fetal testis is a
very active endocrine organ that imposes masculinity against the basic
feminine trend of the body" (9). The introduction of this potent factor has
its first effect on the genital primordia, which at this stage is virtually identical in male and female animals. The presence of testosterone and its metabolic products — in particular, dihydrotestosterone (10) — masculinizes the
internal and then the external genital morphology (6,11). Feminine differentiation proceeds in the absence of hormonal influence and at a later stage in
development. The presence of a masculinizing factor as essential for male
development and the absence of hormonal influence as essential for female
development is the basis of the model for all dimorphic differentiation, be it
morphological, physiological, neural, or behavioral. The majority of the evidence collected to date is based on this understanding and has been studied
mainly in the rodent either by removal of hormonal influences through castration of the male (or chemical interference with the present androgens) or
introduction of exogenous hormones in the androgen-free animal.

Following differentiation of the genitalia and by the identical mechanism,
testosterone seems to have an effect on the CNS. In a series of studies beginning with Pfeiffer in 1936 (12) and continuing in the 1950s (13–15), it was
determined that a part of the CNS was responsible for the cyclic or acyclic
(tonic) pattern of gonadotropin release evident in female and male animals,
respectively. Subsequent research identified the hypothalamus as the structure responsible for the control of gonadatropin release from the pituitary
(4,16–19). It appeared that the hypothalamus, like the genital primordia,
was undifferentiated early in development (20). In the absence of androgen,
the hypothalamus would differentiate so that it would maintain and regulate
the secretion of gonadotropins from the anterior pituitary in a cyclic fashion
and thereby mediate the rhythmic cycle of follicle growth, ovulation, and
corpora lutea formation in the ovary, and control the onset of menstruation
or estrus (4). If androgen were present, the male pattern of tonic gonadotropin release would develop (21); and if ovaries were present during maturity, they would remain in a constant state characterized by the presence of
many follicles without any corpora lutea. Barraclough and Gorski (22,23),
using brain-stimulation techniques, identified the preoptic-suprachiasmic
area as the hypothalamic center responsible for the cyclic control of gonadotropin release; they suggested that it was here that critical period androgen
has its effect, causing a loss of responsiveness to intrinsic or extrinsic stimulation. They placed the tonic control center in the ventromedial arcuate

region. The above observations on the hypothalamus suggested that, just as the genitalia were undifferentiated during early development, so were the areas of the hypothalamus regulating the pattern of gonadotropin release to be manifested dimorphically during adulthood; further, it was the presence or absence of androgen during a set time in development that was responsible for instituting normal adult male and female hypothalamic regulation of the pituitary.

SEXUAL BEHAVIOR

Sexual behavior has also been shown to be affected by the presence or absence of androgen during early development. Many experiments have demonstrated that depriving the male of testosterone perinatally in short-gestation animals—e.g., rats (2-4, 9, 24-30)—or prenatally in species with long gestations—e.g., guinea pigs (31)—will augment the display of female sexual behavior (e.g., lordosis) during adulthood, and in many cases will also reduce male sexual behavior. Conversely, male mating behaviors (e.g., mounting, intromission, ejaculation, and, at least in rats, ultrasonic post-ejaculatory vocalization) can be greatly enhanced by introducing an androgen during the critical period for the development of these behaviors; in these animals, there is a concomitant reduction in female sexual behaviors. This has been demonstrated in rats (3,25,32-39), mice (40), hamsters (41), guinea pigs (2,39,42,43), and rhesus monkeys (2,28,42,44).

It was recently demonstrated that female rats, when treated just after birth with testosterone, "exhibit a temporal patterning of male copulatory behavior identical to that of normal males" (45), even though the frequencies of some aspects of male behavior are lower than in the normal male. However, with testosterone treatment during prenatal and neonatal periods, as well during adulthood, "female rats displayed all the characteristic masculine responses and intervals, including the ultrasonic postejaculatory vocalization." Sexual motivation as measured by "the urge to seek contact with a vigorous male" (46) and sexual attractiveness (8) are dimensions of sexual behavior that also seem to be affected. In general, the presence of testosterone enhances later male behavior, diminishes the display of female behavior, sensitizes the animal's response to androgen during adulthood, and inhibits responsiveness to estrogen and progesterone in later life.

Manipulation of the early hormone environment has further been shown to alter the physiology and physical appearance of treated animals. The presence of androgen during similar stages of early development altered weight, length, and oxygen consumption rate of female guinea pigs in the male direction (28), as well as patterns of steroid metabolism in the liver of rats (47); moreover, male rats, when castrated neonatally, "show female levels of plasma corticosterone following ether stress," an effect which could be reversed by a single injection of androgen at the time of castration (48).

SEXUALLY DIMORPHIC NONMATING BEHAVIOR

Many behaviors dimorphically expressed in male and female animals are not directly related to mating. Investigation into the mechanisms underlying these differences suggests that many dimorphic patterns are influenced by androgen during critical stages in development in a manner similar to that of the differentiation of genital morphology, the hypothalamus, and sexual behavior. Among the behaviors studied were aggression (48–53), open-field behavior (54–56), emergence (8,56,57), running patterns (4), approach and allogrooming (57), avoidance acquisition (55), exploratory behavior (57), sleep patterns (58), saline intake (59), and play and threat (2,28,39,42,60).

Many aspects of maternal behavior have also been investigated. The evidence is somewhat contradictory since both male and nulliparous female rats can be sensitized to show nearly all dimensions of what has generally been defined as maternal behavior if presented with newborn pups for several days (61). However, it has been reported that pup retrieval from a T-maze is affected by the presence of early androgen (62) even when general retrieval is not. In adult mice hormonal induction of construction of a specialized maternal nest seems to depend on lack of androgen during early development (63). Lisk et al. (63) also observed a new type of nest that was unlike the ordinary or maternal nest; it was constructed most frequently by neonatally androgenized females following adult treatment with estrogen and progesterone. Quadagno and co-workers (64–66) obtained conflicting results in their studies and recently suggested that maternal behavior is not influenced by the early hormonal environment. Bridges et al. (61), in a developmental study of maternal responsiveness in rats, found that prepubertal rats are capable of retrieving and grouping pups. He suggested that at least part of the confusion in studies of neonatal androgen and maternal behavior may be explained by the operational definition of these behaviors: "Are the behaviors observed in the sensitized rats really equivalent to those displayed by lactating females? Retrieval *per se* has been suggested to be a poor measure of maternal behavior. Our data now suggest that a sensitized animal may not be displaying maternal behavior, but rather a less specific form of social responsiveness."

Although the behavioral literature, excluding maternal behavior, has been presented thus far as if the results were universally clear, that is not the case. Conflicting results have been obtained over the years on a number of both sexual and nonsexual behaviors (67). There are many possible reasons for the lack of consistency. (a) There is a lack of uniformity of the dosages used in different laboratories. (b) Timing of hormone intervention is crucial to the strength and kind of effect obtained, and the outcome of treatment depends on when (68,69) and for how long (2,3,33,34,36,45,146) androgen is given or deprived, since the critical periods for effects on genital anatomy and the CNS appear to be different. (c) Recently it was noted that the presence of ovaries during development may alter the effect of neonatal androgenization (54,70). If this is true, many of the past experiments need to be re-evaluated since they were carried out with ovaries present. In the past ovarian tenancy was not considered an important factor by many researchers, and therefore the literature contains studies done with ovaries either present or absent. (d) There are differences in the reactions of various strains to testosterone treatment neonatally (53); for example, the work on the rat has been done using Sprague-Dawley, Wistar, and Long Evans rats. No investigation or differential effects of androgens on these animals has been done to my knowledge. (e) There appears to be differential responsiveness to different androgens among species. For example, masculine sexual behavior can be stimulated by dihydrotestosterone propionate in male guinea pigs and rhesus monkeys but not in rats (71). (f) Test situations have not been consistent for studying the same behaviors [see (64) in comparison with Pfaff and Zigmond (8)]. (g) Beach (72) recently remarked that one source of "variability" in covariance between hormones and behavior "is the effect of learning and experience on the development of animals." Sexual and play behavior have been shown to be influenced by rearing conditions in rhesus monkeys (69), and the social environment in which a rat pup is reared seems to have an effect on the development of normal sexual behavior (73). The above is not an exhaustive list of the possible factors affecting the animals reared in different laboratories and tested under different conditions, but it does suggest some of the possible reasons for the conflicting results and variability in the data which have been reported.

CENTRAL NERVOUS SYSTEM

On the assumption that the brain underlies behavior and that the behavioral alteration influenced by the presence or absence of androgen during early development might be associated with alteration of the structure and function of neural tissue, a number of careful studies on the CNS have been executed. The findings are in accord with expectation. Evidence was found of differences between the male and female brain which could be manipulated by early hormone exposure. These include, in relation to the hypothalamus: (a) oxidative metabolism (21,74–79); (b) protein content (80); (c) serotonin levels (81); (d) RNA metabolism (23); (e) nuclear volume (34,82–84); and (f) neural connections (85–87). It had been suggested that hormonal influence on dimorphically expressed brain differences might not be limited only to the hypothalamus, and recently evidence of an effect on the incorporation of ^{3}H-lysine in the cerebellum contributed strong evidence to the notion that hormone influences are not so restricted (88). Further, evidence of a hormone-dependent effect on the reaction to septal lesions and subsequent emotionality implies limbic involvement during the critical period of neural differentiation by androgen (89).

HUMAN RESEARCH

When considering the ethical and methodological problems of designing and implementing human research, the inconsistencies in methodology, design, and approach underlying the conflicting results and problems in interpretation obtained in the animal studies seem minimal. There are a plethora of difficulties awaiting the intrepid researcher who desires to study the effects of the prenatal hormone environment on human development, difficulties which go beyond the obvious ethical restraints and legal injunctions against treating human fetuses for whom such intervention is not clinically indicated. Techniques are either absent, crude, dangerous to the fetus, or so new as to be unavailable to most researchers. It is rare to find and difficult to identify patients whose endocrine status make them likely "experiments in nature" and thereby, at least by analogy, possible subjects for consideration. Rearing conditions, social experience, and learning—which interact with hormone effects even in the lowly rat—have a tremendous influence over the long term of human development. Retrospective treatment data collected from doctor or hospital records are difficult to retrieve, and when finally acquired are frequently imprecise in regard to timing, dosage, and even type of hormone used for therapy. This is but a limited list of the obstacles encountered in trying to study the effects of hormones during early critical stages of human development; yet despite the difficulties, some data are available and can be tentatively interpreted in the light of research on other mammals and nonhuman primates.

Although it seemed clear that testosterone produced by the fetal or neonatal testis was responsible for male differentiation of the genitalia, hypothalamus, and behavior in animals—including monkeys (90,91)—there was little evidence until recently to suggest that the human male fetal testis was capable of synthesizing or forming testosterone at the appropriate time during development. It has now been demonstrated that the human fetal testis is able to form testosterone from C21 precursers as early as 8 to 10 weeks post conception (10). The results of several similar studies (92,93) support these data and together demonstrate that this ability continues to rise to a maximum level at approximately 20 weeks, after which it declines until parturition. Studies of plasma testosterone concentrations have confirmed the *in vitro* evidence (93,95,96). When compared to males, female fetuses show no such concentrations or capability for testosterone formation. Abramovich (94) found the male levels to be nine times those in females during weeks 12 to 18 of gestation and remarked that "it is interesting to note that the levels reached by some of the midpregnancy male fetuses were well within the normal adult male levels measured in this laboratory." These findings have confirmed the presence of testosterone of testicular origin during the period of morphological differentiation of the genitalia in the human, and also well into the stage that has been suggested as the period of hypothalamic differentiation.

There are two populations available for investigating the influence of the prenatal hormone environment on human psychosexual and personality development. The first is patients whose endocrine status necessitates close contact with a hospital center, at least for some time during their life. They include such disorders as adrenogenital syndrome (which produces an excess of adrenal androgen in the fetus) and androgen insensitivity and Turner's syndrome (which result in a deficiency in the hormones available to the fetus). An especially large number of these patients, considering their rarity in the general population, have been evaluated, by Money and his associates at Johns Hopkins Hospital. The second group are patients whose mothers received hormone therapy for the treatment of at-risk pregnancy or toxemia during pregnancy. Many of the hormones used for pregnancy maintenance are synthetic oral progestins, which have some androgenic potential as evidenced by the masculinization of the genitalia in a small percentage of female offspring of these pregnancies (97). Until recently only the masculinized females were available for study (98).

Adrenogenital syndrome is an endocrine anomaly involving excess production of androgen by the fetal adrenals. Just as in the animal experiments when androgen was exogenously introduced, the female with adrenogenital syndrome is born with masculinized genitalia (98) resulting from the higher than normal levels of androgen present during differentiation. A series of studies on these children throughout development has shown that this prenatal androgen not only affects morphology but also influences psychosexual development (100–106). The general conclusions from these investigations are that females with adrenogenital syndrome (a) identified themselves and were identified by their mothers as tomboys during a longer period of childhood as compared to matched controls; (b) evidenced higher levels of energy expenditure; (c) showed stronger interest and participation in outdoor physical activity; (d) reported minimal interest in doll play, dresses, and girls' activities; (e) preferred boys to girls as playmates; and (f) expressed more ambivalence about the advantages and disadvantages of being born a female. Furthermore, there was a difference in the attitudes and concerns with regard to romance, marriage, and maternalism. Ehrhardt (106) reported that these girls had a much lower interest in babies during childhood, had fewer fantasies about romance and marriage, and seemed more concerned about future careers.

The one study on girls treated prenatally with synthetic progestins reports very similar results (98). The 10 subjects were also masculinized physically but not usually to the extent found in adrenogenital syndrome (100). The hormones present during prenatal life were exogenously rather than endogenously produced, and yet the influence on psychosexual development and personality was nearly identical. Although using "convenient" samples such as these precludes true experimental design, the striking similarity between the results of the animal experiments (especially those using androgenized female monkeys) and these human subjects suggests that similar mechanisms are at work.

Patients with androgen insensitivity have endocrine status that allows them to be compared with the experimental animals deprived of hormone influences during critical periods of development. Androgen insensitivity occurs in patients with X,Y chromosomes and involves an inability of the cells to utilize the androgen produced by the testes. This disorder "may be a product of a genetically determined, cellular metabolic barrier in the form of either an additional or missing enzyme" (107). The result is the birth and development of a chromosomal male whose external morphological appearance is female. They are usually reared unconditionally as females and are most likely to come to the attention of the medical community during adolescence when menarche fails to occur. Studies carried out with these patients (108,109) demonstrated that they were very feminine in their attitudes, behavior, and appearance. They were found to have: (a) a configuration of scores on the WISC and WAIS similar to that found in normal female populations, i.e., the sample mean Verbal Score was higher than the sample mean Performance Score; (b) strong interest in marriage and childrearing; (c) minimum interest in outside careers; (d) pleasure in taking care of the home; (e) primarily childhood play with dolls and other girls' toys. Furthermore, none described herself or was described by her mother as having been a tomboy during childhood.

It is evident from these results that chromosomal males deprived of the influence of the androgen during development differentiate as female on many levels, an outcome similar to that seen in animals experimentally deprived of androgens. However, it must be kept in mind when interpreting data from patients with endocrine disorders that there are possible confounding variables, for example, other genetic factors, which could be linked to the primary genetic defect; however, the concordance of the human clinical and the experimental animal data renders this hypothesis unlikely.

One additional and very curious hypothesis was generated from the data collected from the two clinical syndromes characterized by increased amounts of steroid hormones during gestation: Increased amounts of androgen, synthetic progestins, or progesterone during a critical period of prenatal development in humans seemed to raise intelligence. Money and Lewis (110) found that groups of both boys and girls with adrenogenital syndrome evidenced significantly higher IQ scores than the population norm or other groups of endocrine samples tested at Johns Hopkins Hospital. Perlman (111) also found that her sample of adrenogenital patients were of higher IQ than the other samples tested. However, Baker and Ehrhardt

(112) using unaffected siblings and parents as the comparison groups, recently demonstrated that the entire families of adrenogenital patients show higher IQs than expected, even when socioeconomic status was considered (113).

The second group of subjects whose prenatal steroid levels were augmented exogenously and showed IQs elevated from the normal population were children whose mothers received synthetic progestins or naturally occurring progesterone for the maintenance of at-risk pregnancy or toxemia. There are two investigations which, when taken together, strengthened the possibility that a relationship existed between prenatal treatment with steroid homrones and later elevated intelligence. The first was a study by Ehrhardt and Money (98), which included (synthetic) progestin-treated girls. Nine of the 10 subjects were born with masculinized genitalia. The tenth was the treated sister of one of the masculinized subjects. The mean IQ of the group, as measured by the Wechsler Intelligence Scale for Children, was 125 with a standard deviation (SD) of 11.8, compared with the population norm of 100 (SD 15). Sixty percent of the IQs were above 130 when only 2% would be expected in this range from a random sampling of the normal population.

The second study was carried out in England (114). Head teachers rated 29 children whose mothers were treated with progesterone for toxemia during pregnancy and two comparison groups. The comparison groups were comprised of the next child born on the same ward of an uncomplicated pregnancy and the next child delivered of a mother suffering from toxemia who had not been treated with progesterone. The teachers rated all the children as "above standard," "average," or "below standard" on three academic and two nonacademic subjects. The progesterone-treated children were rated significantly higher than the comparison groups on all subjects. The effects were both dose-dependent (subjects receiving 8 g or more during gestation did significantly better than those receiving less than 8 g; and related to time of administration (subjects treated prior to the 16th week received significantly higher ratings than those treated only after the 16th week post conception).

Both these studies, while provocative, are by no means conclusive because of the design problems related to sample selection. In the Money and Ehrhardt study (98), the possibility of sampling bias resulting from the small sample size, and errors accruing from the selection of a convenient sample without a proper comparison group, could have been a source of the main effects obtained. Furthermore, since the testing was carried out by investigators who were aware of the treatment category of the subjects, experimenter bias may also have influenced the results. The Dalton study (113) is flawed by the possibility of selection bias resulting from the differential selection of respondents for the comparison groups, as well as the strategy of matching to achieve pre-experimental equivalence of groups (115).

CALIFORNIA STUDY: A PRELIMINARY REPORT

The California study was designed to minimize the opportunity for confounding the results by sampling bias, selection, and experimenter effects. It employed a larger number of subjects than the two previous investigations and controlled for socioeconomic factors, genetic inheritance, parent education, etc. with a comparison group composed of siblings. As was pointed out by Barker and Edwards (115), since performance on intelligence tests is largely "determined by genetic endowment, prenatal environment, and the physical and social environment after birth, data relating small differences in performance to obstetric events can be interpreted only against a control population matched for similar heredity, prenatal, and postnatal determinants. Therefore unless the differences are large, sibs provide the only possible controls."

Experimenter effects were minimized by employing examiners who were blind to the experimental plan and treatment category of the subjects they tested. Furthermore, all available information from every possible source (doctors, hospitals, and parents) on each pregnancy and delivery was collected — including dosage, duration of treatment, and type of medication as well as pre- and perinatal complications — in order to compile the most complete history possible on the subjects in the treated (experimental) and untreated (comparison) groups.

In order to study the effects of prenatal treatment with steroid hormones on human intelligence and personality using a design which provided the most stringent experimental controls available, 56 families were obtained from two Los Angeles private clinics and from among the cases of a private physician. Each family included at least one offspring whose mother was treated during pregnancy with synthetic progestins and/or estrogen, and at least one sibling whose gestation was untreated. All the offspring available in a family were tested. The smallest sibship included two children, the largest four. Subjects in the experimental group ranged in age from 5 to 17 years and included 25 treated males and 46 treated females. The age range for the comparison group of sibs was 4 to 21 years, with 27 untreated males and 43 untreated females. A total of 84 prenatal and 35 perinatal complications were reported for the treated children as compared to 25 prenatal and 25 perinatal complications for the untreated comparison group.

For the data analysis, subjects were partitioned in two ways: (a) two groups comprised of all treated subjects and all untreated subjects; (b) three treatment subgroups selected by the ratio of progestin/estrogen administered. Table 1 gives the range of the total dose of progestin and estrogen and the progestin/estrogen ratio, which served as the criteria for selection into each subgroup.

Table 2 describes the range, mean, and median of treatment duration and total dosage of progestin and estrogen for the treated group. The mothers

TABLE 1. *Criteria for selection of index cases into three treatment subgroups*

| Treatment subgroups | No. of subjects | Range of total dose (mg) | | Progestin/ estrogen (mg) |
		Progestin	Estrogen	
High estrogen/low progestin	16	478–5,611	3,500–13,905	1:9 to 1:1.5
				>100 to 1
Progestin[a]	26	525–9,890	4–40	100:1 to 358:1
				<100 to 1
High progestin/low estrogen	29	490–10,650	6–1,390	3:1 to 82:1

[a] Seventeen subjects received no estrogen.

of 17 subjects received no estrogen during pregnancy (Tables 1 and 2). These subjects are all found in the progestin group. All mothers of the subjects in both the high-estrogen/low-progestin group and the high-progestin/low-estrogen group received a combination of progestin and estrogen. (For a more detailed description of the sample, complications, and specific hormones administered, see ref. 117.)

The Wechsler Intelligence Scales were chosen for the measurement of IQ and yielded five scores: Full Scale IQ, Verbal IQ, Performance IQ, Verbal Factor, and Performance Factor. The latter two scores were generated according to the factor-analyzed clusters of subtests found by Cohen (118,119). For the measurement of personality, the Cattell series of Personality Questionnaires were administered. Fifteen Primary Factors and four Second Order Factors were available for analysis. The Second Order Factors derived from the Primary Factors include: Anxiety, Extroversion, Tough Poise, and Independence. At this time an analysis of the IQ scores has been completed and analysis of the personality data is underway; some preliminary results are reported here.

A series of three factor analyses of covariance for hormone treatment,

TABLE 2. *Duration and total dosage of progestin and estrogen treatment in treated group*

Drug	No. of subjects	Range	Mean	Median
Progestin				
Duration (weeks)	71	3.97–36.08	17.03	15.95
Dosage (mg)	71	478–10,650	2,779.75	1,857.50
Estrogen				
Duration (weeks)	71	0–34.22	13.36	10.53
	54[a]	0.14–34.22	17.57	16.59
Dosage (mg)	71	0–13,925	1,495.36	58.25
	54[a]	4–13,925	1,966.13	247.5

[a] Data with 17 subjects receiving no estrogen are eliminated.

dose level, and sex, with sibling score and severity of pre- and perinatal influences as covariates were calculated. The first analysis adjusted criterion scores in the treatment group for sibling score. In those instances where a treated subject had two or more untreated siblings, the sibling's average score was entered as a covariate. The correlation (r) of intelligence scores for treated and untreated siblings were all positive and highly significant, correlations varying from $r = .23$ ($F = 8.87$; 1.66; $p < 0.005$) for the verbal factor to $r = .43$ ($F = 17.30$; 1.66; $p < 0.0001$) for Full Scale Score. When intelligence scores were adjusted for the influence of sibling performance, no difference among hormone treatments, dose level, sex, or their interactions for any measure was demonstrable.

Since treated and untreated groups differed markedly in their exposure to complications of pregnancy, and since complications of pregnancy might have an effect on later intelligence, a second analysis adjusted criterion scores for pre- and perinatal influences. Unweighted and weighted scores for pre- and perinatal complications were calculated for subjects within the untreated and treated groups. Because treatment might directly counterbalance the effect of complications within the treated group, a regression equation was calculated for complications and intelligence in the untreated group and applied to the treated groups. The procedure assumes that the best estimate for the effect of pre- and perinatal influences on intelligence is the effect of such influences unconfounded with hormonal influence.

The regression analysis of complications and intelligence in the untreated group revealed significant inverse linear relations between unweighted perinatal complications and full scale ($F = 5.32$; $p < 0.02$), performance ($F = 4.21$; $p < 0.04$), and arithmetic factor scores ($F = 3.99$; $p < 0.04$). However, with these exceptions, pre- and perinatal complications were unrelated to intelligence in either the untreated or treated groups. When intelligence scores were adjusted for complications, once again no significant differences were obtained among hormone treatments, dosage, sex, or their interactions.

In a third analysis the intelligence scores of the treated groups were correlatively adjusted for both the influence of complications and sibling intelligence. The addition of the second covariate, sibling intelligence, did not alter the results. The intelligence of males or females exposed to various hormone treatments at various dosage levels *in utero* appeared unaltered by treatment and largely independent of pre- and perinatal complications. The best predictor for the later intelligence of treated subjects was the intelligence of the subjects' untreated siblings (Table 3).

The negligible relationship between hormone therapy and intelligence (IQ) ought not to be interpreted as an indication that hormone therapy is unrelated to personality, behavior, or achievement. Analysis of the scores from the Cattell Personality Questionnaires, with the same rigorous controls employed in the evaluation of the effect of hormones on intelligence,

TABLE 3. *Full-scale intelligence: observed and predicted group means for two covariates*

Factors and factor levels	X̄ Observed	X̄ Predicted
Treatment		
Estrogen	122.2	121.7
Progestin	121.1	123.1
Mixed	122.3	121.4
Dose		
Low	121.1	121.8
High	122.6	122.4
Sex		
Male	122.9	121.7
Female	121.2	122.3

Predicted means were obtained by a regression analysis of sibling scores and prenatal influences. Since sibling scores alone were significantly related to scores in the treated group, the similarity of observed and predicted means reflect the similarity of scores for treated and untreated groups. Complete observed and predicted means for other measures of intelligence are reported elsewhere (147).

suggests significant, pervasive differences between treated and untreated subjects. For example, in the preliminary analysis undertaken thus far, the relationship of type of treatment and the Cattell Second Order Factor — Tough Poise (Cortertia) — is significant at the $p < 0.0026$ level ($F = 6.56$) when adjusted for sibling scores and prenatal complications. Tough Poise (Cortertia versus Pathemia) is defined by Cattell (120) as follows: "The ratings which go with cortertia are those of cheerfulness, alertness, and readiness to handle problems at the "dry" cognitive, objective level, whereas those at the pathemic pole, as the name indicates, operate at a mood level. . . . Low scoring (pathemic) individuals show a tendency to *feel* rather than *think*."

The results obtained thus far seem to be broadly compatible with the investigations of Ehrhardt and Money (98), Dalton (112), and the extensive animal literature reviewed earlier. Hormone therapy appears to potentiate general dispositional tendencies which relate to the global organization of behavior but only perhaps indirectly to performance on intellectual measures. A detailed analysis and description of the effects of type of treatment, dosage level, and sex of subject will be reported later (Reinisch, van den Daele, and Baker, *in preparation*).

Although the intellectual performance of treated and untreated siblings was subject to a series of statistical evaluations, the results were relatively unequivocal. The intelligence of children on a variety of scores was independent of prenatal hormone treatment, dosage, or sex of subject. This result must be qualified by the general population characteristics of subjects in this investigation. It is plausible that the use of different hormone compounds, dosage levels, or subjects from another population pool might alter results.

Subjects in this study, both treated and untreated siblings, performed unusually well on various measures of intelligence. This result may simply reflect a self-selection bias of patients who seek medical attention and can afford the costs associated with this demand. Socioeconomic status as measured by the Hollingshead rating (113) ranged from I (the highest) to IV, with 43 of the 56 families rated as I or II.

Whatever the ultimate reason for the general elevation of intelligence measures in treated and untreated subjects, the displacement of mean IQs to the superior range of the normal distribution as found in the Money and Ehrhardt patients (98) reduces subject variability and is sufficient to account for an attenuation of correlation between treated and untreated sibling scores. The expected correlation between siblings for full-scale IQ is approximately .50, and the observed correlation in our sample is .46. This attenuation is more marked in less reliable measures such as the verbal factor score.

The displacement of mean IQs to the superior range of the normal distribution may also attenuate treatment effects through a "ceiling effect" on the very superior. Subjects with average or below-average intelligence might plausibly demonstrate treatment effects, but such subjects appear to be unrepresentative of the population who have sought or obtained prenatal hormonal intervention.

Although the results of this study are in disagreement with the hypothesis that an excess of steroids with androgenic properties during appropriate stages of prenatal development have a positive effect on IQ scores later in life, there is perhaps an important difference in the samples used in the Ehrhardt and Money (98) study and this one. Nine of the 10 girls in the Johns Hopkins report were masculinized morphologically by prenatal hormone treatment. None of the subjects in the California sample were so affected. It is possible that the exogenously introduced hormones given to the mothers in both these investigations influenced the mental development only of children who were particularly sensitive to them. This fetal sensitivity might be related to the influence of the hormones on the genitalia. In other words, perhaps IQ is raised only in children whose genitalia are also affected by the hormones, and not in children who show no genital masculinization. The results of the British study (114), aside from its design problems, might be related to two factors. The first is that the hormones affect personality, which in turn relates to school achievement. As suggested earlier, preliminary evidence from the present investigation indicates that this may be the case. The second possible explanation is that treatment with progesterone administered to the pregnant women in the Dalton study (114) is fundamentally different in effect from treatment with synthetic progestins, and that progesterone does affect the CNS, resulting directly in higher mental abilities. However, the results of the present study indicate that synthetic progestins and estrogens administered during pregnancy are neither detrimental nor beneficial to later mental development as measured by standard IQ tests.

CURRENT CONTROVERSIES IN BEHAVIORAL ENDOCRINOLOGY

It seems appropriate to take note of three major controversies currently debated among researchers in behavioral endocrinology or psychoendocrinology. This area of investigation, although relatively new, is "based upon a solid foundation of experimental, clinical, and naturalistic evidence, possesses an impressive armamentarium of techniques, and is beginning to develop a theoretical structure of its own" (72). This theoretical structure, which is in the process of being constructed, is, as would be expected during an early stage of theory formation, confounded by a lack of agreement regarding the model, the mechanism, and the active agent involved in the development of sexually dimorphic behavior.

The model for understanding differentiation of sexually dimorphic behaviors has until recently been held by many researchers to be a unidimensional (Fig. 1A) or polar one (4,43): "for behavior, neural mechanisms normally destined to mediate feminine behavior are differentiated under the

A. UNIDIMENSIONAL

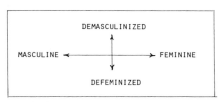

B. ORTHOGONAL

FIG. 1. Models of critical period hormone effects.

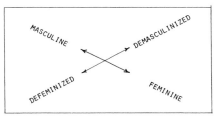

C. OBLIQUE

influence of androgens or other hormones to mediate masculine forms of activity. The differentiating process is viewed as a continuum, and degrees of maleness and femaleness, or degrees of incompleteness of both phenotypes, can be predicted by such a theoretical position. Clearly implied, however, is the corollary that an inverse relationship between the capacities to display masculine and feminine behaviors will normally exist in the adult. That is, feminine characteristics will always be suppressed to an extent which corresponds to the degree of augmentation of masculine characteristics" (42).

This solution proposes that masculinity and femininity lie at opposite ends of a continuum; and therefore as behavior moves from one pole, by definition it moves closer to the opposite end. In other words, less feminine equals more masculine. Whalen (121) recently suggested an orthogonal model (Fig. 1B) as a better explanation of differentiation: "The orthogonal model states that masculinity and femininity are not unitary processes, but reflect many behavioral dimensions that can be independent. The model further states that during development hormones defeminize without masculinizing and masculinize without defeminizing and that hormones can defeminize one behavioral system (e.g., mating) while masculinizing another system. . . . This means that masculinity and femininity need not be opposite ends of the same dimension, but rather can be orthogonal or independent dimensions of the organism's sexuality."

The studies Whalen uses to support this model come primarily from experimental hormone research with rodents. In all the investigations cited, hormones were introduced for a very short time during development, usually as little as 1 day, in a single dose. It is clear from the research carried out with hormones administered over a longer period of development — e.g., the studies of Ward (36) or of Sachs et al. (45) in which rats were given testosterone propionate during the pre- and neonatal periods — that the sexual behavior of the treated females more completely paralleled that of the male than when only one injection was given at any time perinatally. Perhaps masculinity and femininity exist on one dimension only when hormone influence or its absence is extant over the period of development "natural" to the organism's ontogeny. Any point along the continuum would represent a ratio of male/female or female/male behaviors. According to this theory the dimensions would become orthogonal or independent when treatment is nonphysiological or "artificial" to the animal's development in terms of dosage, duration, or timing. In order to encompass both the natural course of events as well as those induced by treatment, an oblique model (Fig. 1C) may be most parsimonious. The oblique model states that masculinization and defeminization, as well as feminization and demasculinization, are correlated, but that one is not necessarily determined, under all circumstances, by the other.

The neuroendocrine mechanism by which the brain mediates behavior is a second point of contention in the growing theory underlying behavioral endocrinology research. Do the presence of steroid hormones during a critical period in development "organize" the CNS to mediate male behavior, or is only the "responsiveness' of neural tissue altered? The position that steroid hormones organize the CNS so that masculine behavior patterns are potentiated later in development was initially proposed by the researchers at the Oregon Regional Primate Center (39,42–44,121); and by Harris (4). Harris expounded the position thusly: "The sex hormones seem to exert a double action on the central nervous system. First, during fetal or neonatal life these hormones seem to act in an inductive way on an undifferentiated brain (as they do on the undifferentiated genital tract), to organize it to a "male type" or "female type" of brain. And second, during adult existence the gonadal hormones act on the central nervous system in an excitatory or inhibitory way, and are thus concerned with the neural regulation of gonadotropic secretion and with the expression of overt patterns of sexual behavior." That is, "prenatal testosterone has an organizing action on the tissues that mediate mating behavior" (44).

The major proponent of the position that exposure to steroid hormones early in development alters the responsivity of the CNS to the hormones available during adulthood, and also the major opponent of the "organization" theory, is Beach (123). "The sex hormones are best regarded not as stimuli or as organizing agents but as chemical sensitizers which alter the stimulability of critical mechanisms within the central nervous system" (123). Beach's view suggests that every adult possesses the neural mechanisms for the expression of both male and female behavior during adulthood, and it is the presence or absence of hormones during the pre- or neonatal period which alters the level at which either system is activated by steroid hormones during later stages of development. Goy (42) does not disagree with the suggestion "that the mammalian nervous system is inherently bisexual;" however, he goes on to point out: "the proposition that early hormones modify thresholds for hormonal activation of behavior in adulthood and do not organize neural structures involved in the mediation of patterns of behavior seems to be only narrowly valid. If the actions of embryonic, foetal, or larval hormones are strictly limited to the role of conditioning thresholds for hormonal activation, then this theoretical position cannot account for psychologically dimorphic behaviors which do not require hormonal activation."

A prime example of nonhormonally activated sexually dimorphic nonmating behaviors was the differences reported in young male and female rhesus monkeys prior to endogenous production of the sex steroids from the gonads (125). These behaviors included rough-and-tumble play, chasing play, threat, and play initiation, all of which were recorded at higher frequencies for young male rhesus than for the females, and were increased in female pseudohermaphrodites whose mothers were treated with testosterone propionate during pregnancy (60).

The final point of controversy centers around the active agent responsible for either "organizing" or "altering the responsiveness" of the appropriate brain centers. The traditional view has implicated fetal testosterone of testicular origin as the hormone which differentiates not only the internal and external genitalia but the brain as well. This view is held by Goy (2), Harris (4), Jost (6,7,9), Phoenix et al. (29,43), and Young (125), to name a few.

However, it has been recognized for some time that exposure to estrogen (estradiol benzoate) neonatally results in alterations of hypothalamic function, sexual behavior, and sexually dimorphic nonmating behavior in mice (127) and rats (4,12,18,20,33,127–129), which resemble those produced by androgen. Recently a number of researchers have suggested that estrogen may be the active agent at the level of the CNS which actually mediates differentiation (130–132). The argument for estrogen as the differentiating hormone is based on the following evidence: (a) It has been shown that nonaromatizable androgens (those that cannot be converted into estrogen) like dihydrotestosterone did not masculinize females (treated neonatally) or elicit masculine mating behavior from adult castrated males (133–135). (b) Estrogen plus dihydrotestosterone was capable of eliciting male behavior as fully as testosterone propionate in castrated males (134,136). (c) Androstenedione (an aromatizable androgen) has been demonstrated to be convertable to estrone by the brain (137,138). Furthermore, Weisz and Gibbs (139) showed that testosterone was converted predominantly to estradiol and androstenedione to estrone in the hypothalamus of 7-day-old rats. Conversion was greater in the infant rat than in the adult. (d) It has been demonstrated that the neonatal rat hypothalamus and limbic system possessed an enzyme system capable of converting androgen to estrogen (132). (e) Rats treated neonatally with testosterone propionate and an estrogen antagonist (MER-25) did not exhibit the anovulatory syndrome typical of androgenized females (131).

One conclusion drawn from the above data is the hypothesis that "the androgen effect on brain differentiation is via estrogenic metabolites formed at the site of action in the brain" (132). Following the investigation of three aromatizable androgens and five which were not, McDonald and Doughty (135) concluded: "If exposure to oestrogen during the neonatal period is responsible for the suppression of female behavior, our results suggest that testosterone is the most effective means of supplying an oestrogenic precursor to the brain."

If estrogen is the active agent at the level of the CNS, it seems clear that it is not supplied in direct form from the periphery since there is ample evidence to suggest that estrogen introduced neonatally in the rat reduced adult male sexual behavior (140,141). This has also been demonstrated in the pig with prenatal treatment (142). In addition, recently Feder et al. (71) found that treatment of experienced adult male castrates with estradiol benzoate and dihydrotestosterone propionate not only elicited masculine mating behavior, but also that large numbers of the treated animals displayed lordosis; 50%, 86%, and 93% of the treated males displayed lordosis, depending on the amount of estrogen administered, in contrast to only 9% of castrated males treated with testosterone propionate and dihydrotestosterone propionate and none of the intact untreated males of the strain. "If male behavior could be activated by an appropriate estrogen treatment *without* concurrent activation of lordosis behavior, this would constitute a stronger argument that the effects of estrogen on male behavior were of physiological significance" (71). It has also been reported that adult castrated male rats treated with testosterone plus antiestrogen are capable of displaying masculine behavior (143), and that females treated with MER-25 (antiestrogen) displayed lower levels of lordosis, but simultaneously the antiestrogen had a facilitatory effect on mounting behavior (144).

Finally, Feder and Wade (145) pointed out that "perinatal hormones decrease responsiveness to estrogen in the adult. It would seem uneconomical for perinatal males to produce substances that would decrease responsiveness to a required hormone in adulthood."

The resolution of the three controversies awaits further research. Data utilized by each side come from studies characterized by the same inconsistencies described earlier. Consistency in the approach to behavioral endocrinology research by the various laboratories concerned may help to clarify and finally resolve the present disagreements concerning the model, mechanism, and agent of critical period steroid hormone effects on the development of later physiological function and behavior.

ACKNOWLEDGMENT

The author wishes to acknowledge Dr. Edward T. Tyler of the Tyler Clinic, Dr. William G. Karow of the Shelton Medical Clinic, and Dr. Paul Steinberg for generously providing the patient records; UCLA School of Medicine, Department of Psychiatry and Dr. Seymour Feshbach of the Department of Psychology for their generosity in providing technical and clerical support and office space; special thanks to Dr. Suzanne Brainard of the National Institute of Education, Dr. Terry Saario of the Ford Foundation, and Ms. Zelda Suplee of the Erickson Educational Foundation for their continued help and interest; Carolyn Kaufman and Helen Gordon for their assistance in patient identification and data collection; Dr. Leland van den Daele for statistical and computer consultation; the late Dr. Herbert G. Birch for his encouragement; and most especially the unstinting support and guidance of Dr. John Money without whom this study would never have been undertaken.

The original research reported here was supported by NIE grant NE–G–00–3–0106, the Ford Foundation, and the Erickson Educational Foundation.

This chapter won the Morton Prince Award of the American Psychopathological Association for the best presentation by a young investigator.

REFERENCES

1. Clark, H. M. (1935): Sex differences in change in potency of anterior hypophysis following bilateral castration in newborn rats. *Anat. Rec.*, 61:193–205.
2. Goy, R. W. (1970): Early hormonal influence on the development of sexual and sex-related behavior. In: *Neurosciences: A Study Program*, edited by G. C. Quarlon, T. Melnechuk, and F. O. Schmitt, pp. 196–207. Rockefeller University Press, New York.
3. Grady, K. L., Phoenis, C. H., and Young, W. C. (1965): Role of the developing rat testis in differentiation of the neural tissues mediating mating behavior. *J. Comp. Physiol. Psychol.*, 59:176–182.
4. Harris, G. W. (1964): Sex hormones, brain development and brain function. *Endocrinology*, 75:627–648.
5. Jost, A. (1953): Problems of fetal endocrinology: The gonadal and hypophyseal hormones. *Recent Prog. Horm. Res.*, 8:379–418.
6. Jost, A. (1972): A new look at the mechanisms controlling sex differentiation in mammals. *Johns Hopkins Med. J.*, 130:38–53.
7. Jost, A. (1973): Development and endocrinological aspects of sex differentiation. In: *Birth Defects*, edited by A. G. Motulsky and W. Lentz, pp. 142–147. Proceedings of the Fourth International Conference, Vienna, Austria.
8. Pfaff, D. W., and Zigmond, R. W. (1971): Neonatal androgen effects of sexual and nonsexual behavior of adult rats tested under various hormone regimes. *Neuroendocrinology*, 7:129–145.
9. Jost, A. (1973): Hormonal effects on fetal development: A survey. *Clin. Pharmacol. Ther.*, 14:714–720.
10. Siiteri, P., and Wilson, J. D. (1974): Testosterone formation and metabolism during male sexual differentiation in the human embryo. *J. Clin. Endocrinol. Metab.*, 38:113–125.
11. Burns, R. K. (1961): Role of hormones in the differentiation of sex. In: *Sex and Internal Secretions*, 3rd edition, edited by W. C. Young, pp. 76–158. Williams & Wilkins, Baltimore.
12. Pfeiffer, C. A. (1936): Sexual differences of the hypophysis and their determination by the gonads. *Am. J. Anat.*, 58:195–225.
13. Harris, G. W., and Jacobsohn, D. (1952): Functional grafts of the anterior pituitary gland. *Proc. R. Soc. Lond. [Biol.]*, 139:263–276.
14. Martinez, C., and Bittner, J. J. (1956): Non-hypohyseal sex differences in estrous behavior of mice bearing pituitary grafts. *Proc. Soc. Exp. Biol. Med.*, 91:506–509.
15. Segal, S. J., and Johnson, D. C. (1959): Absence of steroid hormones on the neural system: Ovulation controlling mechanisms. *Arch. Anat. (Suppl.)*, 48:261–273.
16. Barraclough, C. A. (1961): Production of anovulary, sterile rats by single injections of testosterone propionate. *Endocrinology*, 68:62–67.
17. Barraclough, C. A. (1966): Modification in the CNS regulation of reproduction after exposure of prepubertal rats to steroid hormones. *Recent Prog. Horm. Res.*, 22:503–539.
18. Gorski, R. A., and Wagner, J. W. (1965): Gonadal activity and sexual differentiation of the hypothalamus. *Endocrinology*, 76:226–239.
19. Schally, A. V., and Kastin, A. J. (1969): The present concept of the nature of hypothalamic hormones stimulating and inhibiting the release of pituitary hormones. *Triangle*, 9:19–25.

20. Gorski, R. A. (1973): Perinatal effects of sex steroids on brain development and function. In: *Progress in Brain Research,* edited by E. Zimmerman, W. H. Gispen, B. H. Marks, and D. de Wied, vol. 39, pp. 149–163. Elsevier, Amsterdam.
21. Moguilevsky, J. A., Libertun, C., Schiaffini, O., and Scacchi, P. (1969): Metabolic evidence of the sexual differentiation of hypothalamus. *Neuroendocrinology,* 4:264–269.
22. Barraclough, C. A., and Gorski, R. A. (1961): Evidence that the hypothalamus is responsible for androgen-induced sterility in the female rat. *Endocrinology,* 68:68–79.
23. Gorski, R. A., and Barraclough, C. A. (1963): Effects of low dosages of androgen on the differentiation of hypothalamic regulatory control of ovulation in the rat. *Endocrinology,* 73:210–216.
24. Clayton, R. B., Kogura, J., and Kraemer, H. C. (1970): Sexual differentiation of the brain: Effects of testosterone on brain RNA metabolism in new born female rats. *Nature (Lond.),* 226:810–812.
25. Harris, G. W., and Levine, S. (1965): Sexual differentiation of the brain and its experimental control. *J. Physiol. (Lond.),* 181:379–400.
26. Levine, S. (1966): Sex differences in the brain. *Sci. Am.,* 214:84–90.
27. Nadler, R. D. (1969): Differentiation of the capacity for male sexual behavior in the rat. *Horm. Behav.,* 1:53–56.
28. Neumann, F., Elger, W., and Steinbeck, H. (1970): Antiandrogens and reproductive development. *Philos. Trans. R. Soc. Lond. [Biol. Sci.],* 259:179–184.
29. Phoenix, C. H., Goy, R. W., and Resko, J. A. (1968): Psychosexual differentiation as a function of androgenic stimulation. In: *Perspectives in Reproduction and Sexual Behavior,* edited by M. Diamond, pp. 33–49. Indiana University Press, Bloomington.
30. Ward, I. L. (1972): Female sexual behavior in male rats treated prenatally with an antiandrogen. *Physiol. Behav.,* 8:53–56.
31. Etzel, V., Schenck, B., and Neumann, F. (1974): Influence of cyproterone acetate during pregnancy on the sexual behavior of male guinea-pigs. *J. Reprod. Fertil.,* 37:315–321.
32. Barraclough, C. A., and Gorski, R. A. (1962): Studies on mating behavior in the androgen-sterilized female rat in relation to hypothalamic regulation of sexual behavior. *J. Endocrinol.,* 25:175–182.
33. Dörner, G. (1968): Hormonal induction and prevention of female homosexuality. *J. Endocrinol.,* 42:163–164.
34. Dörner, G., and Staudt, J. (1969): Perinatal structural sex differentiation of the hypothalamus in rats. *Neuroendocrinology,* 5:103–106.
35. Södersten, P. (1973): Increased mounting behavior in the female rat following a single neonatal injection of testosterone propionate. *Horm. Behav.,* 4:1–17.
36. Ward, I. L. (1969): Differential effect of pre- and postnatal androgen on the sexual behavior of intact and spayed female rats. *Horm. Behav.,* 1:25–36.
37. Ward, I. L., and Renz, F. J. (1972): Consequences of perinatal hormone manipulation on the adult sexual behavior of female rats. *J. Comp. Physiol. Psychol.,* 78:349–355.
38. Whalen, R. E., and Rezek, D. L. (1974): Inhibition of lordosis in female rats by subcutaneous implants of testosterone, androstenedione, or dihydrotestosterone in infancy. *Horm. Behav.,* 5:125–128.
39. Young, W. C., Goy, R. W., and Phoenix, C. H. (1964): Hormones and sexual behavior. *Science,* 143:212–218.
40. Manning, A., and McGill, T. E. (1974): Neonatal androgen and sexual behavior in female house mice. *Horm. Behav.,* 5:19–31.
41. Swanson, H. H., and Crossley, D. A. (1971): Sexual behavior in the golden hamster and its modification by neonatal administration of testosterone propionate. In: *Hormones in Behavior,* edited by M. Hamburgh and E. J. Barrington, pp. 677–687. Appleton-Century-Crofts, New York.
42. Goy, R. W. (1970): Experimental control of psychosexuality. *Philos. Trans. R. Soc. Lond. [Biol. Sci.],* 259:149–162.
43. Phoenix, C. H., Goy, R. W., Gerall, A. A., and Young, W. C. (1959): Organizing action of prenatally administered testosterone propionate on the tissues mediating mating behavior in the female guinea pig. *Endocrinology,* 65:369–372.
44. Phoenix, C. H. (1974): Prenatal testosterone in the nonhuman primate and its consequences for behavior. In: *Sex Differences in Behavior,* edited by R. C. Friedman, R. M. Richart, and R. L. Van de Wiele, pp. 19–32. Wiley, New York.

45. Sachs, B. D., Pollak, E. I., Schoelch Kreiger, M., and Barfield, R. J. (1973): Sexual behavior: Normal male patterning in androgenized female rats. *Science*, 181:770–772.
46. Meyerson, B. J., and Lindström, L. (1973): Sexual motivation in the neonatally androgen-treated female rat. *Horm. Brain Function, (Budapest)*, 443–448.
47. Denef, C. (1973): Differentiation of steroid metabolism in the rat and mechanisms of neonatal androgen action. *Enzyme*, 15:254–271.
48. Levine, S. (1970): The influence of hormones in infancy on central nervous system organization. In: *Les Hormones et la Comportement: Problems Actuels d'Endocrinology et de Nutrition*, series 14, edited by H. P. Klotz, pp. 29–43. Expansion Scientifique, Paris.
49. Beanninger, R. (1974): Effects of day 1 castration on aggressive behaviors of rats. *Bull. Psychon. Soc.*, 3:189–190.
50. Edwards, D. A. (1969): Early androgen stimulation and aggressive behavior in male and female mice. *Physiol. Behav.*, 4:333–338.
51. Edwards, D. A. (1970): Post-natal androgenization and adult aggressive behavior in female mice. *Physiol. Behav.*, 5:465–467.
52. Vale, J. R., Ray, D., and Vale, C. A. (1972): The interaction of genotype and exogenous neonatal androgen: Agonistic behavior in female mice. *Behav. Biol.*, 7:321–334.
53. Vale, J. R., Ray, D., and Vale, C. A. (1973): The interaction of genotype and exogenous neonatal androgen and estrogen: Sex behavior in female mice. *Dev. Psychobiol.*, 6:319–327.
54. Blizard, D., and Denef, C. (1973): Neonatal androgen effects on open field activity and sexual behavior in the female rat: The modifying influence of ovarian secretions during development. *Physiol. Behav.*, 11:65–69.
55. Scouten, C. W., Grotelueschen, L. K., and Beatty, W. W. (1975): Androgens and the organization of sex differences in active avoidance behavior in the rat. *J. Comp. Physiol. Psychol.*, 88:264–270.
56. Swanson, H. H. (1967): Alteration of sex-typical behavior of hamsters in open field and emergence tests by neo-natal administration of androgen or oestrogen. *Anim. Behav.*, 15:209–216.
57. Quadagno, D. M., Shryne, J., Anderson, C., and Gorski, R. A. (1972): Influence of gonadal hormones on social, sexual, emergence, and open field behaviour in the rat (Rattus norvegicus). *Anim. Behav.*, 20:732–740.
58. Branchey, L., Branchey, M., and Nadler, R. D. (1973): Effects of sex hormones on sleep patterns of male rats gonadectomized in adulthood and in the neonatal period. *Physiol. Behav.*, 11:609–611.
59. Křeček, J. (1973): Sex differences in salt taste: The effect of testosterone. *Physiol. Behav.*, 10:683–688.
60. Goy, R. W. (1968): Organizing effects on androgen on the behavior of rhesus monkeys. In: *Endocrinology and Human Behavior*, edited by R. P. Michael, pp. 12–31. Oxford University Press, London.
61. Bridges, R. S., Zarrow, M. X., Goldman, B. D., and Denenberg, V. H. (1974): A developmental study of maternal responsiveness in the rat. *Physiol. Behav.*, 12:149–151.
62. Bridges, R. S., Zarrow, M. X., and Denenberg, V. H. (1973): The role of neonatal androgen in the expression of hormonally induced maternal responsiveness in the adult rat. *Horm. Behav.*, 4:315–322.
63. Lisk, R. D., Russell, J. A., Kahler, S. G., and Hanks, J. B. (1973): Regulation of hormonally mediated maternal nest structure in the mouse (Mus musculus) as a function of neonatal hormone manipulation. *Anim. Behav.*, 21:296–301.
64. Quadagno, D. M., and Rockwell, J. (1972): The effect of gonadal hormones in infancy on maternal behavior in the adult rat. *Horm. Behav.*, 3:55–62.
65. Quadagno, D. M., McCullough, J., Kan-hwa Ho, G., and Spevak, A. M. (1973): Neonatal gonadal hormones: Effect on maternal and sexual behavior in the female rat. *Physiol. Behav.*, 11:251–254.
66. Quadagno, D. M., Debold, J. F., Gorzalka, B. B., and Whalen, R. E. (1974): Maternal behavior in the rat: Aspects of concaveation and neonatal androgen treatment. *Physiol. Behav.*, 12:1071–1074.

67. Reinisch, J. M. (1974): Fetal hormones, the brain, and human sex differences: A heuristic integrative review of the recent literature. *Arch. Sex. Behav.,* 3:51–90.
68. Whalen, R. E., Peck, C. K., and LoPiccolo, J. (1966): Virilization of female rats by prenatally administered progestin. *Endocrinology,* 78:965–970.
69. Goy, R. W., and Goldfoot, D. A. (1974): Experiential and hormonal factors influencing development of sexual behavior in the male rhesus monkey. In: *The Neurosciences: Third Study Program,* edited by F. O. Schmitt and F. G. Worden, pp. 571–581. MIT Press, Cambridge, Mass.
70. Hendricks, S. E., and Duffy, J. A. (1974): Ovarian influences on the development of sexual behavior in neonatally androgenized rats. *Dev. Psychobiol.,* 7:297–303.
71. Feder, H. H., Naftolin, F., and Ryan, K. J. (1974): Male and female sexual responses in male rats given estradiol benzoate and 5α-androstan-17β-ol-3-one propionate. *Endocrinology,* 94:136–141.
72. Beach, F. A. (1974): Behavioral endocrinology and the study of reproduction. *Biol. Reprod.,* 10:2–18.
73. Gruendel, A. D., and Arnold, W. J. (1974): Influence of preadolescent experiential factors on the development of sexual behavior in albino rats. *J. Comp. Physiol. Psychol.,* 86:172–178.
74. Moguilevsky, J. A. (1966): Effect of testosterone in vitro on the oxygen uptake of different hypothalamic areas. *Acta Physiol. Lat. Am.,* 16:353–56.
75. Moguilevsky, J. A., and Malinow, R. W. (1964): Endogenous oxygen uptake of the hypothalamus in female rats. *Am. J. Physiol.,* 206:855–857.
76. Moguilevsky, J. A., and Rubinstein, L. (1967): Glycolytic and oxidative metabolism of hypothalamic areas in prepubertal and androgenized rats. *Neuroendocrinology,* 2:213–221.
77. Moguilevsky, J. A., Timiras, P. S., and Geel, S. (1964): Oxygen consumption and carbon dioxide production in the hypothalamus and cerebral cortex. Influence of sexual maturation in male rats. *Acta Physiol. Lat. Am.,* 14:207–210.
78. Moguilevsky, J. A., Schiaffini, O., and Foglia, V. (1966): Effect of castration on the oxygen uptake of different parts of hypothalamus. *Life Sci.,* 5:447–452.
79. Moguilevsky, J. A., Libertun, C., Schiaffini, O., and Szwarcfarb, B. (1968): Sexual differences in hypothalamic metabolism. *Neuroendocrinology,* 3:193–199.
80. Scacchi, P., Moguilevsky, J. A., Libertun, C., and Christot, J. (1970): Sexual differences in protein content of the hypothalamus in rats. *Proc. Soc. Exp. Biol. Med.,* 133:845–848.
81. Ladosky, W., and Gaziri, L. C. J. (1970): Brain serotonin and sexual differentiation of the nervous system. *Neuroendocrinology,* 6:168–174.
82. Dörner, G., and Staudt, J. (1968): Structural changes in the preoptic anterior hypothalamic area of the male rat following neonatal castration and androgen substitution. *Neuroendocrinology,* 3:136–140.
83. Dörner, G., and Staudt, J. (1969): Structural changes in the hypothalamic ventromedial nucleus of the male rat following neonatal castration and androgen treatment. *Neuroendocrinology,* 4:278–281.
84. Staudt, V. J., Dörner, G., Döll, R. and Blöse, J. (1973): Geschlechtsspezifische morphologische unterschiede im nucleus arcuatus und prämamillaris ventralis der ratte. *Endokrinologie,* 62:234–236.
85. Raisman, G. (1974): Evidence for a sex difference in the neuropil of the rat preoptic area and its importance for the study of sexually dimorphic function. *Aggression,* 52:42–51.
86. Raisman, G., and Field, P. M. (1971): Sexual dimorphism in the preoptic area of the rat. *Science,* 173:731–733.
87. Raisman, G., and Field, P. M. (1973): Sexual dimorphism is the neuropil of the preoptic area of the rat and its dependence on neonatal androgen. *Brain Res.,* 54:1–29.
88. Litteria, M., and Thorner, M. W. (1974): Inhibition in the incorporation of (^{3}H) lysine in the Purkinje cells of the adult female rat after neonatal androgenization. *Brain Res.,* 69:170–173.
89. Phillips, A. G., and Deol, G. (1973): Neonatal gonadal hormone manipulation and emotionality following septal lesions in weanling rats. *Brain Res.,* 60:55–64.

90. Resko, J. A. (1974): The relationship between fetal hormones and the differentiation of the central nervous system in primates. In: *Reproductive Behavior,* edited by W. Montagna and W. A. Sadler, pp. 211–222. Plenum Press, New York.
91. Resko, J. A., Malley, A., Begley, D., and Hess, D. L. (1973): Radioimmunoassay of testosterone during fetal development of the rhesus monkey. *Endocrinology,* 93:156–161.
92. Bloch, E. (1964): Metabolism of 4-14C-progesterone by human fetal testes and ovaries. *Endocrinology,* 74:833–845.
93. Taylor, T., Coutts, J. R. T., and MacNaughton, M. C. (1974): Human foetal synthesis of testosterone from perfused progesterone. *J. Endocrinol.,* 60:321–326.
94. Abramovich, R. D. (1974): Human sexual differentiation—in utero influences. *J. Obstet. Gynecol.,* 81:448–453.
95. Diez D'Aux, R. C., and Murphy, B. E. P. (1974): Androgens in the human fetus. *J. Steroid Biochem.,* 5:207–210.
96. Reyes, F. I., Boroditsky, R. S., Winter, J. S. D., and Faiman, C. (1974): Studies on human sexual development. II. Fetal and maternal serum gonadotropin and sex steroid concentrations. *J. Clin. Endocrinol. Metab.,* 38:612–617.
97. Wilkins, L. (1960): Masculinization of female fetus due to use of orally given progestins. *J.A.M.A.,* 172:1028–1032.
98. Ehrhardt, A. A., and Money, J. (1967): Progestin-induced hermaphroditism: I.Q. and psychosexual identity in a study of ten girls. *J. Sex Res.,* 3:83–100.
99. Money, J. (1972): Clinical aspects of prenatal steroidal action on sexually dimorphic behavior. In: *Steroid Hormones and Brain Function,* edited by C. Sawyer and R. A. Gorski, pp. 325–338. University of California Press, Berkeley.
100. Ehrhardt, A. A., Epstein, R., and Money, J. (1968): Fetal androgens and female gender identity in the early-treated adrenogenital syndrome. *Johns Hopkins Med. J.,* 122:160–167.
101. Ehrhardt, A. A., Evers, K., and Money, J. (1968): Influence of androgen and some aspects of sexually dimorphic behavior in women with the late-treated adrenogenital syndrome. *Johns Hopkins Med. J.,* 123:115–122.
102. Ehrhardt, A. A., Greenberg, N., and Money, J. (1970): Female gender identity and absence of fetal gonadal hormones: Turner's syndrome. *Johns Hopkins Med. J.,* 126:237–248.
103. Ehrhardt, A. A., and Baker, S. W. (1974): Fetal androgens, human central nervous system differentiation, and behavior sex differences. In: *Sex Differences in Behavior,* edited by R. C. Friedman, R. M. Richart, and R. L. Van de Wiele, pp. 33–51. Wiley, New York.
104. Money, J. (1965): Influence of hormones on sexual behavior. *Ann. Rev. Med.,* 16:67–82.
105. Money, J., and Meredith, T. (1967): Elevated verbal IQ and idiopathic precocious sexual maturation. *Pediatr. Res.,* 1:59–65.
106. Walker, P. A., and Money, J. (1972): Prenatal androgenization of females. *Hormones,* 3:119–128.
107. Ehrhardt, A. A. (1973): Maternalism in fetal hormonal and related syndromes. In: *Contemporary Sexual Behavior: Critical Issues in the 1970s,* edited by J. Zubin and J. Money, pp. 99–105. Johns Hopkins University Press, Baltimore.
108. Money, J., Ehrhardt, A. A., and Masica, D. N. (1968): Fetal feminization induced by androgen insensitivity in the testicular feminization syndrome: Effect on marriage and maternalism. *Johns Hopkins Med. J.,* 123:105–114.
109. Masica, D. N., Money, J., Ehrhardt, A. A., and Lewis, V. G. (1969): I.Q., fetal sex hormones and cognitive patterns: Studies in the testicular feminizing syndrome of androgen insensitivity. *Johns Hopkins Med. J.,* 124:34–43.
110. Money, J., and Lewis, V. (1966): I.Q., genetic and accelerated growth: Adrenogenital syndrome. *Bull. Hopkins Hosp.,* 118:365–373.
111. Perlman, S. M. (1971): Cognitive function in children with hormone abnormalities. Doctoral dissertation, Northwestern University.
112. Baker, S. W., and Ehrhardt, A. A. (1974): Prenatal androgen, intelligence and cognitive sex differences. In: *Sex Differences in Behavior,* edited by R. C. Friedman, R. N. Richart, and R. L. Van de Wiele, pp. 53–84. Wiley, New York.

113. Hollingshead, A. B. (1957): *Two Factor Index of Social Position.* Yale University, New Haven.
114. Dalton, K. (1968): Ante-natal progesterone and intelligence. *Br. J. Psychiatry,* 114:1377–1383.
115. Campbell, D. T., and Stanley, J. C. (1963): *Experimental and Quasi-Experimental Designs for Research.* Rand McNally, Chicago.
116. Barker, D. J. P., and Edwards, J. H. (1967): Obstetric complications and school performance. *Br. Med. J.,* 2:695–699.
117. Reinisch, J. M., Karow, W. G., Tyler, E. T., and van den Daele, L. (1975): A retrospective study of the effects of hormonal therapy during pregnancy on the intelligence of the offspring: Rationale, sample description, and preliminary results. Presented at The American Fertility Society, Los Angeles, April 3, 1975.
118. Cohen, J. (1957): A factor-analytically based rationale for the Wechsler Adult Intelligence scale. *J. Consult. Psychol.,* 21:451–457.
119. Cohen, J. (1959): The factorial structure of the WISC at ages 7–6, 10–6 and 13–6. *J. Consult. Psychol.,* 23:285–299.
120. Cattell, R. B., Eber, H. W., and Tatsuoka, M. M. (1970): *Handbook for the Sixteen Personality Factor Questionnaire(16PF).* Institute for Personality and Ability Testing, Champaign, Illinois.
121. Whalen, R. E. (1974): Sexual differentiation: Models, methods, and mechanisms. In: *Sex Differences in Behavior,* edited by R. C. Friedman, R. M. Richart, and R. L. Van de Wiele, pp. 467–481. Wiley, New York.
122. Gerall, A. A. (1966): Hormonal factors influencing masculine behavior of female guinea pigs. *Comp. Physiol. Psychol.,* 62:365–369.
123. Beach, F. A. (1971): Hormonal factors controlling the differentiation, development, and display of copulatory behavior in the ramstergig and related species. In: *Biopsychology of Development,* edited by E. Tobach, L. R. Aronson, and F. Shaw, pp. 249–296. Academic Press, New York.
124. Beach, F. A. (1945): Bisexual mating behavior in the male rat: Effects of castration and hormone administration. *Physiol. Zool.,* 18:390–402.
125. Rosenblum, L. (1961): The development of social behavior in the rhesus monkey. Doctoral dissertation, University of Wisconsin Libraries, Madison.
126. Young, W. C. (1967): Prenatal gonadal hormones and behavior in the adult. *Proc. Am. Psychopathol. Assoc.,* 55:173–183.
127. Edwards, D. A., and Herndon, J. (1970): Neonatal estrogen stimulation and aggressive behavior in female mice. *Physiol. Behav.,* 5:993–995.
128. Whalen, R. E. (1968): Differentiation of the neural mechanisms which control gonadotropin secretion and sexual behavior. In: *Perspectives in Reproduction and Sexual Behavior,* edited by M. Diamond, pp. 303–340. Indiana University Press, Bloomington.
129. Whalen, R. E., and Nadler, R. D. (1963): Suppression of the development of female mating behavior by estrogen administered in infancy. *Science,* 141:273–274.
130. Brown-Grant, K. (1973): Recent studies on the sexual differentiation of the brain. In: *Foetal and Neonatal Physiology,* edited by K. S. Comline et al., pp. 527–545. Cambridge University Press, Cambridge.
131. McDonald, P. G., and Doughty, C. (1973/1974): Androgen sterilization in the neonatal female rat and its inhibition by an estrogen antagonist. *Neuroendocrinology,* 13:182–188.
132. Reddy, V. V. R., Naftolin, F., and Ryan, K. J. (1974): Conversion of androstenedione to estrone by neural tissues from fetal and neonatal rats. *Endocrinology,* 94:117–121.
133. Arai, V. (1972): Effect of 5α-dihydrotestosterone on differentiation of masculine pattern of the brain in the rat. *Endocrinol. Jap.,* 19:389–393.
134. Larsson, K., Södersten, P., and Beyer, C. (1973): Sexual behavior in male rats treated with estrogen in combination with dihydrotestosterone. *Horm. Behav.,* 4:289–299.
135. McDonald, P. G., and Doughty, C. (1974): Effect of neonatal administration of different androgens in the female rat: Correlation between aromatization and the induction of sterilization. *J. Endocrinol.,* 61:95–103.
136. Baum, M. J., and Vreeburg, J. T. M. (1973): Copulation in castrated male rats following combined treatment with estradiol and dihydrotestosterone, *Science,* 182:283–285.

137. Naftolin, F., Ryan, K. J., and Petro, Z. (1972): Aromatization of androstenedione by the anterior hypothalamus of adult male and female rats. *Endocrinology*, 90:295–298.
138. Reddy, V. V. R., Naftolin, F., and Ryan, K. J. (1973): Aromatization in the central nervous system of rabbits: Effects of castration and hormone treatment. *Endocrinology*, 92:589–594.
139. Weisz, J., and Gibbs, C. (1974): Conversion of testosterone and androstenedione to estrogens in vitro by the brain of female rats. *Endocrinology*, 94:616–620.
140. Diamond, M., Llacuna, A., and Wond, C. L. (1973): Sex behavior after neonatal progesterone, testosterone, estrogen, or antiandrogens. *Horm. Behav.*, 4:73–88.
141. Gray, J. A., Levine, S., and Broadhurst, P. L. (1965): Gonadal hormone injections in infancy and adult emotional behavior. *Anim. Behav.*, 13:33–45.
142. Schlenker, G., and Hinz, G. (1974): Effect of prenatal treatment with testosterone oenanthate or oestradiol benzoate on gonadal function, sexual behaviour and growth of pigs. *Arch. Exp. Veterinaermed.*, 28:117–133 (English summary).
143. Whalen, R. E., Battie, C., and Luttge, W. G. (1972): Antiestrogen inhibition of androgen induced sexual receptivity in rats. *Behav. Biol.*, 7:311–320.
144. Södersten, P. (1974): Effects of an estrogen antagonist, MER-25, on mounting behavior and lordosis behavior in the female rat. *Horm. Behav.*, 5:111–121.
145. Feder, H. H., and Wade, G. N. (1974): Integrative actions of perinatal hormones on neural tissues mediating adult sexual behavior. In: *The Neurosciences: Third Study Program*, edited by F. O. Schmitt and F. G. Worden, pp. 583–586. MIT Press, Cambridge, Mass.
146. Ward, I. L. (1974): Sexual behavior differentiation: Prenatal hormonal and environmental control. In: *Sex Differences in Behavior*, edited by R. C. Friedman, R. M. Richart, and R. L. Van de Wiele, pp. 3–17. Wiley, New York.
147. Reinisch, J. Machover (1975): The effects on IQ and personality in humans of prenatal treatment with synthetic progestin and estrogen, administered for the maintenance of at-risk pregnancy. Unpublished doctoral dissertation, Columbia University.

Hormones, Behavior, and Psychopathology, edited by
Edward J. Sachar. Raven Press, New York © 1976.

Psychological Correlates of the Menstrual Cycle and Oral Contraceptive Medication

Judith M. Bardwick

Department of Psychology, University of Michigan, Ann Arbor, Michigan 48104

When I began to research relationships between reproductive physiology and psychological correlates in women in 1962, it was considered a bizarre topic — clearly out of the mainstream of legitimate establishment psychology — and only the benign sufferance of my department allowed me to persist in my obvious eccentricity. Why was it important enough that I was willing to risk professional legitimacy? I had spent the preceding 2 years reviewing the psychological and psychiatric literature about women, and at that time the tenets of classic psychoanalysis dominated the psychology of women. It seemed to me lunatic when I read that the fundamental feminine motives derived from the absence of male genitals.

To derive a psychology of women, it seemed necessary to explore normal women's relationships with their own reproductive system. This obvious and simple idea was, in fact, revolutionary. Speaking of revolutionary, I had expected attacks from conservative clinicians, but they did not materialize. I was, instead, attacked as a counterrevolutionary by radical women's liberationists. Radicals seem generally opposed to the idea of physiological determinism and in this case were angered at the idea that physiological differences between the sexes might be significant in accounting for psychological and social differences between women and men. I mention this for a couple of reasons: first to remind everyone that the field of the psychology of women is new,[1] dating back roughly a decade; and second, that this is an area born out of political consciousness where all research has political relevance.

American psychology and psychiatry have paid lip service to the idea that physiological processes are significant variables, but in fact physiology has

---

[1] On the subject of political awareness, I might mention that in this chapter I am restricted to a discussion of the menstrual cycle and effects of oral contraceptives. Obviously this means that only women are discussed. It is true that measures of cyclical changes and endocrine effects are studied far more frequently in reference to women than to men. This is largely because the menstrual cycle is a dramatic and normal event, with men having no comparable obvious cycle, and because the menstrual cycle makes it easy to take measures; this is much more difficult in men. This has the unfortunate effect of giving the naive reader the unacknowledged impression that women are the only ones directly affected by their gonadal steroids. The reader is referred to a paper in which men as well as women are discussed (16).

never been integrated into theories of human behavior. This seems partially due to our uncertainty regarding complex and interactive intrapsychic and interpersonal variables. Thus it is easy to maintain the point of view that if one could understand those variables well enough, we would have a coherent and comprehensive understanding of human motivation and behavior. In addition, physiological research has tended to deal with microvariables that were too specific and too far removed from behavior to seem relevant. I also have the impression that even professionals simplify physiological parameters, attributing to them an absolute determinism that jars badly with our ideal of unlimited progress and change. Still, it seems unlikely that the neck serves as a barrier such that body processes do not affect cerebral processes or that what occurs in the head does not influence what happens in the body. We must attend to physiological processes because we need more valid psychological theory, because physiological variables may be significant in clinical practice, and because physiological differences are likely to be among the significant variables accounting for sex differences.

In this chapter I discuss the gonadal steroids and correlative studies of changes in affect and behavior. What we would like to be able to do is understand steroid effects in the central nervous system. New data suggest that the gonadal steroids influence some parts of the central nervous system by changing transmitter capacities at the synapse. The data I present seem simple, but this is an illusion. In physiological terms we are dealing with a simplistic model of the steroids, synapses, emotions, and behaviors. At the psychological level it is imperative to remember that behavior has many origins. It is easier to discern the effect of physiological states on psychological processes, complicated as these effects are, because we can contain the universe and focus on these variables. Physiological processes, however, are only one of the inputs into behavior; it is much more difficult to contain the universe of significant inputs when attempting to explain behavior.

MENSTRUAL CYCLE DATA

The first study correlating menstrual cycle changes with moods was published in 1942 by Benedek and Rubenstein (2). They did an intensive study of 15 patients who were in psychoanalytical treatment, correlating the therapeutic material with independent assays of menstrual cycle phases. They reported finding women happy, alert, and other- or outward-directed during the first half of the cycle as estrogen levels increased. At ovulation they found a peak in ego integration, the psychological peak correlating with the estrogen peak. After ovulation, in conjunction with the rise in progesterone levels, their patients became more passive, narcissistic, and inward-directed. At premenstruation, when the levels of estrogen and progesterone are both low, they found the women aggressive, tense, and anxious. Since these were studies of a patient population, the question arises as to whether one would find the same mood changes in a nonclinical population. Further studies, some clinical (3) and some based on questionnaire data (1,4), supported the hypothesis that in normal women there are mood shifts in conjunction with (as a result of?) the endocrine changes during the menstrual cycle.

Before discussing our research, it is necessary to make a methodological comment. Most of the studies have asked women to remember their psychological and physical responses during the cycle or after using an oral contraceptive. Memory, as we know, is fallible. Thus measurement and memory confound these data; and, overall, retrospective studies report a greater number of symptoms than do either predictive studies or behavioral measurements. We do not know the extent to which stereotypic beliefs—expectations about what women are supposed to experience during the cycle—affect their self-report regarding symptoms or changes during the cycle (5,6). The most common method of measurement—the retrospective questionnaire—seems to be the worst, although in clinical studies there are problems of expectation and subjective reporting. In addition to studies that report significant correlations, there are reports of no significant changes during the cycle (7). The dependent variable is crucial. It appears that affects are especially sensitive to steroid changes, but cognitive processes seem to be affected much less.

When a standardized personality instrument is used, it is usually designed to measure stable or trait qualities—you are trying to measure changes in state. This means that the investigator is attempting to pick up small and subtle changes—small and subtle enough so there is dispute about their existence—with an instrument designed to be insensitive to transient change. Most researchers have asked their subjects how they felt; this is obviously a self-aware response. Measures yield different results if they tap conscious responses or preconscious processes. We observed the cyclicity of the mood changes most clearly where the response was not self-aware. I think this is because people perceive themselves as more consistent or stable than they are. If, for example, a woman is premenstrual and has an argument, she is far more likely to believe that she had an argument for a real reason, e.g., someone was rude to her. One normally perceives oneself as happy, sad, anxious, or angry because of something specific, and this encourages perceiving the event rather than the precipitating mood. With one exception (8) studies in this area have reported group means, have looked for normative patterns, have not provided any idea of the range or variability of response within the group, and have not reported the percentages within the sample who were severely affected or not affected at all. We do not yet know what the range of normal responses are or the percentages of those insignificantly or severely affected by steroid changes; nor do we know what behaviors would normally change and the percentage for whom behavioral change would be significant.

My students helped conduct the studies. In the first study (9) we asked a group of 26 normal coeds to "tell us about some experience you had." These spontaneous 5-min responses were tape-recorded, transcribed, and scored using Gottschalk and Gleser's Verbal Anxiety Scale (10) for death, mutilation, guilt, shame, and diffuse anxiety. In the next two studies, we also scored for hostility. It is important to note that the only instructions to the subject are to tell about an event. The choice of the event, the emotional tone of the narrative, and the specific words are generated solely by the subject. This scoring technique is literal and not symbolic; this means that when we scored, for example, for mutilation anxiety, the woman had described a body injury (i.e., "I touched the finger and it felt like a nail was going through my hand").

We studied these women during two menstrual cycles, seeing each woman twice at premenstruation and twice at ovulation. The premenstrual anxiety scores were very significantly higher than the ovulation anxiety scores. These data were similar to the observations of Benedek and Rubenstein. We found, for instance, that at ovulation there was a common theme of being able to cope or succeed and feelings of being satisfied with oneself:

> . . . so I was elected chairman. I had to establish with them the fact that I knew what I was doing. I remember one particularly problematic meeting, and afterward, L. came up to me and said, "you really handled the meeting well." In the end it came out the sort of thing that really bolstered my confidence in myself.

Two weeks later, at premenstruation, the same girl spontaneously reported this event:

> They had to teach me how to water-ski. I was so clumsy it was really embarrassing, 'cause it was kind of like saying to yourself you can't do it and the people were about to lose patience with me.

Themes of death and of mutilation were evident at premenstruation:

> I'll tell you about the death of my poor dog. . . . Oh, another memorable event — my grandparents died in a plane crash. That was my first contact with death, and it was very traumatic for me. . . . Then my other grandfather died.

Two weeks later at ovulation, the same young woman told about a different event in her life:

> Well, we just went to Jamaica and it was fantastic. The island is so lush and green and the water is so blue. . . . The place is so fertile and the natives are just so friendly.

This first study strongly supported the hypothesis that the endocrine cycle significantly affects moods — anxiety, hostility, depression, and self-esteem — and if this hypothesis is correct, then there ought to be a change in this affect cycle when women are on the combination pill because the endocrine pattern has been altered. My student Karen Paige (11) replicated the first study, again using the Gottschalk-Gleser Scales, with 38 women not on the pill, 52 on the combination pill, and 12 using a sequential contraceptive. In this study we again found that the anxiety and hostility levels of those not using an oral contraceptive varied significantly during the menstrual cycle with the highest scores on these variables at premenstruation. Women on sequential pills showed the same pattern of affect fluctuations as those who were not on an oral contraceptive. However, women on the combination contraceptive, whose endocrine levels were not only higher than normal but also far more stable than normal, had no significant shift of affects during the four phases of the menstrual cycle that we measured. The levels of anxiety and hostility of women on the combination oral contraceptive remained flat during the month, and the levels were as high as those in women at menstruation.

The first two studies were measures of psychological parameters. In the third study, that by Merilee Oakes (12), we also measured behavior. We used a game called "prisoner's dilemma." In this game the partners can either compete or cooperate with each other. Our subjects unknowingly played against a computer. We were looking for mood changes in women on and off the pill, as we had done before, but now we were also looking for changes in assertiveness, competitiveness, or cooperation as women played the game. All of the women on the oral contraceptive in this study were using a combination pill. We found that in addition to affect changes there were behavioral changes. Again, women on the combination pill showed no significant affect cycle change and they also had a flat pattern in their game playing, suggesting that they played the same way at both midcycle and premenstruation. Women on the pill played less competitively than those not on it. At premenstruation those not on the oral contraceptive played in the same way as women on the pill at both cycle phases. At ovulation women who were not on the pill played significantly more competitively than they did at premenstruation. This competitive game playing did not seem hostile but, instead, seemed to reflect feelings of self-esteem and assertiveness. This peak of self-esteem at ovulation has been a generally consistent finding; there seems to be a heightening of self-esteem in conjunction with a peak in estrogen and no progesterone. Oakes (12) found that personality measures were in accord with the affect measure and the game-playing behavior. At ovulation women not on the pill played competitively and scored high on personality measures of dominance, self-confidence, and aggression. When the psychological measures of women whose pills were estrogen-dominant were compared with those whose pills were progestin-dominant, it was found that those on estrogen-dominant pills described themselves as higher in assertiveness, aggression, and hostility when they were compared with those on progestin-dominant pills. The latter described themselves as high in deference, nurturance, and affiliation.

In addition to estrogen and progesterone, women also produce testosterone, although only approximately one-tenth the amount that men produce. Testosterone seems to peak at ovulation and at premenstruation. Testosterone has long been implicated in studies of hostility, assertiveness, or dominance. The affect need not be thought of as hostile. Assertion and dominance are likely to be experienced by people as a feeling of confidence or self-esteem. It seems plausible to hypothesize that the affect differences between ovulation and premenstruation could reflect an interactive effect of high levels of both estrogen and testosterone at ovulation, increasing the probability of assertive or competitive behavior together with feelings of self-esteem. A high level of testosterone in conjunction with low levels of estrogen and progesterone at premenstruation might increase the probability of aggressive behavior, associated with hostility and low feelings of self-esteem. Why do combination-pill women not show this affect and behavioral cyclicity? Whereas all of the gonadal steroids affect monoamine oxidase (MAO) levels, estrogen seems by far the most potent; and maintaining stable and high levels of estrogen may wash out the effect of a testosterone cycle.

Golub (8) recently studied the size of premenstrual mood changes in a sample of women aged 30 to 45. She compared premenstrual test scores with midcycle and normative data using standardized measures. State anxiety and depression scores were significantly higher premenstrually than at ovulation, but they were much lower than the scores of psychiatric patients. Trait anxiety scores were low and did not relate to premenstrual mood changes. This last point underscores the normalcy of the affect cycle; that is, there was no relationship between a woman's trait or characteristic measure of anxiety or depression and the magnitude of the changes in anxiety or depression scores, or their magnitude, at premenstruation. Three-fourths of her sample had significantly higher depression scores at premenstruation, and two-thirds had significantly higher premenstrual anxiety scores. Although the levels of anxiety and depression were higher at premenstruation, the actual change was generally small and differed markedly from scores of those who were psychiatrically ill or suffering unusual stress.

These data and our own support the hypothesis that menstrual mood changes are a function of physiological changes and perhaps one's sensitivity to them, and are not, in a sample of psychologically healthy women, a function of individual adjustment or psychopathology. Sometimes, it seems, a symptom is less symbolic than metabolic—and normal (13).

GONADAL STEROIDS AND THE CENTRAL NERVOUS SYSTEM

The correlational studies just described are necessarily crude. They tell us that a relationship exists, but we would like to be able to explain the why or the how of the effects we observe. The effects of the gonadal steroids seem wide-ranging, affecting not only the reproductive system but also general body functions. What we do not know seems greater than what we do know; explanation is difficult because the steroid effects are complex, interactive, and systemic, and we are trying to relate these effects to central nervous system functioning—another area of many unknowns. There is currently indirect evidence that the gonadal hormones affect the metabolism of catecholamines or cerebral MAO levels (13). MAO is an enzyme that metabolizes and thus inactivates brain catecholamines, which are either a category of neurotransmitters at the neural synapses or modulators of the synapses. The activity of MAO may be controlled by the gonadal steroids.

In an early study Grant and Pryse-Davies (14) studied the responses of 794 women on a wide variety of oral contraceptives. They concluded that when the level of estrogen in the pill was low, there was a high probability of depression. Thus depression and a low sexual interest was most common where the pill was strongly progestenic, had low levels of estrogen, and thus caused high MAO activity during most of the cycle. They found the lowest depression levels among women whose pills were high in estrogen and who had low MAO levels during most of the cycle. When the pill had high dosages of both estrogen and progestin, the high estrogen level seemed to offset the progestin effect and minimize the depression effect.

When a neuron fires and a catecholamine is released, that amount of the neurotransmitter must be deactivated or metabolized so that the neuron can be activated again. MAO causes the catecholamines to metabolize back to their original state. When there is a high level of MAO at the synapse, there is a low level of the catecholamine at the synapse. This is associated with depression. When there is a high level of estrogen, the level of MAO activity is low. The increase in estrogen may decrease the quantity of MAO or the activity level of MAO, or increase the level of the catecholamine. Both testosterone and estrogen appear to affect MAO levels.

Certain physiological states in women are associated with higher levels of depression: those on combination pills, as well as premenstrual and menopausal women. These are also states associated with high levels of MAO. (Obviously, one ought not to attribute depression solely to a physiological state, although depression seems to be more probable when there is a high MAO level and some forms of depression respond to a decrease in MAO level or activity.) To summarize, depression is likely when there is a high MAO level and a low level of both a catecholamine and estrogen or testosterone.

Klaiber and his associates (15) studied the plasma MAO levels during the pre- and postovulatory phases of the menstrual cycle, in amenorrheic women, in postmenopausal women, and in amenorrheic and postmenopausal women treated with estrogen and a progestin. The plasma level of MAO was significantly higher after ovulation than during the first half of the cycle. MAO levels were also significantly higher in untreated amenorrheic or postmenopausal women, higher than in the postovulatory half of the cycle. When the postmenopausal and amenorrheic women were given estrogen, their MAO levels fell to that of the preovulatory half of the cycle. However, when they were given progestin as well as estrogen, their MAO levels rose, although not as high as before treatment. Women on the combination oral contraceptive who are on a dosage that contains high levels of both estrogen and progestin would be expected, on the basis of Klaiber's data, to maintain a high level of depression throughout the cycle. That is exactly what we found in Paige's data. When women on sequential contraceptives or those not taking the pill are in the first half of the cycle and in the high-estrogen phase, we expect and find that depression levels are very low. When progesterone or progestin levels increase, so does depression. For those not on pills, the highest depression levels are found at premenstruation and secondarily at menstruation, when estrogen levels are low. In our data women on the combination pill maintain an intermediate level of depression during the cycle that is higher than during the estrogen phase but lower than that of non-pill women at premenstruation. Depression levels seem to correlate well with levels of plasma MAO.

We might note that Klaiber observed great differences in plasma MAO levels among individuals and within any one individual under different conditions. In amenorrheic and postmenopausal women, the fluctuations of MAO activity were 18 and 21 times greater, respectively, before treatment with estrogen than after. Even in normal women the fluctuation is significant. The within-individual variance in plasma MAO levels for normal women is three times greater after ovulation than before. If the MAO variability does in fact reflect an affect variability, then one reason women may experience increased levels of depression or anxiety is that they are less stable, less constant, and therefore less predictable to themselves as well as to others.

When estrogen levels are high (testosterone too?) and progesterone is low, plasma MAO levels are low, as are the levels of MAO variability. When plasma progesterone levels are high, so are MAO levels, and MAO variability increases. The addition of progesterone or progestin seems to offset the effect of estrogen or estrogen plus testosterone. This brings us a little closer to understanding our psychoendocrine observations. However, although it does appear that the plasma MAO activity and body tissues appear to be sensitive to changes in the gonadal steroids, we still do not know how the steroids affect MAO levels or activity.

These data support the general hypothesis that physiological variables have a direct influence on the evocation of some moods and behaviors; but this is not to say that, in some simple way, physiology causes psychology. Human behavior is a complex resultant originating in interactions of physiological, psychological, and sociological parameters. The question is posed incorrectly when one asks: "Is this physiological or psychological?" As we progress in the delineation of the psychobiological model, we find that socialization, experience, behavior, or affect may influence the physiology of that organism—and, conversely, the behavior, affects, and anticipations of that organism are influenced by the physiology of that organism. This is an extraordinarily complicated interactive model. I expect that the next decade will confuse us with further complexities, but I am hopeful that if we persist we will begin to perceive the significant variables in elegant simplicity, because that is what usually happens when one finally understands a phenomenon.

REFERENCES

1. Bardwick, J. M. (1971): *Psychology of Women.* Harper & Row, New York.
2. Benedek, T., and Rubenstein, B. B. (1942): The sexual cycle in women: The relation between ovarian function and psychodynamic processes. *Psychon. Med. Monogr.,* 3:vii–307.
3. Shainess, N. (1961): A re-evaluation of some aspects of femininity through a study of menstruation: A preliminary report. *Compr. Psychiatry,* 2:20–26.
4. Coppen, A., and Kessel, N. (1963): Menstruation and personality. *Br. J. Psychiatry,* 109:711–721.
5. Sommer, B. (1973): The effect of menstruation on cognitive and perceptual-motor behavior: A review. *Psychosom. Med.,* 35:515–534.
6. Parlee, M. B. (1974): Stereotypic beliefs about menstruation: A methodological note on the Moos menstrual distress questionnaire and some new data. *Psychosom. Med.,* 36:229–240.
7. Sommer, B. (1972): Menstrual cycle changes and intellectual performance. *Psychosom. Med.,* 34.
8. Golub, S. (1968): The magnitude of pre-menstrual anxiety and depression (*in press*).
9. Ivey, M. E., and Bardwick, J. M. (1968): Patterns of affective fluctuation in the menstrual cycle. *Psychosom. Med.,* 30:336–345.
10. Gottschalk, L. A., and Gleser, G. D. (1969): *The Measurement of Psychological States through the Content Analysis of Verbal Behavior.* University of California Press, Berkeley.
11. Paige, K. E. (1971): The effects of oral contraceptives on affective fluctuations associated with menstrual cycle. *Psychosom. Med.,* 33.
12. Oakes, M. (1970): Pills, periods, and personality. Doctoral dissertation, University of Michigan.
13. Bardwick, J. M. (1973): Psychological factors in the acceptance and use of oral contraceptives. In: *Psychological Perspectives on Population,* edited by J. T. Fawcett, pp. 274–305. Basic Books, New York.
14. Grant, E. C., and Pryse-Davies, J. (1968): Effect of oral contraceptives on depressive mood changes and on endometrical monoamine oxidase and phosphateses. *Br. Med. J.,* 3:777–780.
15. Klaiber, E. L., Kobayashi, Y., Broverman, D. M., and Hall, F. (1971): Plasma monoamine oxidase in regularly menstruating women and in amenorrheic women receiving cyclic treatment with estrogens and a progestin. *J. Clin. Endocrinol. Metab.,* 33.
16. Bardwick, J. M. (1974): The sex hormones, the central nervous system and affect variability in humans. In: *Women in Therapy,* edited by V. Franks and V. Burtle, pp. 27–50. Brunner/Mazel, New York.

Hormones, Behavior, and Psychopathology, edited by
Edward J. Sachar. Raven Press, New York © 1976.

Combined Antiandrogenic and Counseling Program for Treatment of 46,XY and 47,XYY Sex Offenders

John Money, Claus Wiedeking, Paul A. Walker, and Dean Gain

Department of Psychiatry and Behavioral Sciences and Department of Pediatrics, The Johns Hopkins University and Hospital, Baltimore, Maryland 21205

During the last 8 years we have used medroxyprogesterone acetate (Depo-Provera) as an androgen-depleting steroid in combination with counseling therapy in a clinical investigative and treatment program to facilitate self-regulation of behavior in 46,XY men with a history of imprisonable sex offenses and in 47,XYY men with a history of antisocial behavior.

Medroxyprogesterone acetate (MPA) as an androgen-depleting agent for a sex offender was first used in a clinical trial in 1966 (1,2). At that time it was not federally permissible, as it still is not, to use cyproterone acetate, the antiandrogen of choice in Europe (West Germany, Great Britain, and Switzerland).

The purpose of this chapter is to report a pilot study on possible effectiveness of the androgen-depleting agent, MPA, in promoting self-regulation of behavior in 46,XY male sex offenders and in a comparison group of impulsive antisocial 47,XYY males, some of whom were also sex offenders. In each instance the patients were given simultaneous counseling therapy.

SAMPLE SELECTION

46,XY Males, Sex Offenders

The 46,XY sex offender group was composed of 10 men all referred because of a severe paraphiliac problem that threatened eventually to land them in jail. Three were referred by psychiatrists, two by court-appointed psychologists, three by clergymen, and two were self-referred. One of the latter learned of the program through his roommate, who was a participant. Both were homosexual pedophiliacs, as was one other. The other seven diagnoses comprised one bisexual pedophiliac, one heterosexual pedophiliac and voyeur, one homosexual masochist, one homosexual incestuous pedophiliac and transvestite, and three heterosexual exhibitionists. Eight men in this group stayed on hormonal treatment for what was considered an adequate period of time, and two others dropped out prematurely.

47,XYY Males, Impulsive Antisocial

In the cytogenetic laboratory of the Johns Hopkins medical genetics clinic, there are 28 instances of the XYY karyotype on record. In 20 instances the patients were seen in person, and 18 of these were available for behavioral evaluation and follow-up in the present study. Two were too young (aged 10) for inclusion in the present report, and in three others there was no current antisocial, impulsive behavior to warrant treatment with MPA. The remaining sample of 13 is known to be skewed in favor of antisocial, impulsive behavior because cytogenetic screening was done specifically on individuals, many of them institutionalized with known histories of antisocial behavior.

Three XYY cases were ascertained in cytogenetic screening of local penal and detention centers, and two in a private reform school for adolescent boys. The remaining eight cases were referred because of behavior problems in association with tallness by two pediatricians, one neurologist, one endocrinologist, one psychiatrist, one psychologist, one attorney, and one parent.

During the course of treatment, four XYY patients dropped out prematurely. Six others completed an adequate term of treatment, and two of these are now deceased. The remaining three are still on treatment.

Various identifying data in the four groups are given in Table 1. Note that patient 8 is a 46,XY/47,XYY chromosomal mosaic.

Although the sample of XYY men is small in absolute numbers, it is large in terms of the number of patients karyotyped as 47,XYY in any one medical center. It is large also in terms of the amount of time consumed per week in the outpatient clinical care of patients with behavioral disabilities — care which includes social intervention on behalf of the patient in the courts and other agencies of society.

The time available for clinical care limited also the intake of paraphiliac patients. In addition, it is in the nature of paraphilia that the affected male seldom envisions himself as a candidate for medical care, for he does not feel ill. He is most likely to get into treatment under the stress of crisis, as when arrested. Yet the pendulum of the ethics of informed consent has now swung so as to deprive prisoners of any degree of informed consent in certain medical matters: we were expressly forbidden to treat as patients any who were already under arrest or in jail, even if the man himself initiated a request for treatment.

It is fortuitous that the sex offenders are mainly pedophiliacs and exhibitionists. Rapists would have been included had they been referred.

PROCEDURE

The information in this chapter was abstracted from the patients' psychohormonal research records and then tabulated. The records contained

TABLE 1. *Descriptive data of sample*

Patient no.	Age (years completed)	Chromosomal status	Height (cm)	IQ (WAIS; WISC)	Duration of treatment (months)	Weight (kg) Before therapy	Weight (kg) During therapy	Domicile	Referral
1	20	47,XYY	181	92	12	81	86	Local	Geneticist, screening
2	21	47,XYY	188	103	25	85	110	Local	Mother
3	37	47,XYY	205	125	13	111	135	Commuting distance	Physician (psychiatrist)
4	19	47,XYY	176	95	20	69	77	Commuting distance	Geneticist, screening
5	18	47,XYY	185	92	15	74	90	Commuting distance	Physician (neurologist)
6	20	47,XYY	207	121	12	82	100	Local	Psychologist
7	25	47,XYY	202	114	35	91	105	Commuting distance	Attorney
8	16	46,XY/ 47,XYY[a]	186	110	25	78	88	Commuting distance	Physician (pediatrician)
9	22	47,XYY	198	111	34	112	167	Commuting distance	Physician (dermatologist)
10	22	47,XYY	183	82	5	79	95	Local	Geneticist, screening
11	16	47,XYY	196	97	2	92	103	Local	Physician (endocrinologist)
12	28	47,XYY	193	120	4.5	79	—	Commuting distance	Geneticist, screening
13	15	47,XYY	188	116	5	100	100	Long distance	Physician (pediatrician)
14	25	46,XY	178	108	8	65	69	Local	Psychologist (court-appointed)
15	35	46,XY	179	112	12	142	142	Local	Physician (psychiatrist)
16	29	46,XY	180	124	6	78	78	Local	Other patient (roommate), self
17	29	46,XY	185	117	17	64	85	Local	Self
18	42	46,XY	160	111	20	74	68	Local	Minister
19	23	46,XY	169	114	10	58	58	Local	Minister
20	36	46,XY	178	110	8	89	95	Local	Minister
21	24	46,XY	173	79	17	76	78	Commuting distance	Psychologist (court-appointed)
22	22	46,XY	185	100	2	102	102	Local	Physician (psychiatrist)
23	29	46,XY	180	125	1.5	95	97	Local	Physician (psychiatrist)

[a] Chromosomal mosaic with karyotype 46,XY/47,XYY.

data pertaining to chromosomal status, initial and follow-up physical examinations, plasma hormonal determinations, and behavioral histories. The duration of follow-up has varied. The longest follow-up has been since 1966, when MPA was first used in a sex offender case. The shortest follow-up has been since September 1974 (i.e., 6 months), when the final patient of the study was enrolled.

For acceptance into the treatment program patients were required to make a commitment to stay in therapy for at least 1 year, except that three patients were treated for less than 1 year. Two of these were treated for 8 months and one for 6 months. The reasons are that two were taken into the program only recently and are continuing in treatment; the third received a second course of treatment, which after 8 months was sufficient.

No patients have been lost to follow-up. Those no longer under active treatment were recently recalled for behavioral follow-up or else interviewed by long distance telephone.

Chromosomal karyotyping was done in the division of medical genetics, under the supervision of Digamber Borgaonkar, Ph.D. Plasma hormonal determinations were carried out in the laboratory of pediatric endocrinology under the supervision of Claude Migeon, M.D. For the normal adult male, the mean value for plasma testosterone in this laboratory is 575 ± 150 ng/100 ml. For the normal adult female, the mean value is 49 ± 13 ng/100 ml.

Among the XY men of the present study, pretreatment testosterone levels ranged from 212 to 1,170 (median 607 ng/100 ml). Among the XYY men, the range was 384 to 995 (median 568 ng/100 ml).

On treatment, plasma testosterone was lowered in XY men to levels ranging from, at the lowest reported levels, 31 to 196 (median 70 ng/100 ml). In XYY men the corresponding levels were 23 to 236 (median 140 ng/100 ml). These ranges include patients who were sometimes irregular in returning for their injections, but like the others even they showed a dramatic fall in the level of plasma testosterone in response to MPA.

The dosage of MPA administered during the early years of the study was 100 mg (1 cc) a week, or 200 mg every 10 days. Subsequently it was found that the initial dosage, for optimal lowering of plasma testosterone, had to be changed to 200 to 400 mg a week, the exact dosage being dependent on body mass. Behavioral changes were optimized on the higher dosage. Dependent on individual circumstances, after the initial period the dosage could be lowered in progressive increments until a threshold was reached at which testosterone remained lowered. At this threshold level, erection and erotic expression was sometimes possible, although in reduced frequency. Sexual potency was in general more important to XYY men than to sex offenders.

The patients' behavioral records included data from psychological tests and transcribed taped interviews. Interviews were conducted according to a standard schedule of inquiry (3). The records included also data from social agencies and other institutions known to the patient and from interviews with those family members who made themselves available.

From each behavioral record, data were abstracted on large charts under the headings utilized in the figures and findings of this report. Four people did the abstracting, so that at least two people worked on each record, cross-checking with one another. Frequency counts and ratings were tabulated from the abstracts, with a consensus achieved by all four investigators. The criterion of whether to count a type of behavior present or absent was strict rather than lenient; i.e., the evidence had to be unequivocal.

Frequency counts and ratings were made separately for three periods: (a) the 2 years prior to the institution of treatment (except for animal assault that had occurred only during childhood); (b) the period of treatment[1] (Table 1), which ranged from 6 months to 2 years 11 months (median 11 months), except for the six cases of dropout prior to the completion of 6 months of treatment; (c) the period of post-treatment follow-up, gauged as commencing 6 weeks after the last injection and applicable to the six dropouts as well as to nine who satisfactorily completed more than 6 months of treatment, but not to the eight still continuing on treatment. The duration of post-treatment follow-up ranged from 3 months to 4 years 9 months (median 21 months).

FINDINGS

The findings are summarized in Figs. 1 through 4. The numerical data in the figures are self-evident on visual inspection. Repetitious referencing of the figures has been avoided in the sections that follow, since the numerical data and percentages are derived by cross reference as well as directly from Figs. 1 through 4.

Imagistic Eroticism

For each of the three periods (i.e., before, during, and after treatment), a positive rating was made in the category of imagistic eroticism when, regardless of the content or theme of the imagery, a person reported having sexual or erotic sleeping dreams, daydreams, or masturbation or copulation fantasies. During the pretreatment period the presence of such fantasies earned a positive rating, irrespective of individual differences in absolute frequency. During the course of treatment, the rating was changed to nega-

---

[1] Two patients had a history of treatment, relapse, and renewed treatment. In one instance the interval between treatments was 8.5 months, and in the other 3 years 9 months, with the relapse into sex-offending behavior occurring 5.5 years after the patient first begun treatment. Untreated for an additional year, this patient then re-entered the program. Two years later he is still on maintenance therapy. Only the duration of the second treatment period and subsequent follow-up was used in the present calculations.

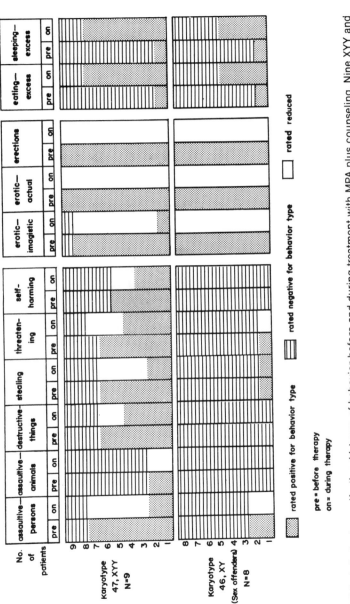

FIG. 1. Patients manifesting 11 types of behavior before and during treatment with MPA plus counseling. Nine XYY and eight XY patients were maintained on treatment without dropping out.

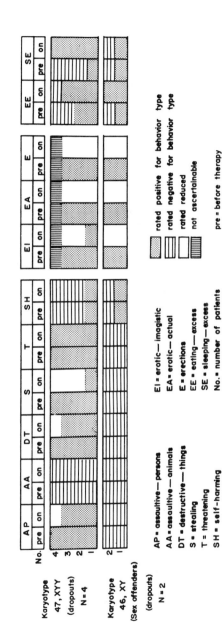

FIG. 2. Patients manifesting 11 types of behavior before and during treatment with MPA and counseling, prior to prematurely dropping out from treatment (N = 6).

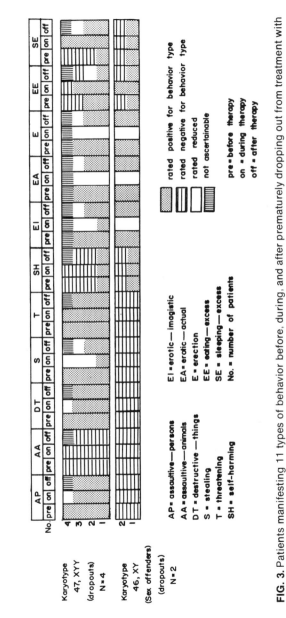

FIG. 3. Patients manifesting 11 types of behavior before, during, and after prematurely dropping out from treatment with MPA plus counseling (N = 6).

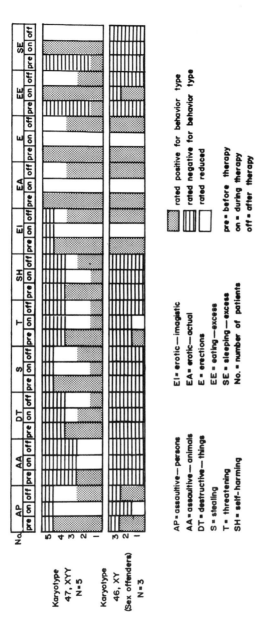

FIG. 4. Patients manifesting 11 types of behavior before, during, and after completing an adequate term of treatment with MPA and counseling (N = 8).

tive if the patient reported a substantial reduction in the frequency of erotic imagery such that the new frequency was self-recognized as unquestionably and significantly decreased. Typically this decrease was paralleled by an increase in erotic inertia and a reduction in frequency of erection and orgasm — in two extreme cases to zero. Total disappearance of imagery, erection, and orgasm was not set as a criterion of effective treatment, since erotic imagery and erection occur during the prepubertal years when plasma androgen levels are low.

46,XY Males, Sex-Offending

In all 10 sex offender cases, erotic imagery was reduced during the period of treatment. In each of the 10 it had been characteristic of the original complaint that the imagery and thoughts of the legally offensive erotic practices had been overly insistent and compulsive. The patients did not complain at the reduction of their erotic imagery but rather experienced it as a relief.

In the five men no longer on treatment (Figs. 3 and 4), the level of erotic imagery was restored on cessation of MPA treatment. Among those who remained on treatment long enough, the content of erotic imagery became reoriented and remained so in three (Table 2). A pedophiliac homosexual youth was able to keep out of legal trouble by being responsive to imagery and actual stimuli of males his own age. A compulsive exhibitionist was no longer bothered by imagery or implementation of his habit, but was erotically content within marriage. A transvestite, pedophiliac father was no longer bothered by imagery or enactment of either transvestism or erotic practices with his juvenile son; he experienced what he defined as a more fulfilling erotic relationship with his wife. The beneficial effect of treatment was repeated 3 years 9 months later, following a relapse, for which reason he is currently still being kept on a low maintenance dosage of MPA.

47,XYY Males, Impulsive Antisocial

Among the 13 patients who had 47,XYY karyotypes and were classified as impulsive antisocial, one adolescent of 17 was electively mute on all matters pertaining to sex and eroticism. Another youth was sexually and erotically inert and indifferent — before as well as on and off treatment — and has remained so until the present. Among the remaining 11, four were assessed as having only the imagery of ordinary heterosexual gender identity, although three youths among them were extremely reticent in dating and romance.

The other seven were varied in their imagery and gender identity. One, while having an isolation psychosis in jail, reported not only ordinary heterosexual imagery but also vengeful heterosexual incest imagery: In jail he serviced homosexual fellators. Another youth claiming to be heterosexual only was known from a parent's report to have engaged in several episodes of fetishistic transvestism, masturbating and ejaculating over his mother's apparel. One man was an exhibitionist, several times arrested, whose imagery and practice was also heterosexual and multiple-partnered. Four others were bisexual. Two of these were predominantly heterosexual with casual homosexual encounters: One had a longstanding history (including arrests) of encounters with young adolescent males until at the time of treatment he underwent a religious conversion and settled down in a heterosexual marriage; and one was predominantly homosexual, with a few trial episodes of being a masochistic "slave." Homeless, this latter youth leapt to his death in the aftermath of a broken homosexual marriage to an older man.

There were only two patients who, despite the antiandrogenic effect of treatment, made claims that erotic imagery and masturbation were not substantially reduced. These two shy youths probably were euphemizing. The electively mute youth continued to be uncommunicative, and the sexually inert youth remained indifferent. The reports of the remaining nine patients all indicated a definite antiandrogenic reduction of erotic imagery and, in the case of the eight older adolescent and adult men, reduction of penile erotic function also (*vide infra*). This dual reduction proved reversible on withdrawal of treatment (Figs. 3 and 4), although two patients dubiously claimed to be sexually more quiescent.

These 13 patients showed an extraordinary pretreatment range in distribution of eroticism from inertia to vigorously active, so that one suspects there may be as great a variability in erotic threshold in XYY men as there is in plasma testosterone level (4). The preliminary evidence from the present sample, however, shows no correlation between testosterone level and either erotic threshold or the varied thematic and paraphiliac content of erotic imagery and practice.

Regardless of erotic type, the eroticism of XYY men was reduced by treatment with MPA, the effect being reversible on cessation of treatment. The reduction effect had a beneficial long-term side effect, in one case helping the man quit a long history of homosexuality with adolescents in favor of a reorganized religious and married heterosexual life, now of 3 years' duration. There was a second beneficial effect — i.e., helping the psychotic man in jail as he recovered to revert to ordinary heterosexual imagery exclusively. In the case of the other paraphilias, change in erotic imagistic content did not occur in connection with treatment (Table 2).

TABLE 2. *Change of erotic imagery from sex-offending to not sex-offending after MPA and counseling*

Sex offenders	Change	No change
XYY (N = 9)	2	7
XY (N = 8)	3	5

Actual Erotic Behavior and Erections

For tabulation purposes, actual erotic behavior and erections were separated, since erection is not imperative to erotic behavior. They are here
presented together, for the findings of the two precisely parallel one another. Ejaculation was not tabulated separately from erection owing to possible uncertainties of ascertainment, should a dry orgasm replace an actual
discharge into the vagina or anus. However, such evidence as appeared
reliable pointed consistently to the fact that, in sufficient amount, MPA
suppressed ejaculation.

It is not necessary here to report the XYY and XY findings separately.
The findings are so uniform that, except for the XYY boy electively mute
on sexual matters, they can be summated: Prior to treatment, all of the
patients were erotically active either in masturbation or some form of
heterosexual or homosexual activity. On treatment with MPA, they all
experienced an unequivocal reduction in the frequency of erections and
actual erotic practice by an estimated minimum rate of at least 50% to a
maximum of 100%.

At the conclusion of treatment (Figs. 3 and 4), return of function was
clearly established in all of the XY sex offenders. The same held true among
the XYY group, except that the two chronically hyposexual individuals and
one sexually inexperienced youth gave responses of doubtful authenticity
and so were rated as reduced in actual eroticism during the post-treatment
period, and the electively mute youth still had nothing to say.

The foregoing findings indicate that the antiandrogenic function of MPA
definitely reduced genital functioning, but that the effect was reversible.
Moreover, in individual cases it was possible to show that by calibrating
the dosage the frequency of ejaculation, erection, and erotic behavior could
also be calibrated. In this way it was possible for a patient to resume his sex
life, although on a reduced scale, while still maintaining some of the desired
suppressive effect on offending erotic imagery and arousal.

None of the patients was personally concerned about a risk of infertility.
Such evidence as is known indicates no loss of fertility from treatment with
MPA. Many XYY patients are idiopathically sterile. One among the present
series, no longer on treatment, has produced one pregnancy.

Nonsexual Offenses

Figures 1 through 4 summarize data not only of sex offenses but also of
behavior that may be loosely categorized as aggressive, i.e., assault against
persons or animals, destructiveness, stealing, threat, and self-harming. The
findings on these variables are presented here only in abbreviated, graphic
form, to be expanded in a separate publication. What these findings show, in
brief, is that antisocial aggression was essentially absent in XY sex of-

fenders but not in the majority of XYY cases. Furthermore, the antiandrogenic effect on aggressive antisocialism in XYY men is not as direct as its effect on sexual behavior. There is, however, a masking effect here from the strict criteria used in making the ratings on aggressive antisocialism. These strict criteria were adhered to because the ideal goal was total remission of penalizable antisocial behavior, and because an attempt to assess degrees of improvement in the absence of 24-hour monitoring was subject to too great a likelihood of false negatives and/or false positives. As a result, the generally negative ratings of the effects of treatment do in fact mask some partial improvements that in at least a few cases made a very great difference in the ability of a younger XYY patient to live with his family or of an older man to live with his community.

By and large, however, these partial improvements were less predictable than improvements in sexual behavior. It is possible that such partial improvements in behavioral aggression as did occur were an indirect rather than a direct effect of hormonal treatment. The very idea of replacing moral judgmentalism by nonjudgmentalism may relieve a person of total personal moral responsibility for socially stigmatized behavior in such a way as to have a paradoxical therapeutic effect. This therapeutic effect is dual — first on the patient himself, and second on the family or other responsible guardians. The power struggle between both parties, replete with antagonistic moralizing and self-righteousness, is replaced by a mutual endeavor to solve a shared problem.

Eating and Sleeping

Figures 1 through 4 include information on eating and sleeping, the reason being that excesses of both were initially suspected as adverse side effects of MPA. The findings do not confirm this suspicion. Such increases as were reported were mild and reversible. The exceptions were patients who were already established as excessive eaters or sleepers prior to treatment.

Subjective Reports

During the course of follow-up interviews, patients were asked to report their own impressions of the effect the medication was having on them. Quotations from their responses are assembled in Table 3.

DISCUSSION

For the purposes of this chapter we set strict rather than lenient criteria for rating behavioral effects of treatment. In this way we avoided the error of a false-positive effect of treatment — but at the expense of creating a false-negative effect. In consequence of this policy, it is possible to accept

TABLE 3. *Patients' personal reports concerning effect of treatment with MPA*

Patient no.	Personal report
1	"The drug has calmed me down, in a way."
2	"I do feel that I am leading a slightly better life."
3	"I have a feeling of relaxation, I don't feel driving tension."
4	Treatment "did help me at the time, put me down in a calm state."
5	"More alertness, I feel a little brighter than before."
6	On a 10-point scale, temper control went from "about an eight point" to "about a two point."
7	"I've not had any disciplinary problems in more than 120 days, which is the longest since I've been in the institution . . . finally subdued my temper, it was a real victory for me."
8	"It slows me down a lot and I conk out a lot faster than anybody else."
9	"When I took the drug regularly I never had an outburst of temper."
10[a]	"When I took the drug I felt much calmer, 100% better."
11[a]	"For a period of time I was very calm, but after 5 weeks it did not have an effect on me any more."
12[a]	"I am less violent, I feel partly cured, after I had an LSD insight."
13[a]	"You know I got mad more."
14	"I have been occupied in so many other things, I don't even think about exposing myself any more."
15	"My mind is at ease, not agitated any more, just like the few minutes of relief I had previously after an ejaculation, now it lasts for months."
16	"It has relieved the sexual tension. It makes being around boys easier."
17	"I feel I am relieved of a great burden. It cut down on the sexual drive."
18	"I feel relaxed, can think before I talk; it is easier to behave."
19	"I can take sex and leave it. Sex is now only one-tenth of my problems."
20	"I have quieted down a lot. I feel normal."
21	"I don't think about sex too much."
22[a]	"I used to think about kids—they are not on my mind."
23[a]	"Less erotically susceptible, but eventually it did nothing to cure my (masochistic) fantasies."

[a] Early dropouts from treatment.

with confidence the finding that the antiandrogenic hormonal effect of MPA is accompanied by a raising or strengthening of the threshold for erotic imagery and practice. This strengthened threshold or barrier is beneficial in the case of men who have no other resource for the reduction or regulation of their sexual imagery and practice if their erotic practice should happen to be so stigmatized by society as to bring them inescapable punishment if apprehended and arrested, or if it should be likely to lead to their own mutilation and death.

In the present study the qualifying cases mirrored the sexual mores and laws of our society, and included exhibitionists, pedophiliacs, and a masochist who risked stage-managing his own murder. Rapists, extreme sadists, and lust murderers could have qualified, but none was referred.

One may argue, without final conclusion, the rights of individuals to their own style of erotic expression. There are no absolute standards. Nonetheless, among the long list of the paraphilias, there are some that are harmless to consenting partners, and some that entail a violation of the rights of one partner by another, without consent. It is here that the cutoff point between "harmless" and "noxious" can be established by social consensus. Individuals on the noxious side of the cutoff point at any given moment in sociolegal history are those who, without the benefits of antiandrogen therapy, suffer extreme and cruel punishment unnecessarily.

The antiandrogenic effect of MPA works not simply by way of the physiological mechanism of depleting plasma androgen nor by a hypothetical direct inhibiting effect on sexual pathways in the brain. It works also, in part, because its use redefines the sex-offending or paraphiliac individual in the eyes of the law, his critics, and himself. It redefines him as an individual whose sexual behavior is not exclusively the province of his own will power, voluntary control, and moral responsibility. At least in part, it is the province also of somatic or physiological variables working within his body which can be externally controlled. In the legal and medical folk philosophy of our times, it is more respectable to have an organic condition than a moral or psychic one, even should it be incurable. It evokes more compassion, less criticism, less punishment, fewer recriminations, less despair, and more collaborative effort toward change.

The androgen-depleting effect of MPA is augmented also by reason of the fact that when the threshold or barrier to sexual arousal is strengthened by the hormone the individual is metaphorically on vacation from the demands and insistence of his sexual drive, and so is able to experience an erotic or psychosexual realignment in conjunction with counseling. The process of realignment is probably similar to that which happens with dramatic suddenness in some cases of religious conversion. Probably also it is expedited if the individual came into treatment because he had reached a crossroads or crisis in his life signifying that maintenance of the paraphiliac pattern of functioning had become categorically impossible.

The most common manifestation of such a crisis is arrest and the prospect or actuality of a long prison sentence. It is a paradox of our times that public concern and professional fear over the rights of informed consent have now reached the point of forbidding antiandrogen therapy to prisoners—at least in Baltimore. Jailed sex offenders have *de facto* been robbed of their rights to treatment and of their rights to informed consent on the basis of the presupposition that a man serving time would sign anything that might lead to a reduction of his sentence. The protective pendulum has swung too far, so far as sex offenders are concerned, for now they are condemned to vindictive punishment without any chance of rehabilitative treatment and full restoration of civil rights. They have only the right to remain incarcerated. There is a moral dilemma here, the solution of which requires better cooperation of medicine and the law.

SUMMARY

Ten 46,XY males with a history of sex offenses and thirteen 47,XYY males with a history of impulsive antisocialism received therapy with antiandrogen (medroxyprogesterone acetate) combined with counseling. Seven of the XYY men were also sex offenders or potentially so. Antiandrogen had a beneficial effect in elevating the threshold for erotic imagery and sexual functioning in all cases. The effect was reversible on discontinuance of treatment. In five cases the sex-offending paraphilia went into remission. The combined program of therapy did not directly affect aggressive behavior in its various guises, but it did have a secondary therapeutic effect in some cases by lessening the pressures of moral blame.

ACKNOWLEDGMENTS

This work was supported by funds from the Grant Foundation, New York; the Upjohn Company, Kalamazoo, Michigan; and Deutsche Forschungsgemeinschaft–Max Kade Foundation, New York.

The authors appreciate the cooperation of the personnel of the pediatric clinical research unit and the cytogenetics laboratory, especially Ms. Sue Blair. Ms. Mary Decker did the art work.

REFERENCES

1. Money, J. (1968): Discussion on hormonal inhibition of libido in male sex offenders. In: *Endocrinology and Human Behavior,* edited by R. P. Michael. Oxford University Press, London.
2. Money, J. (1970): Use of an androgen-depleting hormone in the treatment of male sex offenders. *J. Sex Res.,* 6:165–172.
3. Money, J., and Primrose, C. (1969): Sexual dimorphism and dissociation in the psychology of male transsexuals. In: *Transsexualism and Sex Reassignment,* edited by R. Green and J. Money. Johns Hopkins University Press, Baltimore.
4. Baghdassarian, A., Bayard, F., Borgaonkar, D. S., Arnold, E. A., Solez, K., and Migeon, C. J. (1975): Testicular function in XYY men. *Johns Hopkins Med. J.,* 136:15–24.

Hormones, Behavior, and Psychopathology, edited by
Edward J. Sachar. Raven Press, New York © 1976.

Antiandrogen Therapy of Sex Offenders

Robert M. Rose

*Department of Psychosomatic Medicine, Boston University School of Medicine,
Boston, Massachusetts 02118*

The study reported by Money et al. in the previous chapter (1) is very important for a number of reasons. It argues for a potential treatment for men whose psychological and sexual functioning present problems for society, as well as for themselves. Many of these men are distressed by the intensity of their sexual drive, and are also aware and troubled by the inappropriateness of the object of their sexual fantasies and activities.

Until very recently there has been relatively little available in our psychotherapeutic or psychopharmacological armamentarium to offer these individuals. It now appears that the use of drugs to suppress endogenous secretion of testosterone (e.g., medroxyprogesterone) or such antiandrogens as cyproterone acetate (2) may be quite effective in diminishing the frequency and intensity of their sexual fantasies and behavior. Because the research is new, because of the difficulties with the target population (often coming from our penal system), as well as other problems relating to this aspect of research with humans, there are limitations in interpreting these data, which remain to be clarified.

However, in addition to the finding that there is a very real possibility for treatment, this chapter and others in this volume bring our attention to the area that Beach calls behavioral endocrinology (3). It might be worthwhile to step back for a moment to gain some perspective of our knowledge, or lack of knowledge, in human behavioral endocrinology or psychoendocrinology.

There has been an enormous amount of work reported during the last 20 years on the effects of administration of sex steroids, estrogens, androgens, or progestins on the sexual and aggressive behavior of a variety of animal species (4). Most of this work, however, has been done in the laboratory rat. More recently, workers have been able to correlate changes in the *endogenous* levels of such steroids as testosterone and alterations in sexual and aggressive behavior in animals (5). We certainly cannot review this work here, but it would be fair to say in summary that social and environmental stimuli do influence levels of testosterone, and that changes in plasma levels do appear to affect future behavior (6,7). We now have some data to support this conclusion in animals, but very, very little in man.

There is some work documenting the effect of grossly increased levels of androgens secreted endogenously or administered for therapeutic purposes on sexual behavior in humans (8). Conversely, the effects of castration have been reported somewhat more systematically in recent years by Scandinavian workers, such as Sturup (9) and Bremer (10). There appears to be some consensus that castration in males diminishes sexual libido but does not appear to affect aggressive behavior significantly.

There are serious difficulties, however, with studies on the effects of castration or those involving administration of steroids and reporting changes in behavior. The reporting of diminished sexual drive following castration comes primarily from individuals incarcerated for sexual crimes, and their answers might be biased by their being confined. Many of the studies involving administration of large doses of testosterone were done on patients with serious medical illnesses (11). Studies of the effects of adrenalectomy, which removes the major source of androgens in females, have also been done primarily in medical patients who also have serious medical problems, and the studies were performed with inadequate controls (12). I do not suggest that these studies are not useful. They are, but they fall short of documenting the magnitude of influence or degree of therapeutic effect that we have come to expect from research in psychopharmacology.

Perhaps an even more serious problem arises in our lack of knowledge of the relationship between alterations in levels of circulating steroids that occur naturally and their potential effect on behavior. Some argue that aggressive and sexual behavior are learned in man, a function of expectation or role (13). Others argue that sexual or aggressive behavior in man is much more clearly structured or determined by hormonal influences as observed in some animals. There is no doubt that the answer lies somewhere in between, since even in a variety of animal species, endocrine influences are seen to interact significantly with the experience of the individual animals.

Nevertheless, the gaps in our knowledge of human behavioral endocrinology are conspicuous. We do not know if variations in the normal level of testosterone in the male or female fetus contribute significantly to altered behavioral propensity that may emerge or be expressed later in life. We have no studies on the possible behavioral correlates of adrenarche in young girls, i.e., when there is a significant increase in adrenal secretion of androgens before menarche. In an even more obvious area, we do not know how the behavior of adolescent males changes concomitantly with the dramatic increases in plasma testosterone, which may rise three- to fivefold within 18 to 24 months. We have been able to document the precise endocrinological events that characterize the menstrual cycle for the last 5 to 10 years; yet I know of no studies that report changes in mood, percept, or activity in women during the menstrual cycle in which the individuals being studied have had estrogen, progesterone, or testosterone measured concurrently.

Perhaps the most dramatic endocrinological event, at least quantitatively, is the drop in progesterone levels that accompanies parturition and delivery of the placenta (14). A 300-fold drop occurs during this time. No study of parallel observations of progesterone levels and behavioral data during this postpartum period has yet been reported.

What might be some of the reasons for these large gaps in the literature? Endocrinological techniques have changed rapidly only during the last 10 years, so as to permit both efficient as well as reliable measurement of various sex steroids. Taboos on meaningful research of sexual behavior have been diminishing rapidly. Excellent articles are appearing in a variety of journals, as well as the fairly recently established *Archives of Sexual Behavior*. We are also beginning to overcome conceptual and methodological problems in measuring assertive or aggressive behavior in man. However, these approaches are as yet unproved. Now a new problem seems to be emerging to squelch research in this area. This relates to what its advocates argue as a resurgence of medical ethics. One translation of this is that people must be protected from knowledge about themselves — knowledge that others judge to be too difficult for them to handle or assimilate. This recent form of anti-intellectualism is carried under the banner of protection of human subjects, which in itself is indeed a valid and important undertaking. There have been excesses in research, as well as a lack of sufficient sensitivity on the part of some researchers to the issues of informed consent. However, it is unwise and unethical to be so certain one knows what is good or bad for others.

In summary, research in human behavioral endocrinology is really only beginning. We should be able to conduct controlled studies of possible treatment techniques for sexual offenders. We should begin research looking for the parallels between psychological and endocrinological events in human maturation. Research in this area is essential for us to judge how hormones may affect human sexual and aggressive behavior. Whereas conduct of this research was previously hampered by technological shortcomings and social stricture or taboo, it now appears to be threatened by a new form of political pressure and anti-intellectualism.

REFERENCES

1. Money, M., Wiedeking, C., and Walker, P. (1975): Anti-androgen therapy of sex offenders. *This volume.*
2. Cooper, A. J., Ismail, A. A. A., Phanjoo, A. L., and Lore, D. L. (1972): Antiandrogen (cyproterone acetate) therapy in deviant hypersexuality. *Br. J. Psychiatry,* 120:58–64.
3. Beach, F. A. (1975): Behavioral endocrinology: An emerging discipline. *Am. Sci.,* 63:178–187.
4. Beach, F. A., editor (1965): *Sex and Behavior.* Wiley, New York.
5. Rose, R. M., Bernstein, I. S., Gordon, T. P., and Catlin, S. F. (1974): Androgens and aggressive behavior: a review and recent findings in primates. In: *Primate Aggression, Territoriality, and Xenophobia,* edited by R. L. Holloway, pp. 275–304. Academic Press, New York.
6. Rose, R. M., Gordon, T. P., and Bernstein, I. S. (1972): Plasma testosterone levels in the male rhesus: Influences of sexual and social stimuli. *Science,* 178:643–645.

7. Rose, R. M., Bernstein, I. S., and Gordon, T. P. (1975): Consequences of social conflict on plasma testosterone levels in rhesus monkeys. *Psychosom. Med.,* 37:50–61.
8. Rose, R. M. (1972): The psychological effects of androgens and estrogens: a review. In: *Psychiatric Complications of Medical Drugs,* edited by R. I. Shader, pp. 251–293. Raven Press, New York.
9. Sturup, G. K. (1968): Treatment of sexual offenders in Herstedvester Denmark: The rapists. *Acta Psychiatr. Scand. [Suppl.],* 204:5–62.
10. Bremer, J. (1959): *Asexualization. A Follow-Up Study of 244 Cases.* Macmillan, New York.
11. Foss, G. L. (1951): The influence of androgens on sexuality in women. *Lancet,* 1:667–669.
12. Waxenburg, S. E., Drellich, M. G., and Sutherland, A. M. (1959): The role of hormones in human behavior. I. Changes in female sexuality after adrenalectomy. *J. Clin. Endocrinol.,* 19:193–202.
13. Montagu, M. F., editor (1968): *Man and Aggression.* Oxford University Press, New York.
14. Yannone, M. E., Mueller, J. R., and Osborn, R. H. (1969): Protein binding of progesterone in the peripheral plasma during pregnancy and labor. *Steroids,* 13:773–781.

Hormones, Behavior, and Psychopathology, edited by
Edward J. Sachar. Raven Press, New York © 1976.

Psychological Disturbances Associated with Endocrine Disease and Hormone Therapy

Peter C. Whybrow and Thomas Hurwitz

Department of Psychiatry, Dartmouth Medical School, Hanover, New Hampshire 03755

In 1786 in Bath, England, Caleb Parry unknowingly presided over the birth of endocrinological psychiatry. In a paper published only after his death, he noted with curiosity a young woman who had developed a highly nervous condition after falling from a wheelchair. Her distressed mental state was clearly associated with a large swelling on the neck (1). More than 40 years later, Robert Graves described the same syndrome, one which has subsequently been given his name (2).

The endocrine system may be viewed as an evolution of the simplest form of communication between cells in a multicellular organism. The chemical substance liberated from one cell arrives at the surface of a second cell and modifies the behavior of the latter. Despite its evolution, the nervous system retains this basic principle of communication. Although there has been considerable adaptation of cell shape and function, promoting intimate physical contact between cells, the actual transfer of information from one cell body to another remains dependent on the release of a chemical transmitter passing across the cell gap or synapse. It is not surprising then to find that several neurotransmitters are also numbered among the "chemical messengers" of the endocrine system (e.g., norepinephrine). Furthermore, although both are self-regulating or homeostatic dynamic systems, each has the capacity to influence the behavior of the other. Indeed, we may place the endocrine and nervous systems on a conceptual continuum.

A disturbance of endocrine function thus represents a fascinating experiment of nature—one that Parry noticed in 1786. Since his time the accumulated literature has provided us with some understanding of the complex relationship between hormones and behavior. It is, however, a literature replete with anecdote and imprecise description, leaving many questions unanswered. It is the task of this chapter to describe some of the significant trends in the exploration of these phenomena and through a critical appraisal to offer some suggestions for their future study.

WHY AN INTEREST IN PSYCHOENDOCRINOLOGY?

It was obvious to early workers that an understanding of the mental symptoms associated with the endocrinopathies was important in their differential diagnosis from the psychoses and organic dementias. Only too frequently an endocrine dysfunction presenting with bizarre psychiatric symptoms was missed and the individual relegated to institutional life. Sensitive and specific laboratory tests now aid the clinician but in no sense remove the importance of this area of clinical understanding. An area of interest closely related to and largely stimulated by these clinical observations is the time-honored quest for a greater understanding of the biological underpinnings of mental function.

That behavior might be explained on the basis of endocrine function has long been an attractive notion. Cushing (3) in 1913 wrote: "It is quite probable that the psychopathology of everyday life hinges largely upon the effect of the ductless glands' discharge upon the nervous system." In 1922 Tucker (4) noted: "It has been observed that a large number of delinquent girls have hypothyroidism. Hyperpituitary cases are frequently called incorrigible, opinionated and hypersexual, while the hypo- cases may be lazy, truant and ungrateful and deficient in integrity. Hyperadrenal cases may be cowardly and easily led to fear. Hypergonads show sexual irregularity and go to extremes in dress and action, and hypogonad cases may be perverts, asocial, cunning, cruel or cringing."

These colorful speculations imply a specificity of behavior relating to a particular endocrinopathy. When combined with the need for accurate differential diagnosis, this question becomes one of the most pressing in the study of psychoendocrinology. Are the mental symptoms in the endocrinopathies specific to the syndrome itself or are they general, reflecting a final common path of central nervous system (CNS) dysfunction?

METHOD OF REVIEW

The major literature in psychoendocrinology was reviewed for the preparation of this chapter. An attempt to present it all would be impossible in the space available and would provide little insight for future advance. To those interested, two recent and fairly comprehensive reviews are recommended (5,6). We have chosen to confine our discussion to three areas of endocrine function—the adrenal cortex and the parathyroid and thyroid systems—using these to illustrate our present understanding in the field and to highlight some of the difficulties in conducting research in psychoendocrinology.

In review of the first two, the major literature is summarized in a series of diagrams. The cases presented in each reference were closely scrutinized for psychiatric pathology reported by the original authors, and where it was identified, a category of disturbance was assigned based on the following criteria:

1. *Psychosis:* Bizarre behavior or utterings in the presence of a clear sensorium were noted by the author.

2. *Disturbed cognition:* A confused sensorium, which might or might not have been associated with bizarre thought or behavior, was reported.

3. *Disturbed affect:* A clear disturbance of mood (euphoria or depression) was noted by the author.

Many of the studies present in the literature are collections of case histories. Frequently these patients presented primarily with psychiatric symptomatology, and hence a very biased sample is reviewed. It is difficult to draw a conclusion as to the prevalence of psychiatric disorder in such studies. In other instances there has been no such selection for psychiatric disability, and the percentage of individuals developing overt psychopathology may be roughly estimated. In reviewing this material we attempt to note this basic difference in methodology.

No quantification of the degree of endocrine change is presented in older reviews and is frequently absent also in more recent studies; hence it is impossible to draw conclusions regarding the relationship of the degree of psychopathology and of endocrine disturbance. The difficulty is further compounded in some disorders where a traditional syndrome is now known to be potentially one of several endocrinopathies. This confounds any attempt to answer questions regarding specificity of psychopathology to particular endocrine disturbance.

ADRENAL CORTEX

Addison's Syndrome

In 1868 Thomas Addison (7) noted evidence of cognitive dysfunction in three individuals. He considered the symptomatology to be indicative of disturbed cerebral circulation, advance in the disease being characterized by insidious and increasing "languor" and a wandering mind. This final dysfunction usually occurred just prior to death.

These observations are reinforced by the major descriptive review by Engel and Margolin in 1942 (8). In a review of an unselected sample of 25 cases of Addison's disease, they reported 16 with neuropsychiatric symptomatology. Reviewing the study, it is obvious that, of these, two had delirium, one had psychotic dysfunction in a clear sensorium, and five had depressive symptomatology. Electroencephalographic (EEG) studies were abnormal in five of eight individuals investigated, but no clinical cognitive assessment was reported. Engel and Margolin proposed that the disturbance of cerebral function was secondary to hypoglycemia, a concept supported by Sorkin (9).

The psychopathology present in Addison's disease is given crude representation in Fig. 1. It may be seen that there is little to add to the picture first formed by Addison himself — one of confusion and depression in a small percentage of cases.

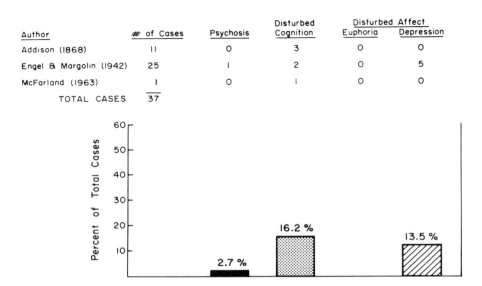

Author	# of Cases	Psychosis	Disturbed Cognition	Disturbed Affect	
				Euphoria	Depression
Addison (1868)	11	0	3	0	0
Engel & Margolin (1942)	25	1	2	0	5
McFarland (1963)	1	0	1	0	0
TOTAL CASES	37				

FIG. 1. Summary of literature review on Addison's disease and mental change.

Adrenocortical Hyperfunction (Cushing's Syndrome)

Cushing's syndrome is an example of a situation in which the originally noted syndrome is now known to be a collection of endocrinopathies. In general, however, this complication is not stressed in the psychoendocrinology literature, the basic trends of which are diagramed in Fig. 2.

Cushing's zeal for psychopathology appears to have waned by 1932 (10). Of the 12 cases described, only passing reference is made to the mental state of four individuals. Two clearly experienced confusion, one perceptual distortion, and a fourth fatigue, forgetfulness, and impotence associated with fits of unnatural irritability and depression.

This sample, unselected for mental dysfunction, is in contrast to the seven cases reported by Spillane in 1951 (11). Among this group of young individuals (all aged 20 to 30), poor memory and concentration was evident in three, and a clear-cut depressive syndrome in three others. After reviewing 50 case reports from the literature, in addition to those seven individuals presented, Spillane concluded that depressive syndromes predominate in Cushing's syndrome. Trethowen and Cobb (12) and Gifford et al. (13) support these earlier findings. The latter study represents a biased sample of individuals presenting with psychiatric disability, whereas that of Trethowen and Cobb reports an unselected population. Even in the former study, however, 15 of the 25 cases demonstrated clear evidence of psychopathology.

The information available in the literature thus suggests that delirium is associated with both hyper- and hypofunction of the adrenal cortex. Mood

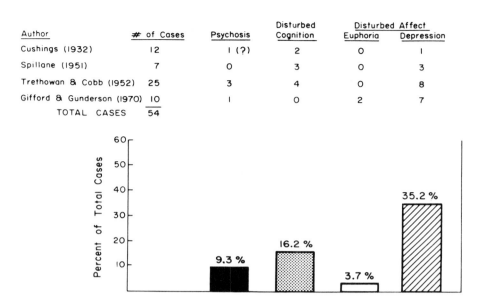

Author	# of Cases	Psychosis	Disturbed Cognition	Disturbed Affect Euphoria	Depression
Cushings (1932)	12	1 (?)	2	0	1
Spillane (1951)	7	0	3	0	3
Trethowan & Cobb (1952)	25	3	4	0	8
Gifford & Gunderson (1970)	10	1	0	2	7
TOTAL CASES	54				

FIG. 2. Summary of literature review on Cushing's syndrome and mental change.

disturbance, predominantly of a depressive nature, is also apparent. Such conclusions are similar to those drawn by Carpenter et al. (14) who reviewed the literature with the specific purpose of clarifying the relationship of the depressive syndrome to the disorders of steroid metabolism. They suggest that both hyper- and hypofunction of the adrenal cortex impair cognitive, perceptual, and affective ability. They review the role of cortisol in mediating the electrolyte balance and influencing catechol and indoleamine metabolism.

Administered Steroids

During the early 1950s the introduction of cortisone and adrenocorticotropic hormone (ACTH) as therapeutic agents spurred investigation of the effects of exogenous steroids on mentation (Fig. 3). Furthermore, it provided an opportunity for methodological advance in that individuals could be studied both during and after treatment. Clark et al. (15) reported eight cases of mental disturbance in individuals treated with ACTH and in two who received cortisone. Evidence of a mild delirium with decreased concentration and greater difficulty in organization of thought, together with the subjective sense of "fogginess" in the head, was apparent in two individuals. Two others experienced mild affective change with symptoms of euphoria and depression. In six others there was major psychiatric disorder, including paranoid thoughts, visual and auditory hallucination, and gross confusion including a disruption of recent memory. Complete recovery oc-

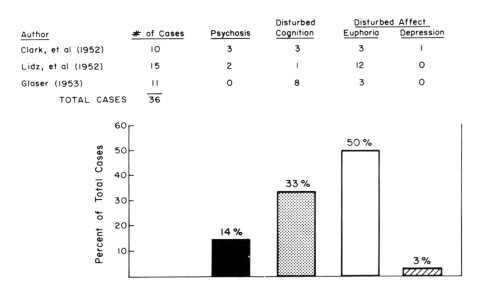

Author	# of Cases	Psychosis	Disturbed Cognition	Disturbed Affect	
				Euphoria	Depression
Clark, et al (1952)	10	3	3	3	1
Lidz, et al (1952)	15	2	1	12	0
Glaser (1953)	11	0	8	3	0
TOTAL CASES	36				

FIG. 3. Summary of literature review relating to steroid administration and mental change.

curred in all cases on withdrawal of hormone therapy. Glaser (16) reported similar findings and Lidz et al. (17) improved on the previous studies by introducing objective measurements. Using interviews, Kohs Block test, the Rorschach test, and EEG, they showed clear evidence of euphoria in 12 of the 15 individuals receiving exogenous steroids. Whereas Kohs Block test did not, interestingly enough, reveal a deficit in cognition, in almost all cases EEG changes were shown to be compatible with a mild delirium (a slow-wave pattern). When disturbance seen in the hyperadrenocortical syndrome is compared with that of exogenous steroids, it is apparent that literature concerning the latter emphasizes the presence of euphoria, whereas depressive affect predominates in the former. Lidz explained this phenomenon on the basis of improvement in the physical symptoms experienced by those receiving exogenous steroids. However, it occurs to us — especially in light of the presence of euphoria in some individuals with Cushing's syndrome (Fig. 2) — that the phenomenon is perhaps time-related. Obviously those presenting with Cushing's or other syndromes of increased adrenocortical secretion have had the disability for some time, whereas those observed while receiving steroids (frequently in large doses) are experiencing an acute effect. This hypothesis is obviously open to testing, and we return to it later when proposing future research design.

A review of this literature highlights some of the difficulties in traditional psychoendocrine research. Collecting large numbers of individuals with psychoendocrinopathy is difficult. Correlation of the mental testing with an endocrine assessment demands precise collaboration between behavioral scientists and internists. Furthermore, if the ideal is sought and the individual is retested after recovery from the endocrinopathy, the period of study is necessarily extended. All these factors conspire against a definitive study with a large group of individuals. Finally, in the case of a group of hormones whose influence is as pervasive as that of the steroids, the biological fulcrum for any mental changes found is unclear. Hence it comes as no surprise that precise information regarding the relationship of psychopathology to steroid metabolism is not available.

Other areas, however, offer a more fertile ground for optimism. One such area is the parathyroid system.

PARATHYROID SYSTEM

Psychoendocrine research on the parathyroid system has one major advantage over studies of the adrenal cortex. The predominant change occurring in the disturbances of parathyroid secretion at the cellular level is a change in the circulating available calcium ion, the level of which is easily measured. This, together with the diligence of two or three workers, has led to a more precise understanding of the relationship of mental function to parathormone change.

Hypoparathyroidism

In 1920 Barrett (18) reported a toxic mental disorder associated with tetany. Retrospectively, it is probable that these individuals did not have hypoparathyroidism as the basic cause of the tetany. Greene and Swanson in 1941 (19) reported on 18 individuals with specific hypoparathyroid tetany, five of whom had evidence of a psychosis associated with a "toxic delirium." However, the definitive study on the mental changes associated with hypoparathyroidism is clearly that of Denko and Kaelbling (20) published in 1962 (Fig. 4). Their exhaustive review of the literature was prompted by the study of a 45-year-old woman who had presented with epilepsy associated with dementia and hypoglycemia, and whose mental function improved totally when the abnormal calcium levels were rectified.

The review presented by Denko and Kaelbling (20) is exemplary in many ways. First, all available studies in the literature were reviewed. Unfortunately, however, only those where psychiatric descriptions were available were retained. This introduced a bias in the sample and made it difficult to determine the true prevalence of psychopathology in hypoparathyroidism. Offsetting this bias, however, was the fact that the sample was cleared of any individuals known to have a disorder capable of producing psychiatric symptoms or who had had psychiatric disability in the past. The authors also categorized the syndrome of hypoparathyroidism into its four major subgroups (compare with the literature on Cushing's syndrome), and each was

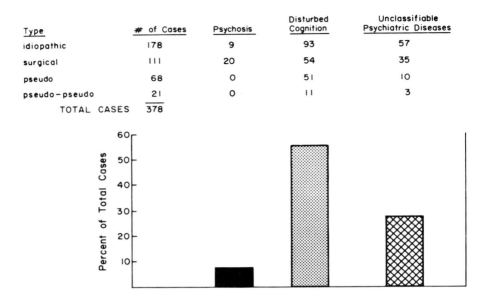

Type	# of Cases	Psychosis	Disturbed Cognition	Unclassifiable Psychiatric Diseases
idiopathic	178	9	93	57
surgical	111	20	54	35
pseudo	68	0	51	10
pseudo-pseudo	21	0	11	3
TOTAL CASES	378			

FIG. 4. Mental changes in hypoparathyroidism. (After Denko and Kaelbling, ref. 20.)

then reviewed separately. These investigators concluded that idiopathic hypoparathyroidism may show dementia, delirium, and all varieties of psychiatric disorder, some of each category reversing with treatment of the disease. Surgical hypoparathyroidism produces a similar variety of psychic conditions, but organic brain syndrome clearly predominates and in most instances responds to treatment. Perhaps here there is again a reflection of the time element in the production of psychiatric symptomatology. Pseudo- and pseudopseudohypoparathyroidism also show intellectual dysfunction, with some of the former individuals responding to treatment but none in the latter category. The study in general exemplifies how a detailed and exhaustive literature review of one syndrome can assist in our understanding of psychoendocrinology.

Hyperparathyroidism

Figure 5 summarizes the major studies available in the psychoendocrinology of hyperparathyroidism. Two studies stand out: that of Karpati and Frame in 1964 (21) and that of Petersen in 1968 (22).

Karpati and Frame (21) report mental changes observed in 33 consecutive cases of primary hyperparathyroidism, excluding those individuals with renal, electrolyte, or other abnormalities that would bias observation. Psychiatric symptoms such as nervousness, tension, and irritability are reported in one-third of these cases. Major psychiatric disturbances, however, are

Author	# of Cases	Psychosis	Disturbed Cognition	Disturbed Affect Euphoria	Depression
Eitinger (1942)	1	0	0	--	(?)
Nielsen (1955)	5	0	2	--	--
Fitz & Hallman (1952)	2	--	2	--	--
Karpati & Frame (1964)	33	--	1	--	2
Agras & Oliveau (1964)	1	--	--	--	1
Petersen (1968)	54	5	12	--	36
TOTAL CASES	96				

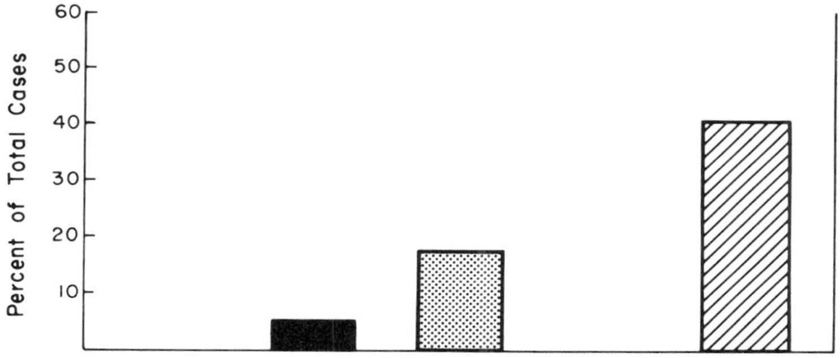

FIG. 5. Summary of literature review relating to hyperparathyroidism and mental change.

slight, with evidence of cognitive disturbance in only one individual and depression of a severe nature in two others. The study suggests that the mental changes are associated with the serum calcium level, but as no quantification of the psychopathology is offered this is difficult to interpret.

Petersen (22) also avoided merely reviewing those individuals presenting with psychiatric disability and, in addition, studied the mental state of individuals with Boeck's sarcoid and vitamin D intoxification in an effort to clarify whether the mental dysfunction was related to parathormone or circulating calcium levels. The only drawback in this study was that no objective testing of the psychopathology was employed. However, by using a crude rating scale, Petersen clearly demonstrated that the degree of psychic disturbance increased with an increasing serum calcium level and not with parathormone. He concluded that affective and drive disorders are associated with the milder endocrinopathy, and with increasing disturbance an acute organic psychosis occurs. He clearly established a linear relationship between impairment of cognition and elevated serum calcium concentrations (Fig. 6). Again, one is struck by the considerable clarification offered by one detailed and carefully executed study.

It is apparent from the previous discussion that despite the difficulties of research in psychoendocrinology certain factors in the study design can increase the chance of obtaining meaningful information.

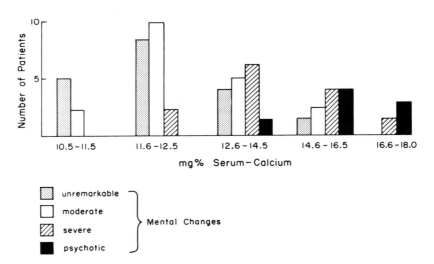

FIG. 6. Mental changes relating to serum calcium levels in 59 patients. (After Petersen, ref. 22.)

1. The study must be prospective, screening all consecutive admissions with the diagnosis of the endocrine disorder under study.

2. All those individuals with a prior history of psychiatric disorder must be removed from the sample.

3. All those with proved epilepsy, arteriosclerotic disease, or other disability potentially disruptive of CNS function must be removed from the sample.

4. Objective tests for psychopathology must be employed.

5. These must be correlated with biological measures of endocrine dysfunction.

6. The individual must be retested after treatment, again correlating the tests of psychopathology with further biological measures. This takes advantage of the usual course of the endocrine disorder when treatment is instigated and therefore uses the individual as his own control.

THYROID SYSTEM

It was with the above factors in mind that in 1966, in collaboration with colleagues at the University of North Carolina, we undertook a study of the mental changes occurring in thyroid dysfunction. All persons between the ages of 16 and 65 who were admitted to the North Carolina Memorial Hospital during 1966 with a suspected disturbance of thyroid function were screened. Those individuals with a history of previous disturbance unassociated with diagnosed thyroid illness, epilepsy, mental deficiency, arteriosclerosis, or other cardiovascular disease, plus individuals with fewer than 8 years of formal education, were excluded from the study. Finally, a total of 17 persons were evaluated, 10 with confirmed hyperthyroidism (mean protein-bound iodine, or PBI, 14.6 μg/100 ml) and seven with confirmed hypothyroidism (mean PBI 1.85 μg/100 ml).

After identification, a careful evaluation of the individual's mental status was undertaken. A subjective test of mood (Clyde Mood Scale) was followed by a clinical interview of approximately 1 hour during which details of previous life adjustment, current life circumstance, and current psychiatric symptoms were obtained. Subsequently, tests of cognitive function were administered (the Porteus Maze Test and the Trailmaking Test of Reitan). The patients also completed the MMPI, and the interviewer completed the Brief Psychiatric Rating Scale (BPRS) after the interview. This comprehensive series of tests was repeated after euthyroidism had been re-established. The average interval between these test periods was 10.5 months (range 4 to 15 months).

The results of the investigation demonstrated a disturbance of mental function in both the hyper- and hypothyroid groups—a finding grossly consistent with the consensus of previous reports. The mean rating on the BPRS in Fig. 7 illustrates this. The disturbance was profound in seven hyperthyroid and six hypothyroid patients (76% of the group).

The mental dysfunction observed falls essentially under two headings. One is an impairment of cognitive function, the other a disturbance of affect. In both instances the hypothyroid group seemed to be more grossly disturbed than the hyperthyroid group, but the impairment was considerable in both.

Impairment of Cognitive Function

All 10 individuals in the hyperthyroid group had noticed increased difficulty in concentration and some impairment of recent memory. Another individual close to thyroid storm denied any difficulty but reported a loss of visual and auditory acuity. All of these patients and one who had subjectively noticed no change in cognitive function had difficulty with simple arithmetic. The Trailmaking Test of Reitan confirmed this impairment, the mean of the hyperthyroid group (192.4 sec) being well above the limit established by Reitan for hospitalized individuals (145 sec). This disturbance, which fell to the upper limits of Reitan's group after treatment, is illustrated in Fig. 8.

The disturbance was even more profound in the hypothyroid group. Six of the seven individuals had noticed deterioration of recent memory and difficulty in concentration. They had difficulty with simple dollars-and-cents arithmetic during the interview, and several of the women complained that they could no longer remember cooking recipes without constant reference

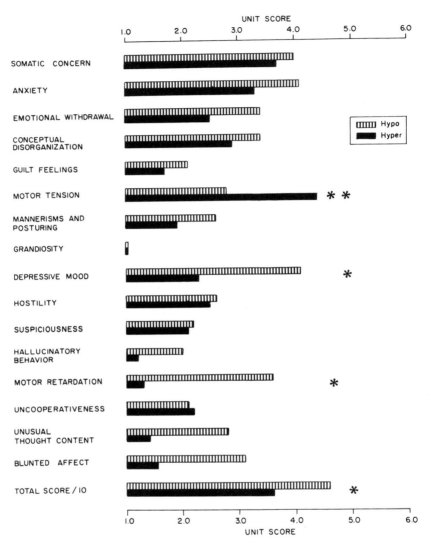

FIG. 7. Mean ratings on BPRS of hypothyroid and hyperthyroid groups. *, $p < 0.05$. **, $p < 0.01$.

to the book. Another had come to rely on her children to remember where she had placed things in the house. This profound disturbance was again reflected in the objective testing. The performance on the hypothyroid group was far worse than that of the hyperthyroid group, falling within the range of that considered by Reitan to be evidence of brain damage. Five of the seven hypothyroid individuals failed to complete the test in 300 sec. Furthermore with euthyroidism established there was evidence on retesting

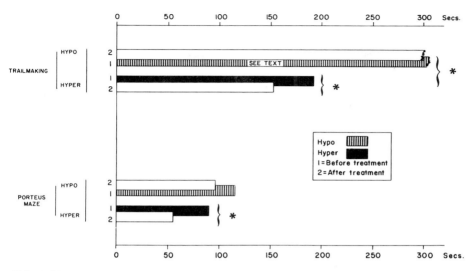

FIG. 8. Mean performance of hyperthyroid and hypothyroid groups on Trailmaking and Porteus Maze Tests. *, $p < 0.05$.

that individuals who had had prolonged hypothyroidism had residual impairment of cognition.

Hence both hyper- and hypothyroid persons suffered disturbance of recent memory and confusion of intellectual function. The same change in mental function thus appears to occur at the extremes of thyroid function in a fashion similar to the biphasic effect on protein metabolism. Tata (23) suggested that this biphasic effect is more likely "to be observed in activities whose manifestations depend upon a delicate balance between synthetic and degradative metabolic process of the cell." We found suggestions that the CNS is extremely sensitive to the effects of thyroid hormone. It was the first organ system to show improvement when thyroid medication was begun in a patient with a hypothyroid psychosis (24); a similar sensitivity was noted by Libow and Durell (25) in a detailed case study. Parenthetically these gross clinical observations suggested to us that the currently held view that the adult brain is refractory to the influence of thyroid hormone may need revision. Chronic deficiency appears to leave a residual deficit.

It was also apparent that the confusional state was associated with "psychotic" symptomatology in some persons. One individual, whom we reported previously (24), had a classic myxedema psychosis.

Others, predominantly but not exclusively in the hypothyroid group, experienced disturbances of perception (auditory and visual) as well as delusional and hallucinatory phenomena, these being broadly reflected in the F-scale of the MMPI. [This scale has 64 items, many of which deal with peculiar thoughts and beliefs. The content of the questions is undisguised,

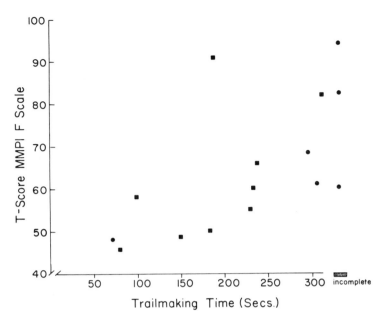

FIG. 9. Relationship between T-scores on MMPI F-scale and Trailmaking Test performance. ●, Hypothyroidism. ■, hyperthyroidism.

and many are frank statements of psychiatric symptoms (26).] Figure 9 shows the relationship between each individual's T-score on the F-scale and his performance on the Trailmaking Test. The scatter suggests that with increasing confusion of intellectual function there was an associated increase in the reporting of bizarre thoughts and beliefs. The organization of cognitive function and the integration of perceptual input appear to decline together, suggesting (as in previous clinical impressions) that the "psychoses" occurring in disordered thyroid states are the result of an increasing delirium.

Disturbance of Affect

In contrast to the cognitive disturbance, where a departure from euthyroidism in either direction appeared to result in a generalized disruption of intellectual function, affective disturbance appeared to be intimately and specifically linked to the prevailing thyroid state. Persons with hyperthyroidism complained of subjective anxiety, fatigue, and irritability. Paranoid symptoms occurred in two persons, one of whom had hallucinatory visual disturbance in which swarms of bees appeared to be flying toward her. These latter symptoms disappeared when the thyrotoxicosis was treated. Although motor tension was considerably increased, there was no euphoria;

and the hyperthyroid persons were not hypomanic in the generally accepted psychiatric sense. After treatment the combination of motor tremor and subjective anxiety described as nervousness by most of the patients showed early improvement. This was reflected subjectively on both the Clyde Mood and the MMPI Scales as well as in the clinical examination.

In the hypothyroid group the predominant affectual disturbance was a marked depression of mood. One individual was frankly psychotic, with depression and paranoid features. Others had considerable symptoms that were at times indistinguishable from the symptoms of profound melancholia. One patient, for example, considered herself a burden to her family. She had frequent thoughts of suicide and had become preoccupied with memories of her son, who had died in an automobile accident some years previously. She wished she had been the one killed. She dreamed of digging him from his grave with her bare hands, and she heard his voice calling her during the day. She felt she had been saved from a previous illness to be punished by her present one.

Another individual had experienced increasing insomnia with vivid morbid dreams, the content of which she at times believed with an almost delusional intensity. She dreamed of her only son's death and of mutilation of other family members. She was increasingly concerned that she was "losing her mind" and had transient ideas that someone was "bugging" her hospital room. An EEG showed bilateral slow waves consistent with a diffuse organic impairment. A third individual was preoccupied with impending disaster and was tearful throughout the clinical interview, sometimes crying, although the discussion had no obvious depressive connotation.

After euthyroidism had been re-established, the testing procedure was repeated, and a return to normal functioning was noted in general. The hyperthyroid group showed little residual impairment. The affective disturbance in the hypothyroid group returned to euthymia, but there was some evidence of a continuing deficit in cognitive function in this group. Figure 10 summarizes these findings, the details of which have been reported elsewhere (27).

Probably one of the most intriguing findings of this study was the striking difference in the predominant mood of the two groups. The hyperthyroid persons complained of subjective anxiety, fatigue, and irritability. No severe depressive affect was present, despite a gross disturbance of bodily function in several of the individuals. The findings of Artunkal and Togrol (28), who also used the MMPI to evaluate 20 thyrotoxic women, are in general agreement. These investigators found only mild depression in their group, with little change after treatment. Lidz (29) and Kleinschmidt and Waxenberg (30) noted, however, that a profound depressive state frequently precedes the clinical onset of thyrotoxicosis.

In contrast, all but one of the patients in the hypothyroid group reported depression of mood, and the MMPI profile of the group was very similar to

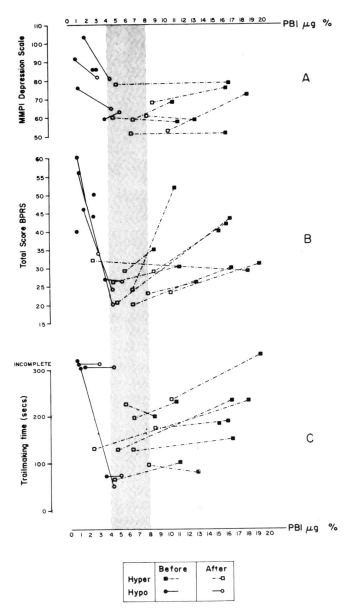

FIG. 10. Change occurring in hyperthyroid and hypothyroid individuals on return to euthyroid status.

those individuals diagnosed as psychotically depressed (31). The number of case reports of hypothyroidism in which depression affect was a leading presenting symptom is considerable and suggests that this is not an isolated finding (27).

This disparity of affect between the two groups is even more notable when one considers that all individuals were suffering from an illness that produces serious physical disability. Several of the hyperthyroid patients had mild myopathies, two had cardiac arrhythmias, and all complained of physical weakness. In the hypothyroid individuals physical fatigue was again a common complaint, as was mental confusion. The disturbance of body image, however, resulting from weight gain and change in body habitus was selected as a predominant problem by only one individual. In others the changes had occurred slowly and were attributed to overeating or increasing age. A psychological reaction to the illness itself, although undoubtedly playing a part, appears to be an insufficient explanation for either the profound depression seen in those persons with hypothyroidism or the difference in affect between the hyper- and hypothyroid groups.

CONCLUSIONS

We must return now to our original question regarding the specificity of the psychopathology associated with particular endocrine dysfunction. First, it is apparent from our three examples that as any endocrinopathy becomes severe a cognitive clouding ensues. In addition to the evidence of poor concentration and memory, perceptual distortion and psychotic disturbance may develop. Hence with increasing severity, all endocrinopathies appear to promote a final common path of psychopathology — that of a delirium.

In addition, however, there is evidence that hormonal balance may contribute in a special way to some elements of mental dysfunction. It seems, for example, that a major decrease in circulating thyroid hormone fosters depressive affect and that a rise in circulating thyroid hormone protects the individual from that affect (27). Similarly, euphoria is associated with a rising level of corticosteroids (15–17), whereas depressive affect appears as these levels are maintained (12,13). We submit that it may be reasonably concluded that endocrine function does have a relationship to affect and to the lability of that affect.

Beyond these global statements there is little to glean from the literature available. Despite its initial appeal, the study of endocrinological psychiatry is complex and difficult, providing a ready snare for the unwary. The collection of a large number of individuals with a uniform disability has become an increasing problem as laboratory tests now allow earlier diagnosis and therapeutic intervention. Indeed much is to be said, in our opinion, for discarding the efforts to quantify the psychic disturbance experienced by groups of individuals with a particular endocrinopathy, turning instead to the psychobiological model of investigation in single persons. Close measurement and analysis of psychic and physiological parameters, followed over time and through the course of the natural experiment of the endocrinopathy, potentially offer far more information than the traditional method. Other experimental approaches also may be of value.

Assessing the effect of the exogenous hormone — as in situations produced by surgical operation or where hormones are used therapeutically (e.g., inflammatory disease, contraceptive efforts) — may provide useful information. The advantage here is that the number of variables being manipulated at any one time can potentially be controlled. This is especially true, of course, where replacement therapy is concerned.

Other information may come from closer scrutiny of the effects of pharmacological intervention, such as the efforts by Greer et al. (32) using propranolol in thyrotoxic anxiety. The use of calcium lactate in hypoparathyroidism with close monitoring of the physical and mental states as the lactate is given, is typical of another potentially fruitful approach.

The ultimate goal is to define a covariation between endocrinological and behavioral events. Of itself, endocrinological psychiatry can continue to provide some clinical insight, but it is obvious that information gathered in this arena alone will provide answers to only a small part of the total jigsaw.

ACKNOWLEDGMENT

The authors wish to thank Mrs. Laurie Jordan for her unfailing assistance during the preparation of this manuscript.

REFERENCES

1. Parry, C. H. (1940): Diseases of the heart. *Med. Classics*, 5:8–20.
2. Graves, R. J. (1940): Clinical lectures, lecture XII. *Med. Classics*, 5:25–43.
3. Cushing, H. (1913): Psychic disturbances associated with disorders of the ductless glands. *Am. J. Insan.*, 69:965.
4. Tucker, B. R. (1922): The internal secretions in their relationship to mental disturbance. *Am. J. Psychiatry*, 2:259.
5. Smith, C. K., Barish, J., Correa, J., and Williams, R. H. (1972): Psychiatric disturbance in endocrinologic disease. *Psychosom. Med.*, 34:69.
6. Williams, R. H. (1970): Metabolism and mentation. *J. Clin. Endocrinol.*, 31:461.
7. Addison, T. (1868): Disease of the supra-renal capsules. In: *Collection of the Published Writings of the Late Thomas Addison*, pp. 209–239. MDA, New Sydenham Society, London.
8. Engel, G. L., and Margolin, S. G. (1942): Neuropsychiatric disturbances in internal disease. *Arch. Intern. Med.*, 70:236.
9. Sorkin, S. Z. (1949): Addison's disease. *Malacologia*, 28:371.
10. Cushing, H. (1932): The basophil adenomas of the pituitary body and their clinical manifestations (pituitary basophilism). *Bull Hopkins Hosp.*, 50:137–195.
11. Spillane, J. D. (1951): Nervous and mental disorders in Cushing's syndrome. *Brain*, 74:72.
12. Trethowan, W. H., and Cobb, S. (1952): Neuropsychiatric aspects of Cushing's syndrome. *Arch. Neurol. Psychiatry*, 67:283.
13. Gifford, S., and Gunderson, J. G. (1970): Cushing's disease as a psychosomatic disorder:

A selective review of the clinical and experimental literature and a report of ten cases. *Medicine (Balt.)*, 49:397–409.

14. Carpenter, W. T., Jr., Strauss, J. S., and Bunney, W. E., Jr. (1972): The psychobiology of cortisol metabolism: clinical and theoretical implications. In: *Psychiatric Complications of Medical Drugs,* edited by R. I. Shader. Raven, New York.

15. Clark, L. D., Bauer, W., and Cobb, S. (1952): Preliminary observations on mental disturbances occurring in patients under therapy with cortisone and ACTH. *N. Engl. J Med.,* 246:206.

16. Glaser, G. H. (1953): Psychotic reactions induced by corticotropin (ACTH) and cortisone. *Psychosom. Med.,* 15:280.

17. Lidz, T., Carter, J. D., Lewis, B. I., and Surratt, C. (1952): Effects of ACTH and cortisone on mood and mentation. *Psychosom. Med.,* 14:363.

18. Barrett, A. M. (1920): Psychosis associated with tetany. *Am. J. Insan.,* 76:273.

19. Greene, J. A., and Swanson, L. W. (1941): Psychosis in hypoparathyroidism with a report of five cases. *Ann. Intern. Med.,* 14:1233.

20. Denko, J. D., and Kaelbling, R. (1962): The psychiatric aspects of hypoparathyroidism: Review of the literature and case reports. *Acta Psychiatr. Scand. [Suppl.]* 38:7.

21. Karpati, G., and Frame, B. (1964): Neuropsychiatric disorders in primary hyperparathyroidism. *Arch. Neurol.,* 10:388.

22. Petersen, P. (1968): Psychiatric disorders in primary hyperparathyroidism. *J. Clin. Endocrinol. Metab.,* 28:1491.

23. Tata, J. R. (1964): Biological action of the thyroid hormones at the cellular and molecular levels. In: *Actions of Hormones on Molecular Processes,* edited by G. Litwack and D. Kritschevsky. Wiley, New York.

24. Treadway, C. R., Prange, A. J., Doehne, E. F., Edens, C. H., and Whybrow, P. C. (1967): Myxedema psychosis: Clinical and biochemical changes during recovery. *J. Psychiatr. Res.,* 5:289–296.

25. Libow, L. S., and Durell, J. (1965): Clinical studies on the relationships between psychosis and the regulation of thyroid gland activity. *Psychosom. Med.,* 28:377–382.

26. Dahlstrom, W. G., and Welsh, G. S. (1960): *An MMPI Handbook: A Guide to Use in Clinical Practice and Research.* University of Minnesota Press, Minneapolis.

27. Whybrow, P. C., Prange, A. J., and Treadway, C. R. (1969): Mental changes accompanying thyroid gland dysfunction. *Arch. Gen. Psychiatry,* 20:48–63.

28. Artunkal, S., and Togrol, B. (1964): Psychological studies in hyperthyroidism. In: *Brain-Thyroid Relationships,* edited by M. P. Cameron and M. O'Connor. Little, Brown, Boston.

29. Lidz, T. (1949): Emotional factors in the etiology of hyperthyroidism. *Psychosom. Med.,* 2:2–8.

30. Kleinschmidt, H. J., and Waxenberg, S. E. (1956): Psychophysiology and psychiatric management of thyrotoxicosis: A two year follow-up study. *J. Mt. Sinai Hosp.,* 23:131–153.

31. Dahlstrom, W. G., and Prange, A. J. (1960): Characteristics of depressive and paranoid schizophrenic reactions on the Minnesota Multiphasic Personality Inventory. *J. Nerv. Ment. Dis.,* 131:513–522.

32. Ramsay, I., Greer, S., and Bagley, C. (1973): Propranolol in neurotic and thyrotoxic anxiety. *Br. J. Psychiatry,* 122:555–560.

Hormones, Behavior, and Psychopathology, edited by
Edward J. Sachar. Raven Press, New York © 1976.

Neuroendocrine Effects of Psychotropic Drugs

Richard J. Wurtman and John D. Fernstrom

*Department of Nutrition and Food Science, Massachusetts Institute of Technology,
Cambridge, Massachusetts 02139*

Virtually all psychotropic drugs can modify the concentrations of particular hormones in the blood of humans and experimental animals. These
modifications sometimes result from direct actions of the drugs on neuroendocrine transducer cells (described below) or in other cases on loci which
have not yet been identified but which possibly include monoaminergic
brain neurons. Although the neuroendocrine effects of psychotropic drugs
generally attract little clinical interest, these effects occasionally are sufficiently troublesome to require changes in medication (e.g., the lactation
sometimes caused by phenothiazines or reserpine).

This chapter describes the neuroendocrine transducer cells, which link
brain function to plasma hormone levels, and summarizes the evidence
showing how these hormone-secreting cells are controlled by the same monoaminergic synapses that react to the presence of psychotropic drugs. It
also reviews the best-characterized neuroendocrine effect of psychotropic
drugs, i.e., the change in prolactin secretion caused by their interactions
with dopamine receptors in the brain.

NEUROENDOCRINE TRANSDUCER CELLS

Mammalian organs utilize two types of signals to communicate with each
other: hormones and neurotransmitters (1,2). Hormones include members
of various chemical families: They can be water-soluble (adrenocorti-
cotropin or epinephrine) or lipid-soluble (the steroids or melatonin); they
can range in size from simple amino acids (thyroxine) to peptides (thyrotropin-releasing factor), proteins (insulin), or glycoproteins (follicle-
stimulating hormone). In contrast, all known neurotransmitters are low-
molecular-weight, water-soluble amines or amino acids that are ionized at
the pH of body fluids. Hormones are distributed throughout the body by the
circulation, thus bathing virtually every cell in the body. Their messages
remain "private," however, because only tissue-specific receptors can "de-
code" them; hence only cells of the thyroid contain the decoding apparatus
needed to respond to circulating thyroid-stimulating hormone (TSH).
Neurotransmitters, however, impinge on only a very small number of cells —

those that receive synapses from neurons that release them; "privacy" in neuronal communications thus derives from anatomic specificity.

Theoretically, four types of specialized communications cells could exist, differentiated by their use of hormones or neurotransmitters as input and output signals. These are: (a) endocrine cells, which both respond to and emit hormonal signals (e.g., thyroid, gonadal, and adrenocortical cells); (b) true neurons, which both respond to and emit neurotransmitters; (c) neuro-endocrine transducer cells, which respond to neurotransmitter inputs by releasing hormones into the circulation (3,4); and (d) an as yet unnamed category of cells which respond to circulating hormones by releasing a neuro-transmitter at synapses. No examples of this last category have yet been positively identified, but such cells presumably do exist, inasmuch as the brain can be shown experimentally to contain "sensors" (probably neurons) that respond to changes in the plasma concentrations of hormones such as cortisol or testosterone by increasing or decreasing secretion of the hypo-thalamic releasing factors (*vide infra*). The carotid body may also contain such "humoral-neural" transducer cells.

Identification of a cell or organ as a neuroendocrine transducer requires two kinds of evidence: First, electron microscopic techniques must show that the cell does receive synapses; second, experimental data must demon-strate that the cell's capacity to secrete appropriate quantities of its hormone disappears when its innervation is interrupted. (For example, the chromaffin cells of the adrenal medulla fail to secrete epinephrine in response to insulin-induced hypoglycemia when the adrenal is transplanted or when its pre-ganglionic cholinergic nerves are transected.)

The family of neuroendocrine transducers includes cells both within and outside the brain. Examples within the brain include: (a) the modified hypothalamic neurons, which secrete releasing factors into the portal blood vessels that supply the anterior pituitary; and (b) the magnocellular hypo-thalamic neurons of the supraoptic and paraventricular nuclei, which liberate the polypeptides oxytocin and vasopressin into the portal circula-tion (5) and, from terminals within the posterior pituitary, directly into the systemic bloodstream. Examples of neuroendocrine transducer cells out-side the brain include: (a) the pineal organ, which synthesizes and secretes its hormone (melatonin) in response to the release of norepinephrine from sympathetic neurons (4,6); (b) the adrenal medulla, which secretes epineph-rine in response to acetylcholine released from preganglionic cholinergic neurons; (c) the renal juxtaglomerular cells, which secrete renin in response to norepinephrine released from sympathetic neurons; and perhaps (d) the β-cells of the pancreas, whose secretion of insulin in response to circulating glucose is suppressed by norepinephrine released from sympathetic neurons. It should be noted that all of these neuroendocrine transducer cells can be influenced by hormonal as well as neural inputs. Thus epinephrine synthesis is controlled by glucocorticoid hormones (7); insulin secretion depends on circulating glucose; hyperkalemia stimulates renin secretion; and some steroid hormones probably feed back directly on hypothalamic releasing-factor cells. What makes these cells unique, however, is their capacity to transduce synaptic to humoral signals.

The hypothalamic neuroendocrine transducer cells which secrete releasing factors have been exceedingly difficult to study; only very recently has it become possible to identify some of these cells by anatomic techniques (i.e., immunohistofluorescence methods for labeling cells containing peptide-releasing factors of known structure). It cannot be assumed that such releasing-factor cells contain distinct neurosecretory granules (as do those of the supraoptic and paraventricular nuclei of the hypothalamus, or of the anterior pituitary) or, conversely, that cells in the basal hypothalamus which contain such granules are releasing-factor cells.

Physiologic approaches have demonstrated that the hypothalamus contains compounds which affect secretion from the anterior pituitary. Such experiments have shown that: (a) hypothalamic extracts could cause (or, in the case of prolactin, inhibit) pituitary secretion; (b) hypothalamic lesions could interfere with pituitary function; and (c) the placement of small amounts of hormones at specific loci within the hypothalamus could suppress secretion of these hormones from their glands of origin. During the past few years, major advances have been made in the chemical characterization of hypothalamic releasing factors. For example, considerable evidence now indicates that the hypothalamic compound which causes the pituitary to release TSH is a specific tripeptide: pyroglutamyl-histidyl-proline amide. The structures of some other releasing factors have also been determined, including the luteinizing hormone-releasing factor (a decapeptide) and a hypothalamic compound that suppresses growth hormone secretion (somatostatin, a 14 amino acid peptide). The substance secreted by the hypothalamus and inhibiting the secretion of prolactin from the pituitary appears to be a catecholamine (dopamine), and not a peptide (*vide infra*). The structure of the hypothalamic constituent (corticotropin-releasing factor) that controls ACTH secretion remains unknown.

MONOAMINERGIC SYNAPSES AND NEUROENDOCRINE FUNCTION

There is abundant evidence that the monoamine neurotransmitters dopamine, norepinephrine, epinephrine, and serotonin mediate the actions of many if not most psychotropic drugs. These same neurotransmitters appear to be of special significance in controlling the secretory activity of neuroendocrine transducer cells. All such cells outside the brain either: (a) secrete catecholamines (the adrenal medullary cells); (b) are caused to secrete their hormones by norepinephrine liberated from sympathetic nerve terminals (the pineal and renal juxtaglomerular cells); or (c) have their secretory activity suppressed by circulating epinephrine or sympathetic neuronal norepinephrine (the pancreatic β-cells). The hypothalami of virtually all mammals studied also contain relatively large concentrations of all four monoamine neurotransmitters, localized within the terminals of neurons whose cell bodies are located in the upper brainstem, or in the case of dopamine within neurons located entirely within the hypothalamus.

Unfortunately it has not yet been possible to determine the anatomic relationships between these monoamine-containing neurons and the cells that secrete releasing factors, primarily because of the lack, until recently, of any anatomic method for identifying the particular cells that contain a given releasing factor. However, considerable experimental evidence shows that the administration of drugs which modify the release or metabolism of hypothalamic monoamines can affect anterior pituitary function: Thus reserpine and monoamine oxidase inhibitors can block ovulation; drugs that increase hypothalamic serotonin content can stimulate the secretion of growth hormone and suppress ACTH secretion (paradoxically, 5-hydroxytryptophan stimulates adrenocortical secretion); L-DOPA administration accelerates growth hormone secretion while suppressing that of prolactin; and phenothiazines and other dopamine receptor-blocking agents can raise serum prolactin levels to the point of facilitating lactation (*vide infra*). On the basis of such observations, it seems likely that hypothalamic monoamines exert major influences on the secretion of releasing factors and thereby on anterior pituitary function.

There are a number of loci at which monoamine neurotransmitters might affect secretion of a particular releasing factor:

1. The releasing factor itself might be a monoamine. It seems likely that the hypothalamic factor that inhibits prolactin secretion is dopamine itself (*vide infra*).

2. The releasing factor and a monoamine might be synthesized and stored within the same cell, and the monoamine might participate in the mechanism causing secretion of the releasing factor. A number of peptide-synthesizing cells also contain relatively large amounts of serotonin and possibly other tryptophan derivatives including tryptophanyl peptides. However, there is currently no direct evidence implicating these compounds in pituitary hormone secretion.

3. The neuroendocrine transducer cell that secretes the releasing factor could receive a primary synaptic input from monoamine-containing neurons.

4. Monoaminergic neurons outside the hypothalamus could influence the secretion of releasing factors via multisynaptic mechanisms.

5. The release of monoamines from peripheral neurons could affect anterior pituitary function directly (by means of circulating epinephrine secreted from the adrenal medulla) or indirectly (by causing hyper- or hypotension, with consequent changes in hormone secretion).

All of these possible loci should be considered when attempting to explain the mechanism by which a psychotropic drug, acting on monoaminergic synapses, affects anterior pituitary function.

Psychotropic drugs may also act on monoaminergic synapses to modify the secretion of hormones from the posterior pituitary. Detectable levels of dopamine, norepinephrine, and serotonin are found in both the supraoptic and paraventricular nuclei of the hypothalamus (8,9), regions which contain the cell bodies of neuroendocrine transducers that secrete oxytocin and vasopressin from their terminals in the posterior pituitary. Bridges and Thorn (10) demonstrated that pretreatment with phenoxybenzamine prevents the increase in vasopressin release that follows injection of a hyperosmolar solution. The same response has been obtained by pretreating animals with reserpine. At the present time there is no information on the influence of serotonin-containing neurons on posterior pituitary secretion.

Just as monoaminergic neurons appear to influence the secretion of releasing factors from the hypothalamus, and consequently hormones from the pituitary and target glands, these hormones in turn affect the biochemical activity of monoamine neurons in the brain. Thus hyperthyroidism reportedly enhances the sensitivity of dopaminergic receptors (11); ovariectomy (12) or the proestrous phase of the rat estrous cycle (13) accelerates brain norepinephrine synthesis; glucocorticoids enhance the activity of the epinephrine-forming enzyme phenylethanolamine-N-methyl transferase (PNMT) in rat brain (14); and behaviorally active peptides that are fragments of the ACTH molecule stimulate brain catechol synthesis (D. Versteeg and R. J. Wurtman, *unpublished observations*).

PSYCHOTROPIC DRUGS AND PROLACTIN SECRETION

Probably the best-characterized neuroendocrine effect of psychotropic drugs in humans is the change in mammary gland function that follows the administration of phenothiazines and other agents that block dopamine receptors. Patients so treated frequently develop galactorrhea, which when tested is usually found to be associated with elevated serum prolactin levels. The chronic administration of either chlorpromazine (15) or reserpine (16) — which depletes tissues of dopamine — can be associated with galactorrhea and elevated serum prolactin levels; moreover, reserpine administration may also increase the incidence of mammary carcinoma (17).

In contrast, the administration of L-DOPA (18), which elevates tissue and circulating dopamine levels, suppresses prolactin release and elevates serum growth hormone levels (19); in rats the dopamine agonists apomorphine (20) and ergocornine (21) suppress prolactin secretion. It has been suggested that dopamine agonists may have a place in the treatment of advanced hormone-dependent mammary carcinomas (22).

The following evidence suggests that a catecholamine (possibly dopamine) is *the,* or perhaps *a,* hypothalamic hormone which suppresses prolactin release when secreted into the pituitary portal system:

1. The prolactin-inhibitory activity of particular hypothalamic extracts varies in proportion to their catecholamine content (23). Such activity de-

creases with treatments that decrease hypothalamic catecholamines *in vivo* (e.g., reserpine administration) or that remove the catecholamines present in extracts, e.g., alumina adsorption or incubation with purified monoamine oxidase (24).

2. Injection of dopamine into the third ventricle (25) or into the pituitary portal veins (26) suppresses prolactin release.

3. Both dopamine and norepinephrine inhibit prolactin release from cultured pituitaries (24).

4. Injections of L-DOPA (which elevate tissue and blood dopamine levels) or of apomorphine (a dopamine agonist) decrease serum prolactin levels in rats (20,27), whereas administration of α-methyldopa (which interferes with the synthesis and storage of dopamine) or dopamine receptor blockers, (e.g., pimozide and haloperidol) increases serum prolactin (28–30).

Unfortunately, technical limitations have so far precluded measuring the dopamine concentrations in pituitary portal blood, and thus it remains to be determined whether physiologic increases or decreases in prolactin secretion are associated with inverse changes in portal venous dopamine concentrations.

SUMMARY

Abundant evidence associates monoamine neurotransmitters with the control of neuroendocrine functions and also with the mechanisms of action for numerous psychotropic drugs. Hence finding that administration of psychotropic drugs can modify plasma hormone levels is not surprising and may point to a need for limiting the use of these drugs.

REFERENCES

1. Wurtman, R. J. (1973): Biogenic amines and endocrine function; Introduction: neuroendocrine transducers and monoamines. *Fed. Proc.*, 32:1769–1771.
2. Wurtman, R. J. (1970): Neuroendocrine transducer cells in mammals. In: *The Neurosciences: Second Study Program,* edited by F. O. Schmitt, pp. 530–538. Rockefeller University Press, New York.
3. Wurtman, R. J., and Anton-Tay, F. (1969): The mammalian pineal as a neuroendocrine transducer. *Recent Prog. Horm. Res.*, 25:493–522.
4. Wurtman, R. J., and Axelrod, J. (1961): The pineal gland. *Sci. Am.*, 213:493–522.
5. Zimmerman, E. A., Carmel, P. W., Husain, M. K., Ferin, M., Tannenbaum, M., Frantz, A. G., and Robinson, A. G. (1973): Vasopressin and neurophysin: High concentrations in monkey hypophyseal portal blood. *Science,* 182:925–927.
6. Wurtman, R. J., Axelrod, J., and Kelly, D. E. (1968): *The Pineal.* Academic Press, New York.
7. Wurtman, R. J., Pohorecky, L. A., and Baliga, B. S. (1972): Adrenocortical control of the biosynthesis of epinephrine and proteins in the adrenal medulla. *Pharmacol. Rev.*, 24:411–426.
8. Saavedra, J. M., Palkovits, M., Brownstein, M. J., and Axelrod, J. (1974): Serotonin distribution in the nuclei of rat hypothalamus and preoptic region. *Brain Res.*, 77:157–165.
9. Palkovits, M., Brownstein, M. J., Saavedra, J. M., and Axelrod, J. (1974): Norepinephrine and dopamine content of hypothalamic nuclei of the rat. *Brain Res.*, 77:137–149.

10. Bridges, T. E., and Thorn, N. A. (1974): The effect of autonomic blocking agents on vaso-pressin release in vivo induced by osmoreceptor stimulation. *J. Endocrinol.*, 48:265–276.
11. Klawans, H. L., Jr., Coetz, C., and Weiner, W. J. (1973): Dopamine receptor site sensi-tivity in hyperthyroid guinea pigs: A possible model of hyperthyroid chorea. *J. Neural Transm.*, 34:187–193.
12. Anton-Tay, F., and Wurtman, R. J. (1971): Brain monoamines and endocrine function. In: *Frontiers in Neuroendocrinology,* edited by L. Martini and W. F. Ganong, pp. 45–66. Oxford University Press, New York.
13. Zschaeck, L. L., and Wurtman, R. J. (1973): Brain ^{3}H-catechol synthesis and the vaginal estrous cycle. *Neuroendocrinology,* 11:144–149.
14. Pohorecky, L. A., Zigmond, M., Karten, H., and Wurtman, R. J. (1969): Enzymatic con-version of norepinephrine to epinephrine by the brain. *J. Pharmacol. Exp. Ther.*, 165:190–195.
15. Frantz, A. (1973): Catecholamines and the control of prolactin secretion in humans. *Prog. Brain Res.*, 39:34–322.
16. Somlyo, A. P., and Waye, J. D. (1960): Abnormal lactation: Report of case induced by reserpine and brief review of the subject. *J. Mt. Sinai Hosp.*, 27:5–9.
17. Reserpine and breast cancer. (1974): Report from the Boston Collaborative Drug Sur-veillance Program, Boston University Medical Center. *Lancet*, 2:669–671.
18. Kleinberg, D. L., Noel, G. L., and Frantz, A. G. (1971): Chlorpromazine stimulation and L-dopa suppression of plasma prolactin in man. *J. Clin. Endocrinol.*, 33:873–876.
19. Brown, W. A., Van Woert, M. H., and Ambani, L. M. (1973): Effect of apomorphine on growth hormone release in humans. *J. Clin. Endocrinol. Metab.*, 37:463–465.
20. Smalstig, E. B., Sawyer, B. D., and Clemens, J. A. (1974): Inhibition of rat prolactin re-lease by apomorphine in vivo and in vitro. *Endocrinology*, 95:123–129.
21. Shaar, C. J., and Clemens, J. A. (1972): Inhibition of lactation and prolactin secretion in rats by ergot alkaloids. *Endocrinology*, 90:285–288.
22. Frantz, A. G., Habif, D. V., Hyman, G. A., and Suh, H. K. (1972): Remission of metastatic breast cancer after reduction of circulating prolactin in patients treated with L-dopa. *Clin. Res.*, 20:864.
23. Schally, A. V., Arimura, A., Takahara, J., Redding, T. W., and Dupont, A. (1974): Inhibi-tion of prolactin release in vitro and in vivo by catecholamines. *Fed. Proc.*, 33:237 (Abstr.).
24. Shaar, C. J., and Clemens, J. A. (1974): The role of catecholamines in the release of an-terior pituitary prolactin in vitro. *Endocrinology*, 95:1202–1212.
25. Kamberi, I. A., Mical, R. S., and Porter, J. C. (1971): Effect of anterior pituitary perfusion and intraventricular injection of catecholamines in prolactin release. *Endocrinology*, 88:1012–1020.
26. Takahara, J., Arimura, A., and Schally, A. V. (1974): Suppression of prolactin release by a purified porcine PIF preparation and catecholamine infused into a rat hypophyseal portal vessel. *Endocrinology*, 95:462–465.
27. Euker, J. S., Shaar, C. J., and Riegle, G. D. (1973): Prolactin and LH responses to LRH, L-dopa, and methyldopa in young and aged male rats. *Fed. Proc.*, 32:307 (Abstr.).
28. Lu, K.-H., and Meites, J. (1971): Inhibition by L-dopa and monoamine oxidase inhibitors of pituitary prolactin release: Stimulation by methyldopa and d-amphetamine. *Proc. Soc. Exp. Biol. Med.*, 137:480–483.
29. Ojeda, S. R., Harms, P. G., and McCann, S. M. (1974): Effect of blockade of dopaminergic receptors on prolactin and LH release: Median eminence and pituitary sites of action. *Endocrinology*, 94:1650–1657.
30. Dickerman, S., Kledzik, G., Gelato, M., Chen, H. J., and Meites, J. (1974): Effects of haloperidol on serum and pituitary prolactin, LH and FSH, and hypothalamic PIF and LRF. *Neuroendocrinology*, 15:10–20.

Hormones, Behavior, and Psychopathology, edited by
Edward J. Sachar. Raven Press, New York © 1976.

Chemical Lesioning of the Hypothalamus as a Means of Studying Neuroendocrine Function

John W. Olney, Bruce Schainker, and Vesela Rhee

*Department of Psychiatry, Washington University School of Medicine,
St. Louis, Missouri 63110*

In this chapter, we raise some questions — more questions than answers we fear — about the origins of the tuberinfundibular tract. Students of neuro-endocrinology have focused increasing attention in recent years on this tract as the apparent final common fiber pathway through which the hypo-thalamus delivers several releasing factors to the pituitary portal vascular system. Parallel attention has been focused on a particular hypothalamic nucleus, the arcuate nucleus, which is believed from early Golgi studies (1) to contain all or nearly all of the cell bodies giving rise to this important fiber pathway. The arcuate-tuberinfundibular system has been a source of additional intrigue recently because of evidence from fluorescence histo-chemical studies (2) that the putative neurotransmitter dopamine (DA) is present both in the cell bodies of arcuate neurons and in their presumptive terminals in the external mantle region (neurohemal contact zone) of the median eminence.

Several years ago we reported that the subcutaneous or oral administra-tion of monosodium glutamate to experimental animals selectively destroys arcuate neurons (3–5). In addition to its simplicity, a major advantage of this chemical approach over electrolytic or surgical techniques for lesioning the arcuate nucleus is the selectivity of glutamate's toxic action for neuronal dendrites and cell bodies. Other tissue components in the arcuate region are spared, including axons not originating from arcuate neurons. Endocrine disturbances resulting from glutamate treatment therefore can be inter-preted with some confidence as stemming specifically from the loss of arcuate neurons. This chemical lesioning technique also provides a unique tool for testing the assumption that arcuate neurons richly supply the ex-ternal mantle region of the median eminence with axon terminals. If this assumption is correct, a large glutamate-induced lesion of the arcuate nucleus should result in a substantial loss of axon terminals from the external mantle region of the median eminence. The findings presented here suggest that a re-evaluation of this assumption may be in order.

Our methods of administering glutamate and histologically evaluating its effects on arcuate neurons have been described in detail elsewhere (3,4,6).

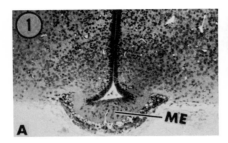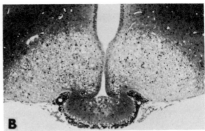

FIG. 1. A: Arcuate nucleus of the hypothalamus of a normal 10-day-old mouse. Arcuate nucleus is comprised of small neurons that cluster bilaterally in arch-like configuration over the median eminence (ME). **B:** Acute lesion induced in the brain of a littermate by a subcutaneous injection of glutamate gives the arcuate region a rarefied appearance. The margins of the lesion accurately demarcate the anatomical boundaries of the arcuate nucleus. ×75.

The arcuate nucleus of the hypothalamus of a normal young mouse not treated with glutamate is depicted in Fig. 1A. The same region of the brain from a littermate, 6 hours following treatment with glutamate, is illustrated in Fig. 1B. An acute necrotizing process affecting the arcuate nucleus but not other regions of the hypothalamus is evident.

The hypothalamus of an adult mouse that received a series of daily treatments with glutamate during infancy, resulting in the loss of essentially all of the neurons of the arcuate nucleus, is seen in Fig. 2A. Despite elimination of arcuate neurons from this brain, the external mantle region of the median eminence is difficult to distinguish from normal by either light microscopy (Fig. 2) or electron microscopy (Fig. 3).

We also examined the ultrastructure of the median eminence at various acute post-treatment intervals in an effort to garner evidence that at least some terminal degeneration does occur there secondary to destruction of arcuate neurons. The survey electron micrograph in Fig. 4 shows an acute arcuate lesion 6 hours following glutamate treatment with a substantial portion of the median eminence also present. Note the striking contrast between the lesioned area, which contains numerous necrotic arcuate neurons, and the external mantle region of the median eminence, which appears normal. Absence of degenerating terminals in the external mantle of the median eminence at this post-treatment interval was confirmed by careful ultra-structural examination at higher magnifications. Following the same basic approach, we studied both infant and adult mice at various post-treatment intervals. Between approximately 10 and 20 hours post treatment we identified a few degenerate terminals in the median eminence region (Fig. 5); however, these were exceedingly rare—perhaps in the range of 1% of the terminals populating the external mantle of the median eminence.

Although these findings suggest that the axonal input of arcuate neurons to the median eminence may be quite limited, there is ample evidence from

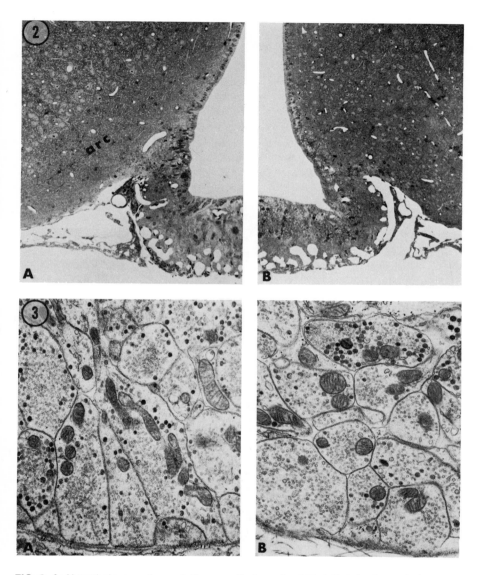

FIG. 2. A: Hypothalamus of an adult mouse that received 10 daily injections of glutamate during infancy. The arcuate region (arc) is devoid of neurons, and the ventricle is widened compared to the hypothalamus of a normal adult **(B).** Note that there is no appreciable difference between the median eminence of the brain that has no arcuate nucleus **(A)** and the one with a full complement of arcuate neurons **(B).** ×250.

FIG. 3. Ultrastructural views of the neurohemal contact zone from the brains in Figs. 2A and B are illustrated here in **A** and **B,** respectively. There are more dense core vesicles in the boutons of the glutamate-treated animal **(A),** but this is merely a normal sampling variation. Both scenes are consistent with the normal appearance of the external mantle region of the adult mouse median eminence. ×12,000.

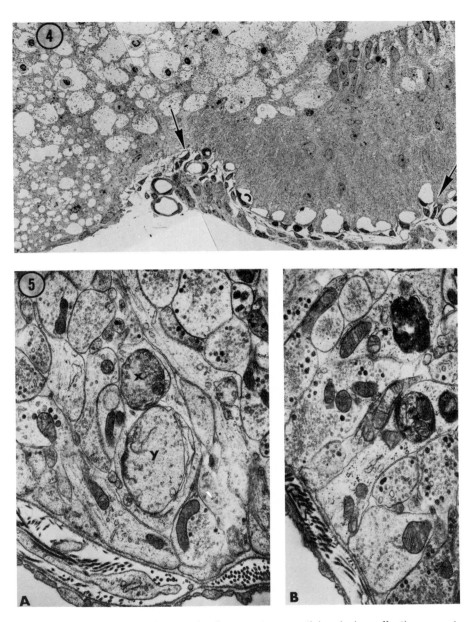

FIG. 4. Survey electron micrograph of an acute necrotizing lesion affecting arcuate neurons (*left* and *upper central*) 6 hours after glutamate treatment, but there is no discernible reaction or secondary degeneration occurring in the external mantle region of the median eminence (between *arrows*). ×500.

FIG. 5. A: Neurohemal contact zone of an adult mouse 12 hours following glutamate administration. Two structures (x, y) that may be degenerating axon terminals appear to be surrounded by and partially enwrapped in tanycyte processes. **B:** Another site from the neurohemal contact zone of the same animal reveals dense bodies within tanycyte processes. These may be the remnants of degenerate axonal boutons in that the internal content includes mitochondria and possibly some synaptic vesicles undergoing degradation. ×12,000.

other glutamate studies that arcuate neurons do play important roles in neuroendocrine regulatory function. When arcuate neurons are eliminated from the infant rodent brain by treatment with glutamate, treated animals manifest multiple endocrine disturbances as adults, including skeletal stunting (3,7,8); obesity (3,7–12); female sterility (3,8); reduced mass of the anterior pituitary (3,7,9,13), ovaries (7,8,13), and testes (7,13); reduced pituitary content of growth hormone (7) and luteinizing hormone (7); and interference in prolactin secretion (9).

Others (8,14) recently reported findings similar to ours; i.e., they were able to lesion the arcuate nucleus with glutamate but were unable to detect an accompanying degeneration of terminals in the external mantle region of the median eminence. Scott and colleagues (15) argued from these findings that arcuate neurons may secrete their bioactive substances into the third ventricle from whence they are taken up and transported by tanycytes to the portal vascular system. Hökfelt and Fuxe (2) suggested that DA-containing arcuate neurons, instead of secreting releasing factors, might terminate on other axons in the external zone of the median eminence to regulate secretion of releasing factors by the latter. This possibility is not absolutely ruled out by our findings, since arcuate neurons, by maintaining only a relatively small fiber input to the median eminence region, could possibly exert significant dopaminergic influence over the secretory activity of other nerve terminals. These authors (2) estimated, however, that over 50% of the axons terminating in the external mantle of the median eminence arise from DA-arcuate neurons. Our findings tend to contradict this view and suggest that the vast majority of such fibers must still be considered of undetermined origin.

To trace further the origins of boutons which populate the external mantle region of the median eminence, it is suggested that DL-homocysteic acid, one of glutamate's more potent excitotoxic analogues (16), be introduced directly into various hypothalamic regions. By a direct action on neural dendrites and cell bodies, the neurons of the region (but not other tissue components) are destroyed. The median eminence can be systematically examined for bouton degeneration after each such lesion placement to establish the extent to which any given group of hypothalamic neurons sends axons to the external mantle region.

ACKNOWLEDGMENT

This study was supported in part by grants NS-09156 and DA-00259 from the United States Public Health Service. Dr. Olney is the recipient of USPHS Career Development Award MH-38894.

REFERENCES

1. Szenthágothai, J., Flerko, B., and Halász, B. (1962): *Hypothalamic Control of the Anterior Pituitary*. Akademiai Kiadó, Budapest.

2. Hökfelt, T., and Fuxe, K. (1972): On the morphology and the neuroendocrine role of the hypothalamic catecholamine neurons. In: *Brain-Endocrine Interaction,* edited by K. M. Knigge, D. E. Scott, and A. Weindl. Karger, Basel.
3. Olney, J. W. (1969): Brain lesions, obesity and other disturbances in mice treated with monosodium glutamate. *Science,* 164:719.
4. Olney, J. W., and Ho, O.-L. (1970): Brain damage in infant mice following oral intake of glutamate, aspartate or cysteine. *Nature (Lond.),* 227:609.
5. Olney, J. W., Sharpe, L. G., and Feigin, R. D. (1972): Glutamate-induced brain damage in infant primates. *J. Neuropathol. Exp. Neurol.,* 31:464–488.
6. Olney, J. W. (1971): Glutamate-induced neuronal necrosis in the infant mouse hypothalamus: An electron microscopic study. *J. Neuropathol. Exp. Neurol.,* 30:75.
7. Redding, T. W., and Schalley, A. V. (1970): The effects of MSG on the endocrine axis in rats. *Fed. Proc.,* 29:755.
8. Holzwarth, M. A., and Hurst, E. M. (1974): Manifestations of monosodium glutamate (MSG) induced lesions of the arcuate nucleus of the mouse. *Anat. Rec.,* 178:378.
9. Nagasawa, H., Yanai, R., and Kikuyama, S. (1974): Irreversible inhibition of pituitary prolactin and growth hormone secretion and of mammary gland development in mice by monosodium glutamate administered neonatally. *Acta Endocrinol. (Kbh.),* 75:249–259.
10. Knittle, J. L., and Ginsberg-Feller, F. (1970): Cellular and metabolic alterations in obese rats treated with monosodium glutamate during the neonatal period. *Bulletin: American Pediatrics Society General Meeting, Program Abstracts,* p. 6.
11. Djazayery, A., and Miller, D. S. (1973): The use of gold-thioglucose and monosodium glutamate to induce obesity in mice. *Proc. Nutr. Soc.,* 32:30A.
12. Matsuyan, S. (1970): Studies on experimental obesity in mice treated with MSG. *Jap. J. Vet. Sci.,* 32:206.
13. Trentini, G. P., Botticelli, A., and Botticelli, C. (1974): Effect of monosodium glutamate on the endocrine glands and reproductive function of the rat. *Fertil. Steril.,* 25:478–483.
14. Paul, W. K., and Lechan, R. (1974): The median eminence of mice with a MSG induced lesion. *Anat. Rec.,* 178:436.
15. Scott, D. E., Dudley, G., and Knigge, K. (1974): The ventricular system in neuroendocrine mechanisms. II. In vivo monoamine transport by ependyma of the median eminence. *J. Cell Tissue Res.* 154:1–16.
16. Olney, J. W., Misra, C. H., and de Gubareff, T. (1975): Cysteine-S-sulfate: Brain damaging metabolite in sulfite oxidase deficiency. *J. Neuropathol. Exp. Neurol.,* 34:167–177.

Hormones, Behavior, and Psychopathology, edited by
Edward J. Sachar. Raven Press, New York © 1976.

Prostaglandins

Howard A. Gross and Ronald Fieve

*Department of Mental Hygiene, New York State Psychiatric Institute,
New York, New York 10032*

Prostaglandins (PGs) constitute a group of neuroregulatory agents. They are a class of hormone-like substances comprising over 20 distinct compounds. They probably function as modulators or mediators of humoral and other stimuli to neurons. The PGs were first isolated from seminal vesicles and plasma by Goldblatt and Von Euler during the early 1930s. The chemical structure and biosynthesis of the PGs were elucidated by Bergström et al. (1) during the later 1950s and early 1960s. Their immediate precursors are the dietary essential polyunsaturated fatty acids, of which arachidonic acid is the most abundant representative.

Certain PGs are found diffusely throughout central nervous tissue, which is endowed with complete systems capable of synthesizing and metabolizing distinct PG types. PG actions are varied and include effects on behavior, regulation of food intake, body temperature, and cardioregulatory and motor functions. Depending on the tissue involved, PGs can inhibit or stimulate the formation of cyclic adenosine monophosphate (cyclic AMP) and, through this substance, the responsiveness of the neuronal membrane to neurotransmitter substances.

One model of PG action proposes that prostaglandin E (PGE) formed within the postsynaptic junction region as a result of transmitter action diffuses into the synaptic cleft and then acts at prejunctional sites to curtail the further release of transmitter. It is also postulated that PGE has a secondary action at postjunctional sites, changing the response to transmitters.

In contrast, it is thought that PGF, preferentially formed after stimulation of certain other neuronal pathways, although not affecting the cyclic AMP system, may stimulate the cyclic guanosine $3',5'$-monophosphate (cyclic GMP) formation, thereby altering the cyclic AMP effects.

As a consequence of the use of pharmacological agents in man that preferentially affect the cholinergic or sympathetic pathways, one might anticipate finding differences in the levels of PGs in cerebrospinal fluid (CSF). We are currently studying a number of hospitalized depressed patients under conditions of metabolic control, using radioimmunoassay techniques to measure PGE and PGF in CSF. Our results are still at a preliminary stage and so I cannot yet present the data.

I have chosen two sets of data from other laboratories, however, to discuss the types of neuroregulatory function in which the PGs have been shown to be of significance. Ojeda et al. (2) showed that PGE_2, when injected into the third ventricle of male rats, induces a 40-fold increase of plasma luteinizing hormone (LH) at 15 min. PGE_1 induced a much smaller increase in LH. Neither $PGF_{1\alpha}$ nor $PGF_{2\alpha}$ altered the plasma LH after their injection into the third ventricle. We can see that the response is specific to a unique class of PGs and of a physiologically significant magnitude.

The second set of data is from Yue et al. (3). These investigators measured serum prolactin increases in response to intra-amniotic $PGF_{2\alpha}$ during a second-trimester abortion. They reported no deviation in prolactin with intra-amniotic hypertonic saline infusion from normal second-trimester pregnancy.

From these two examples (among many appearing in the current literature), we can see that the PGs play a significant role in neurophysiology and that it is of great interest to pursue studies of their significant relationship to biological psychiatry.

REFERENCES

1. Bergström, S., Danielsson, H., and Samuelsson, B. (1964): The enzymatic formation of prostaglandin E_2 from arachidonic acid. *Biochim. Biophys. Acta,* 90:207–210.
2. Ojeda, S. R., Harma, P. G., and McCann, S. M. (1974): Effect of third ventricular injections of prostaglandins on gonadotropin release in conscious free moving male rats. *Prostaglandins,* 8:545–552.
3. Yue, D. K., Smith, I. D., Turtle, J. R., and Shearman, R. P. (1975): Effect of prostaglandin $F_{2\alpha}$ on the secretion of human prolactin. *Prostaglandins,* 8:387–395.

Hormones, Behavior, and Psychopathology, edited by
Edward J. Sachar. Raven Press, New York © 1976.

Use of Neuroendocrine Techniques in Psychopharmacological Research

Edward J. Sachar, Peter H. Gruen, Norman Altman,
Frieda S. Halpern, and *Andrew G. Frantz

*Department of Psychiatry, Albert Einstein College of Medicine, Bronx, New York 10461;
and *Department of Medicine, Columbia University College of Physicians and Surgeons,
New York, New York 10032*

Each of the topical papers in this volume has implicitly offered another insight into the way endocrinological techniques can serve as research strategies to illuminate problems in clinical psychiatry and psychobiology — through the study of hormonal effects on brain and behavior, or by measuring hormonal changes resulting from alteration in brain function and psychological state. In this chapter we focus on another of these psychoendocrine research strategies — the use of neuroendocrine techniques in the study of problems in neuropharmacology. Dr. Wurtman's chapter gives an overview of neurotransmitter regulation of the secretion of hypothalamic releasing factors. Administration of neuropharmacological agents with known neurotransmitter effects have helped clarify the neurotransmitter effects of psychotropic drugs.

EFFECT OF AMPHETAMINE ON NEUROENDOCRINE FUNCTION

Our first example involves studies of the effects of d- and l-amphetamine on neuroendocrine function. While it is agreed that both these isomers significantly stimulate brain catecholaminergic activity, there is controversy in the literature about their relative effects on noradrenergic and dopaminergic neurons in various regions of the brain (1). This is an important psychopathological question, since both d- and l-amphetamine are of roughly comparable potency in man in their ability to induce psychotic states resembling acute paranoid schizophrenia or to exacerbate existing psychoses.

One group concluded that d- and l-amphetamine were equipotent in their effects on dopaminergic synaptosomes derived from rat striatum, whereas the d- form was 10 times more potent than the l-form on noradrenergic synaptosomes derived from cerebral cortex (5). However, in contrast, three other groups reported that the d-isomer was four to five times as potent as the l-isomer on dopaminergic striatal synaptosomes, whereas d- and l-isomers were equipotent in nonstriatal synaptosomes (6–8). Studies of the effects of these isomers of amphetamine *in vitro* on a variety of motor and self-stimulating systems in animals have also yielded conflicting reports (9–17) and confusion regarding their relative effects on dopaminergic and noradrenergic neurons in different brain regions. In particular, the net physiological effects of these drugs on the intact limbic system has remained obscure.

Approaching this problem neuroendocrinologically, we noted that catecholamines mediate certain secretory responses of ACTH and growth hormone (18–21). It thus seemed possible to study the effects of *d*- and *l*-amphetamine on the limbic system through their relative effects on these hormones.

Ganong, Van Loon, Scapagnini, and others have presented considerable evidence for a noradrenergic system which normally *inhibits* (i.e., reduces) ACTH and cortisol secretion (22–25). (These observations, by the way, make the observation of hypersecretion of cortisol in depression of particular relevance to the norepinephrine hypothesis of affective disorders.) Similarly, there is much evidence that both dopamine and norepinephrine (26–34) stimulate growth hormone (GH) secretion, and that the GH responses to L-DOPA (33) and hypoglycemia (34) are catecholaminergically mediated, although the relative roles of dopamine and norepinephrine in these responses are still unclear.

My colleagues (Drs. Robert Marantz, Elliot Weitzman, Jon Sassin) and I decided to administer equivalent doses of *d*- and *l*-amphetamine to monkeys, and to determine their acute effects on the secretion of cortisol and GH (35,36). Figure 1 indicates that both *d*- and *l*-amphetamine equally sup-

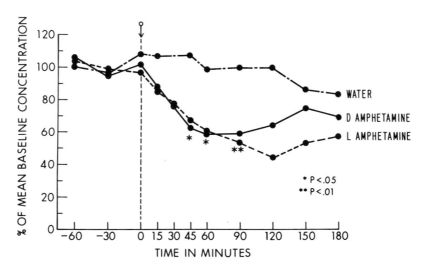

FIG. 1. Mean plasma cortisol response to *d*- and *l*-amphetamine in six studies in three monkeys.

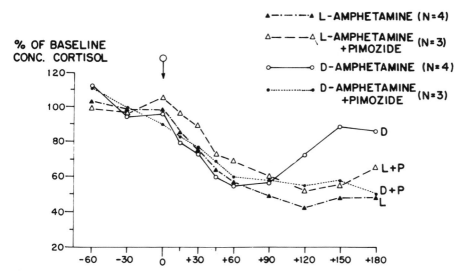

FIG. 2. Mean plasma cortisol responses to *d*- and *l*-amphetamine before and after pimozide in monkeys.

pressed cortisol secretion, with plasma cortisol concentration falling to approximately 60% of control values 60 to 90 min after injection. To be sure that the effect on cortisol was caused by amphetamine's noradrenergic and not its dopaminergic effects, we repeated the experiment after pretreating the animals with a substantial dose of pimozide (37). Pimozide specifically blocks dopamine receptors and should eliminate any dopaminergic effect of amphetamine (38). Figure 2 shows that the cortisol responses to amphetamine after pimozide were identical to those before, and that *d*- and *l*-amphetamine continued to suppress cortisol secretion equally.

Turning now to the GH responses (Fig. 3), we see that, before pimozide, *d*- and *l*-amphetamine equally stimulated GH secretion. There is a lot of variability, as is typical of GH secretion in the monkey (36), but the mean responses are very similar. Again, after pimozide pretreatment (Fig. 4) the GH responses are completely preserved, and the effects of *d*- and *l*-amphetamine remain identical.

We conclude therefore that: (a) amphetamine inhibits cortisol secretion and stimulates GH by its noradrenergic effects; and (b) in the intact limbic system, *d*- and *l*-amphetamine have identical noradrenergic effects on neuroendocrine tracts.

BRAIN DOPAMINERGIC SYSTEM

Let us turn now to the study of a brain dopaminergic system, the tubero-infundibular system, which regulates the secretion of prolactin inhibitory

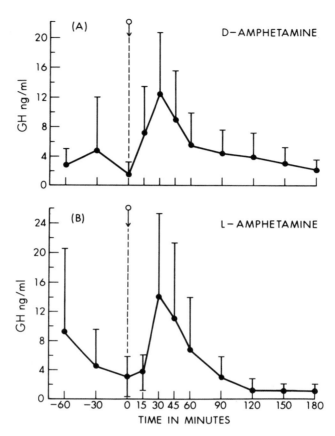

FIG. 3. Mean plasma growth hormone responses to *d*- and *l*-amphetamine in six studies in three monkeys.

factor from the hypothalamus. This factor, which Wurtman suggested may be dopamine itself, regulates prolactin secretion from the pituitary by varying degrees of inhibition. Dopaminergic stimulation thus leads to a suppression of prolactin secretion, whereas dopaminergic blockade results in stimulation of prolactin secretion. As Fig. 5 shows, ingestion of 500 mg L-DOPA, a dopamine precursor, typically leads to a *fall* in plasma prolactin concentration—to less than 50% of control levels. Conversely (Fig. 6), intramuscular injection of 25 mg chlorpromazine markedly stimulates prolactin secretion. It is interesting to note that the peak prolactin response to intramuscular chlorpromazine is achieved at approximately 1 hour; the blood level of chlorpromazine also reaches its peak between 30 min and 1 hour after an intramuscular dose (39,40). This prolactin response, by the way, is responsible for galactorrhea, an occasional side effect of phenothiazines. These prolactin effects of chlorpromazine seem to be

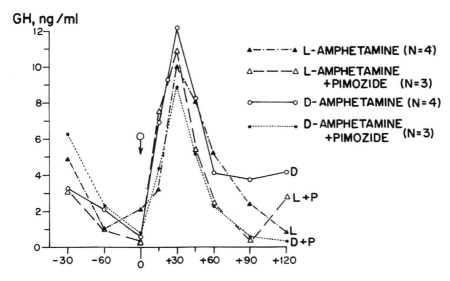

FIG. 4. Mean plasma growth hormone responses to *d*- and *l*-amphetamine in monkeys before and after pimozide.

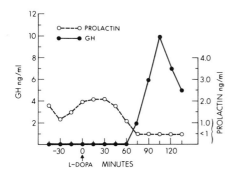

FIG. 5. Prolactin and growth hormone responses to ingestion of L-DOPA 500 mg by a normal 60-year-old woman.

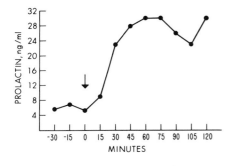

FIG. 6. Plasma prolactin response to chlorpromazine 25 mg i.m. in a 25-year-old schizophrenic man.

exerted at the level of the central nervous system (CNS) rather than directly on the pituitary, since women with pituitary stalk sections do not have a prolactin response to chlorpromazine (41). Similarly, haloperidol stimulates prolactin in rats but not on pituitary tissue *in vitro* (42).

How might the psychopharmacologist make use of this tuberoinfundibular prolactin system? For one thing, he could use it to test the hypothesis that all effective antipsychotic drugs have in common the ability to block dopamine receptors or to reduce dopaminergic activity. If so, they should all stimulate prolactin secretion. We have seen that chlorpromazine, the prototypic phenothiazine, has the expected effect on prolactin. From studies done with Meltzer and Frantz (43), we have evidence that the piperidine phenothiazines, fluphenazine, and trifluoperazine (Prolixin and Stelazine) also stimulate prolactin in schizophrenic patients.

With colleagues at Albert Einstein we have been testing the effects of other classes of antipsychotic drugs in man. In patients and normal subjects, haloperidol (a butyrophenone) also markedly stimulates prolactin (Fig. 7). We note in passing that for several of the subjects a 1-mg dose of haloperidol stimulated as large an increase in prolactin as did 25 mg chlorpromazine given intramuscularly (Fig. 6). This is approximately in accord with the relative clinical potencies of the two drugs and suggests that the

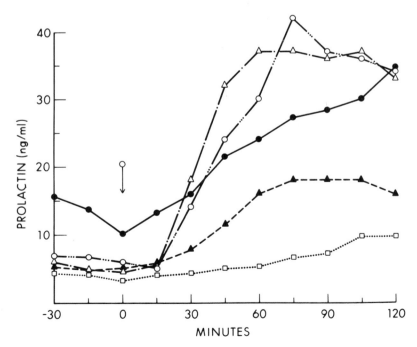

FIG. 7. Plasma prolactin responses to haloperidol 1 mg i.m. in five normal young men.

prolactin response may reflect not only clinical efficacy but also clinical potency.

Pimozide, to which we have already referred, is a diphenylbutylpiperidine antipsychotic. Figure 8 shows that 1 mg pimozide taken orally by normal subjects 12 hours before testing and another 1 mg 3 hours before testing was sufficient to double the plasma prolactin concentration over control levels. Whereas unmedicated subjects showed a 65% suppression of prolactin concentration after L-DOPA, after pimozide ingestion the mean L-DOPA-induced fall was only 20%. This supports the idea that pimozide blocks dopamine receptors and attenuates the effects of dopamine synthesized from L-DOPA.

So far, however, the results would please but not surprise the neuropharmacologists, since all of these drugs have been shown in animals to be potent dopamine antagonists in a variety of brain systems—e.g., their effects on HVA production and neuronal firing in the caudate, their ability to block amphetamine-induced stereotyped motor behavior, to block apomorphine-induced emesis, or to block dopaminergic effects on dopamine-sensitive adenyl cyclase. One might argue that a simple blood test readily applied to humans has some special clinical advantage, but truthfully there has not been much advancement in neuropharmacological *theory* with these

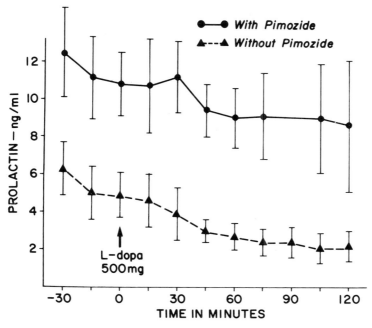

FIG. 8. Mean plasma prolactin responses to L-DOPA alone and after pretreatment with pimozide in five normal young men.

studies. What about thioridazine (Mellaril)? This drug is a very effective antipsychotic—milligram for milligram as potent as chlorpromazine. Yet in a variety of systems, particularly the caudate, it appears to exert only *slight* dopamine antagonism (44,45). Figure 9 shows that thioridazine taken orally markedly stimulated prolactin secretion in normal human subjects within 2 hours; in fact, the oral preparation seemed to act faster even than chlorpromazine in the same subjects. This study confirms our observations on prolactin secretion in schizophrenic patients receiving thioridazine treatment (43). It strongly suggests that in the tuberoinfundibular system thioridazine, like the other antipsychotic drugs, is a potent dopamine antagonist.

Why does thioridazine successfully antagonize dopamine in this neural system but not in the corpus striatum? There are several possibilities; one for which there is some good evidence is that since cholinergic influences antagonize dopaminergic influences a drug which has both potent anticholinergic and antidopaminergic effects would yield little net effect on dopamine activity (46). The two properties of the drug would cancel each other out in a system like the extrapyramidal one, which has *both* cholinergic and dopaminergic input. Thioridazine is such a drug—a far more potent anticholinergic agent than any of the other phenothiazines. This might explain its lack of antidopaminergic effects in the extrapyramidal system, but why is it apparently such a potent dopamine antagonist in the tuberoinfundibular system? Perhaps the tuberoinfundibular tract, in contrast to the caudate nucleus, has little cholinergic input, so that thioridazine exerts

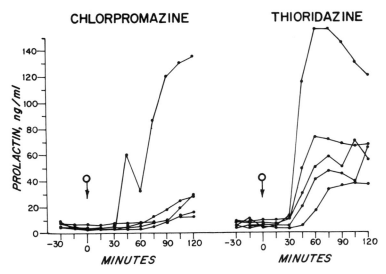

FIG. 9. Plasma prolactin responses to 50-mg doses of chlorpromazine and thioridazine concentrates in five normal men.

its effect on dopamine unopposed. To test this possibility, we studied the effect of physostigmine administration (1 mg i.m.) on prolactin secretion. If there is substantial cholinergic input to the tuberoinfundibular system, physostigmine should stimulate cholinergic activity, thereby antagonizing dopamine and releasing prolactin. If, however, there is no significant cholinergic input into the tuberoinfundibular system, physostigmine would have no effect on prolactin.

This is not an easy study to do. Physostigmine has powerful peripheral cholinergic effects that make people nauseated and sick. Prolactin is a stress-sensitive hormone, and sick subjects could have a stress-induced rise in prolactin unrelated to cholinergic antagonism of dopamine. We therefore decided to block the peripheral effects of physostigmine with methscopolamine, an anticholinergic agent which does not cross the blood-brain barrier. However, we underestimated the dose of methscopolamine we would need for adequate blockade, and our first few subjects became sick anyway. In these subjects the effects of the stress on plasma cortisol, GH, and prolactin all markedly increased (Table 1). We then increased the dose of methscopolamine, and the next group of subjects felt reasonably well. Note that there was also no effect on cortisol, GH, or prolactin. This lack of a prolactin response to physostigmine supports the view that cholinergic input to the tuberoinfundibular system is slight, and it may account for thioridazine's effects in this system—unantagonized dopamine blockade.

Now the picture changes. The tuberoinfundibular prolactin system, because of its properties, may be superior to the customarily studied dopamine system of the neostriatum when studying antipsychotic drug effects. Indeed few people really think that schizophrenia "resides" in the extrapyramidal system or that this is the site of the therapeutic effect of antipsychotic drugs. The extrapyramidal effects of neuroleptic drugs wear off quickly or can be counteracted, although the antipsychotic effects of the

TABLE 1. *Prolactin responses to physostigmine in normal men*

Subject	Meth-scopolamine (mg)	Discomfort	Cortisol ($\Delta\mu$g%)	HGH (Δng/ml)	Prolactin (Δng/ml)
High stress					
A	0.25	2+	+13.3	+10	+19.5
B	0.25	3+	+12.2	+ 2.5	+41.2
C	0.25	3+	+10.1	+10.6	+23.7
D	0.4	2+	+11.2	+10.9	+ 6.0
Mean		2.5	+11.7	+ 8.5	+22.6
Low stress					
E	0.4	1+	− 0.7	0	− 1.3
F	0.4	1+	+10.6	− 0.4	− 4.6
G	0.4	0	− 9.3	+ 0.1	+ 0.6
H	0.4	0	− 0.6	− 0.6	− 0.7
Mean		0.5+	0	− 0.2	− 1.5

drugs persist. It has been proposed that one likely neural system for the "locus" of schizophrenia is the mesolimbic dopamine system (47). It is relevant that the tuberoinfundibular and mesolimbic systems are closely related both anatomically and physiologically (48,49). In any case, the *tuberoinfundibular system may be a better model of the system where antipsychotic drugs exert their therapeutic effect.*

Could this system also provide a solution to the problem of clozapine? Clozapine is a dibenzoazapine widely used in Europe as an effective antipsychotic. Like thioridazine, it has very few extrapyramidal side effects and is a poor dopamine antagonist in the caudate (50–54). Like thioridazine, it has powerful anticholinergic effects. However, it does not have the typical neuropharmacological profile of effective antipsychotic drugs (52–54). We decided to try clozapine in our system. Figure 10 shows that 12.5 mg clozapine taken orally had no effect on prolactin secretion in young normals. Although this was a small dose, it markedly sedated our subjects, and in two schizophrenic patients receiving 100 mg/day there was also no prolactin elevation. The prolactin response test, then, does not elucidate the mystery of how clozapine exerts its antipsychotic effect.

What about phenothiazines which do not have antipsychotic efficacy, drugs like promazine (Sparine) and promethazine (Phenergan)? Figure 11 shows that 25 mg promazine given intramuscularly has no effect on prolactin secretion, consistent with its lack of ability to block dopamine or relieve psychosis. The same is true of promethazine (Fig. 12).

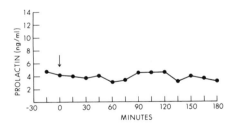

FIG. 10. Plasma prolactin response to ingestion of clozapine 12.5 mg in a normal 22-year-old man.

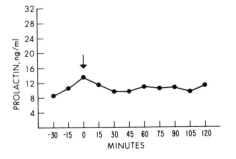

FIG. 11. Plasma prolactin response to promazine 25 mg i.m. in a normal 25-year-old man.

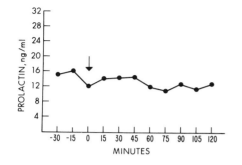

FIG. 12. Plasma prolactin response to promethazine 25 mg i.m. in a normal 23-year-old man.

Thus far, then, with the exception of clozapine, all the drugs we tested that are proved antipsychotics stimulate prolactin. Drugs of similar structure which are clinically ineffective fail to stimulate prolactin.

We turn now to a drug which has yet to be tested as an antipsychotic: thiethylperazine (Torecan) widely used as an antiemetic. Matthysse showed it to be a powerful dopamine antagonist in the caudate of monkeys, markedly increasing dopamine turnover (44). It blocks amphetamine-induced stereotypy (45) and stimulates prolactin secretion in rats (55). The manufacturer says it is not an effective antipsychotic, but in fact it has had only one poorly controlled clinical trial (in Czechoslovakia), and those psychiatrists report that two-thirds of their schizophrenics improved markedly or moderately on 20 mg/day (56). Figure 13 shows that 10 mg of this drug markedly stimulated prolactin secretion. We predict that this drug would be an effective antipsychotic, and suggest that the *prolactin response test might be useful as a predictor of the antipsychotic potential of neuroleptic drugs.*

We suggest yet another possible clinical application for prolactin measures. Figure 9 shows the individual prolactin responses to oral concentrates of chlorpromazine and thioridazine. Note the large individual differences in response, amounting to a 10-fold variation. A similar range of prolactin

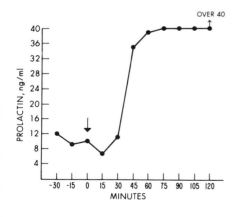

FIG. 13. Prolactin response to thiethylperazine 10 mg i.m. in a normal 27-year-old man.

responses were seen after identical doses of haloperidol (Fig. 7). Note that the biggest responder to chlorpromazine is also the biggest responder to thioridazine, and that the prolactin response to oral chlorpromazine is much slower than is seen in response to an intramuscular dose (Fig. 6). The point is that if one did not know that the vertical axis represented prolactin values, it would be assumed that these curves represented *drug* blood levels after a standard dose (levels which are known to vary widely from person to person), the levels rising more acutely after intramuscular administration than after oral ingestion.

The importance of drug blood levels in clinical psychopharmacology is now well known. Many patients fail to respond clinically to drugs because of idiosyncracies in their metabolism leading to inadequate blood levels and presumably inadequate brain effects. We suggest that if prolactin responses can be shown roughly to correlate, within limits, with drug blood levels, prolactin measures might replace drug blood levels, providing a physiological index of the actual degree of brain dopaminergic antagonism achieved in individual patients.

That such quantitative measures may well be possible is apparent in Fig. 14, which shows the prolactin response to 12.5, 25, and 37.5 mg in the same normal subject. There is a prolactin response to all three doses, but the response is not all-or-nothing but graded according to dose — and presumably to the blood level of the drug.

Presumably there would eventually be a plateau in prolactin concentration; once all dopamine receptors are blocked, a further increase in drug dose or blood levels should have no more effect on prolactin secretion, but the point at which this plateau occurs in patients — in terms of drug dose, drug blood level, prolactin concentration, and clinical response — should deserve careful analysis.

Indeed one may have in the prolactin response a pharmacotherapeutic measure to aim for in the schizophrenic patient: prolactin levels high enough to indicate adequate brain dopaminergic blockade. Levels below that concentration may well be associated with therapeutic failure.

One final application of the prolactin response to clinical pharmacology is illustrated in Fig. 15. This schizoaffective patient had a sustained prolactin elevation for approximately 2 weeks after a single intramuscular dose

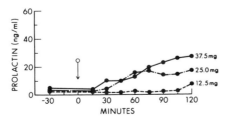

FIG. 14. Plasma prolactin response to intramuscular injections of 12.5 mg, 25 mg, and 37.5 mg chlorpromazine on different occasions in a normal 33-year-old man.

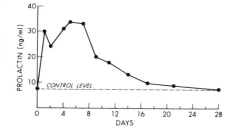

FIG. 15. Prolonged prolactin response to a single injection of depot fluphenazine enanthate 10 mg i.m. in a schizoaffective woman.

of depot fluphenazine enanthate. This prolonged prolactin response is in accord with the prolonged clinical effects claimed for this depot preparation. Thus the prolactin response may be useful in determining the duration of action of neuroleptic drugs or even as a way of determining if the patient is taking his medication.

SUMMARY

The prolactin response to neuroleptic drugs, as a reflection of dopaminergic blockade in the CNS, may have great value to the psychopharmacologist as a means of testing the dopamine theory of antipsychotic drugs, as a screening test in humans for the efficacy and potency of neuroleptics, as a semiquantitative measure in medicated patients of the actual degree of CNS dopamine blockade achieved by the drug regimen (which can be correlated with therapeutic response), as an index of the duration of action of drugs, and as a way of monitoring drug compliance in outpatients. If these possibilities are borne out in future research, this neuroendocrine strategy can become one of the most powerful tools in psychopharmacology.

ACKNOWLEDGMENTS

This research was supported in part by grants from the U. S. Public Health Service MH 25133–02 (Sachar); 5 MO1-RR-50 (Clinical Center); Scottish Rite Freemasonry, Northern Jurisdiction, United States (Gruen); AM-11294; CA 11704; and American Heart Association 68–111 (Frantz).
Dr. Marvin Liefer assisted in the studies of prolactin responses to neuroleptics.

REFERENCES

1. Matthysse, S. (1974): Dopamine and the pharmacology of schizophrenia: The state of the evidence. *J. Psychiatr. Res.,* 11:107–113.
2. Snyder, S. H. (1973): Amphetamine psychosis: A "model" schizophrenia mediated by catecholamines. *Am. J. Psychiatry,* 130:61–67.
3. Angrist, B. M., Shopsin, B., and Gershon, S. (1971): Comparative psychotomimetic effects of stereoisomers of amphetamine. *Nature (Lond.),* 234:152–154.

4. Davis, J. M. (1975): A two factor theory of schizophrenia. *J. Psychiatr. Res.,* 11:25–29.
5. Coyle, J. T., and Snyder, S. H. (1969): Catecholamine uptake by synaptosomes in homogenates of rat brain: Stereospecificity in different areas. *J. Pharmacol. Exp. Ther.,* 170:221–231.
6. Ferris, R. M., Tang, F., and Maxwell, R. A. (1972): A comparison of the capacities of isomers of amphetamine, deoxypipradiol and methylphenidate to inhibit the uptake of tritiated catecholamines into rat cerebral cortex slices, synaptosomal preparation of rat cerebral cortex, hypothalamus and striatum and into adrenergic nerves of rabbit aorta. *J. Pharmacol. Exp. Ther.,* 181:407–416.
7. Harris, J. E., and Baldessarini, R. J. (1973): Uptake of [³H]-catecholamines by homogenates of rat corpus striatum and cerebral cortex: Effects of amphetamine. *Neuropharmacology,* 12:669–679.
8. Thornburg, J. E., and Moore, K. E. (1973): Dopamine and norepinephrine uptake by rat brain synaptosomes: Relative inhibitory potencies of l- and d-amphetamine and amantadine. *Res. Commun. Chem. Pathol. Pharmacol.,* 5:81–89.
9. Carr, L. A., and Moore, K. E. (1970): Effects of amphetamine on the contents of norepinephrine and its metabolites in the effluent of perfused cerebral ventricles of the cat. *Biochem. Pharmacol.,* 19:2361–2374.
10. Taylor, K. M., and Snyder, S. H. (1971): Differential effects of d- and l-amphetamine on behavior and on catecholamine disposition in dopamine and norepinephrine containing neurons of rat brain. *Brain Res.,* 28:295–309.
11. Scheel-Krüger, J. (1972): Behavioral and biochemical comparison of amphetamine derivatives, cocaine, benztropine and tricyclic antidepressant drugs. *Eur. J. Pharmacol.,* 18:63–73.
12. Svensson, T. H. (1971): Functional and biochemical effects of d- and l-amphetamine on central catecholamine neurons. *Naunyn Schmiedebergs Arch. Pharmacol.,* 271:170–180.
13. Chiueh, C. C., and Moore, K. E. (1974): Relative potencies of d- and l-amphetamine on the release of dopamine from cat brain in vivo. *Res. Commun. Chem. Pathol. Pharmacol.,* 7:189–199.
14. Von Voigtlander, P. F., and Moore, K. E. (1973): Involvement of nigro-striatal neurons in the in vivo release of dopamine by amphetamine, amantadine and tyramine. *J. Pharmacol. Exp. Ther.,* 184:542–552.
15. Phillips, A. G., and Fibiger, H. C. (1973): Dopaminergic and noradrenergic substrates of positive reinforcement: Differential effects of d- and l-amphetamine. *Science,* 179:575–576.
16. Christie, J. E., and Crow, T. J. (1971): Turning behavior as an index of the action of amphetamines and ephedrines on central dopamine containing neurones. *Br. J. Pharmacol.,* 43:658–661.
17. Bunney, B. S. (1973): Reported at the Symposium on Catecholamines and Their Enzymes in the Neuropathology of Schizophrenia, May 18–21, 1973, Strasbourg, France. Cf. *J. Psychiatr. Res.,* 11:55, 1974.
18. Hökfelt, T., and Fuxe, K. (1972): On the morphology and the neuroendocrine role of the hypothalamic catechoamine neurons. In: *Brain-Endocrine Interaction: Median Eminence Structure and Function,* pp. 181–223. Karger, Basel.
19. Wurtman, R. J. (1974): Possible sites of action of monamines in hypothalamohypophyseal function. *Neurosci. Res. Prog. Bull.,* 9:214–217.
20. Brown, G. M., and Martin, J. B. (1973): Neuroendocrine relationships. *Prog. Neurol. Psychiatry,* 28:193–240.
21. Ganong, W. F. (1974): Minireview: The role of catecholamines and acetylcholine in the regulation of endocrine function. *Life Sci.,* 15:1401–1414.
22. Ganong, W. F. (1972): Evidence for a central noradrenergic system that inhibits ACTH secretion. In: *Brain-Endocrine Interaction: Median Eminence Structure and Function,* pp. 254–266. Karger, Basel.
23. Van Loon, G. R. (1973): Brain catecholamines and ACTH secretion. In: *Frontiers in Neuroendocrinology,* edited by F. W. Ganong and L. Martini, pp. 209–247. Oxford University Press, New York.
24. Scapagnini, U. (1973): Effect of drugs acting on brain monoamines and control of adrenocortical function. In: *Endocrinology,* edited by R. A. Scow, pp. 125–130. Excerpta Medica, Amsterdam.

25. Marks, B. H., Hall, M. M., and Bhattacharya, A. N. (1970): Psychopharmacological effects and pituitary-adrenal activity. *Prog. Brain Res.,* 32:58–70.
26. Lal, S., DeLaVega, C. E., Sourkes, T. L., and Friesen, H. G. (1973): Effect of apomorphine on growth hormone, prolactin, luteinizing hormone and follicle-stimulating hormone levels in human serum. *J. Clin. Endocrinol. Metab.,* 37:719–724.
27. Brown, W. A., VanWoert, M. A., and Ambani, L. M. (1973): Effect of apomorphine on growth hormone release in humans. *J. Clin. Endocrinol. Metab.,* 37:463–465.
28. Müller, E. E. (1973): Nervous control of growth hormone secretion. *Neuroendocrinology,* 11:338–369.
29. Müller, E. E., Pecile, A., Felici, M., and Cocchi, D. (1973): Norepinephrine and dopamine injection into lateral brain ventricle of the rat and growth hormone releasing activity in the hypothalamus and plasma. *Endocrinology,* 86:1376–1382.
30. Müller, E. E., Dal Pra, P., and Pecile, A. (1968): Influence of brain neurohormones injected into the lateral ventricle of the rat on growth hormone release. *Endocrinology,* 83: 893–896.
31. Müller, E. E., Sawano, S., Arimura, A., and Schally, A. V. (1967): Blockade of release of growth hormone by brain norepinephrine depletors. *Endocrinology,* 80:471–476.
32. Toivala, P. T. K., and Gale, C. C. (1972): Stimulation of growth hormone release by microinjection of norepinephrine into hypothalamus of baboons. *Endocrinology,* 90:895–902.
33. Boyd, A. E., Lebovitz, H. E., and Pfieffer, J. B. (1975): Stimulation of human growth hormone secretion by L-dopa. *N. Engl. J. Med.,* 238:1425–1429.
34. Blackard, W. G., and Heidingsfelder, S. A. (1968): Adrenergic receptor control mechanism for growth hormone secretion. *J. Clin. Invest.,* 47:1400–1414.
35. Besser, G. M., Butler, P. W. P., Landon, J., and Rees, L. (1969): Influence of amphetamines on plasma corticosteroid and growth hormone levels in man. *Br. Med. J.,* 4:528–530.
36. Jacoby, J. H., Greenstein, J. F., and Weitzman, E. D. (1974): The effect of monamine precursors on the release of growth hormone in the rhesus monkey. *Neuroendocrinology,* 14:95.
37. Janssen, P. A. J., Niemegeers, C. J. E., Schellekens, K. H. L., Dresse, A., Lenaerts, F. M., Pinchard, A., Schaper, W. K. A., Van Nueten, J. M., and Verbruggen, F. J. (1968): Pimozide, a chemically novel, highly potent and orally long-acting neuroleptic drug. *Arzneim. Forsch.,* 18:261–279.
38. Anden, N.-E., Butcher, S. G., Corrodi, H., Fuxe, K., and Ungerstedt, U. (1970): Receptor activity and the turnover of dopamine and noradrenaline after neuroleptics. *Eur. J. Pharmacol.,* 11:303–314.
39. Hollister, L. E., Curry, S. H., Derr, J. E., and Kanter, S. L. (1970): V. Plasma levels and urinary excretion of four different dosage forms of chlorpromazine. *Clin. Pharmacol. Ther.,* 11:49–59.
40. Curry, S. H., Marshall, J. H. L., Davis, J. M., and Janowsky, D. S. (1970): Chlorpromazine plasma levels and effects. *Arch. Gen. Psychiatry,* 22:289–296.
41. Lister, R. C., Underwood, L. E., Marshall, R. N., Friesen, H. G., and Van Wyk, J. J. (1974): Evidence for a direct effect of thyrotropin-releasing hormone (TRH) on prolactin release in humans. *J. Clin. Endocrinol. Metab.,* 39:1148–1150.
42. Dickerman, S., Kledzik, G., Gelato, M., Chen, H. J., and Meites, J. (1974): Effects of haloperidol on serum and pituitary prolactin LH and FSH, and hypothalamic PIF and LRF. *Neuroendocrinology,* 15:10–20.
43. Meltzer, H. Y., Sachar, E. J., and Frantz, A. G. (1974): Serum prolactin levels in unmedicated schizophrenic patients. *Arch. Gen. Psychiatry,* 31:564–569.
44. Matthysse, S. (1973): Antipsychotic drug actions: A clue to the neuropathology of schizophrenia. *Fed. Proc.,* 32:200–205.
45. Janssen, P. A. J., Niemegeers, C. J. E., Schellekens, H. K. L., and Lanaerts, F. M. (1967): Is it possible to predict the clinical effects of neuroleptic drugs (major tranquilizers) from animal data? *Arzneim. Forsch.,* 17:841–854.
46. Snyder, S., Greenberg, D., and Yamamura, H. I. (1974): Antischizophrenic drugs and brain cholinergic receptors. *Arch. Gen. Psychiatry,* 31:58–61.
47. Stevens, J. R. (1973): An anatomy of schizophrenia. *Arch. Gen. Psychiatry,* 29:177–189.
48. Nauta, W.: Personal communication.
49. Lichtensteiger, W. (1973): Changes in hypothalamic monoamines in relation to endocrine

states: functional characteristics of tuberoinfundibular dopamine neurons. In: *Endocrinology*, edited by R. O. Scow, pp. 131–137. American Elsevier, New York.

50. Simpson, G. M., and Varga, E. (1974): Clozapine—a new antipsychotic agent. *Curr. Ther. Res.*, 16:679–686.

51. Anden, N.-E., and Stock, G. (1973): Effect of clozapine on dopamine turnover in the corpus striatum and limbic system. *J. Pharm. Pharmacol.*, 25:346–348.

52. Bürki, H. R., Ruch, W., and Asper, H. (1975): Effects of clozapine, thioridazine, pulapine, and haloperidol on the metabolism of the biogenic amines in the brain of the rat. *Psychopharmacologia*, 41:27–33.

53. Bartholini, G., Haefely, W., Jaffre, M., Keller, H. H., and Pletscher, A. (1972): Effects of clozapine on cerebral catecholaminergic neurone systems. *Br. J. Pharmacol.*, 46:736–740.

54. Sedvall, G., and Nybäck, A. (1973): Effect of clozapine and some other antipsychotic agents on synthesis and turnover of dopamine formed from ^{14}C-tyrosine in mouse brain. *Isr. J. Med. Sci.*, 9 (Suppl.):24–30.

55. Clemens, J. A., Smalstig, E. B., and Sawyer, B. D. (1974): Antipsychotic drugs stimulate prolactin release. *Psychopharmacologia*, 40:123–127.

56. Vencovsky, E. (1967): Antisychoticke pusobeni thiethylperazinu a klinicke zkusenosti s jeho aplikaci v psychiatrii. (The antipsychotic effect of thiethylperazine and clinical findings—with its application in psychiatry.) *Cesk. Psychiatr.*, 63:1–8.

Hormones, Behavior, and Psychopathology, edited by
Edward J. Sachar. Raven Press, New York © 1976.

Serum Prolactin Levels in Schizophrenia—Effect of Antipsychotic Drugs: A Preliminary Report

H. Y. Meltzer and *V. S. Fang

*Department of Psychiatry, University of Chicago Pritzker School of Medicine, and Illinois
State Psychiatric Institute, Chicago, Illinois 60612; and *Department of Medicine,
University of Chicago Pritzker School of Medicine, Chicago, Illinois 60637*

The relationships between brain neurotransmitters and pituitary hormones have been a major focus of recent neuroendocrine research (1–3). Sachar pioneered in utilizing these relationships as probes to test current theories of neurotransmitter deficiencies or excesses in the affective psychoses and the schizophrenias (4–6), as well as the mechanisms of action of neuroleptic drugs (7). With his direct encouragement and suggestions of important areas for investigation, we have started to investigate dopaminergic activity in the schizophrenias with neuroendocrine techniques.

In this chapter we can report only on the serum prolactin levels in stored frozen samples that had been collected for a study of serum enzyme levels in psychiatric patients (8). Because prolactin is quite stable in frozen samples, there is no analytical problem as a result of even prolonged storage. However, we have not yet collected samples following a rigorous experimental design appropriate to these studies. Clearly this will be necessary to utilize neuroendocrine techniques fully in psychiatric biological and psychopharmacological research. Nevertheless, there are a number of interesting directions for further research which we believe emerge from the data already in hand, and we therefore think it is justified to report them at the current time.

First, a brief word about methodology. The patients employed in this study have already been described in reports on serum creatine phosphokinase activity in psychiatric patients (8). Our use of neuroleptic drugs is important to summarize since it is critical to the interpretation of some of the results. During the first week of hospitalization, all patients are generally given placebo. Thereafter the patients diagnosed as schizophrenic or acutely psychotic are given chlorpromazine or thioridazine (100 mg p.o. b.i.d.) or trifluoperazine (5 mg p.o. b.i.d.), all of which are considered equivalent in dosage. The dosage of the medication given the patient is increased by the equivalent of chlorpromazine (100 mg p.o. b.i.d.) until optimal clinical response is achieved and it is felt further increase is unlikely to produce additional benefit. Most patients who receive trifluoperazine also receive trihexyphenidyl (5 mg p.o. q.i.d. or b.i.d.). Medication is given at 9 A.M. and 9 P.M. Serum samples are obtained Monday through Friday at 8 A.M. throughout hospitalization. Serum prolactin levels are determined by a double-antibody radioimmunoassay method (9). A highly purified human prolactin (72–11–23) and a highly specific rabbit antiprolactin antiserum (65–5) were kindly supplied by Dr. Henry Friesen. A less pure human prolactin preparation (Friesen 73–4–27) was used as standard reference. The assay sensitivity is 0.05 ng of the weight of the standard or a concentration of 1 ng/ml. The intra-assay variation is less than 5%, and the interassay variation is less than 15% with samples containing prolactin ranging from 2 to 200 ng/ml.

SERUM PROLACTIN LEVELS IN NEWLY ADMITTED PSYCHIATRIC PATIENTS

Secretion of prolactin by the pituitary in both man and laboratory animals can be influenced by a number of factors, such as administration of serotonin precursors (10), anticholinergic drugs (11), cholinomimetic drugs (12), antiadrenergic drugs (13), prostaglandins (14), thyrotropin-releasing hormone (15), the gonadal hormones estradiol and testosterone (16), and amino acids such as arginine, leucine, and phenylalanine (17). However, the major influence on prolactin levels in serum appears to be inhibition of prolactin secretion from the pituitary by dopamine. Evidence for this has been summarized elsewhere (2,3,7,18–21), as well as by Wurtman and Fernstrom in this volume. At one time it was widely held that dopamine from tuberoinfundibular (TI) dopamine neurons controlled the release of prolactin-inhibiting factor (PIF), which inhibited prolactin secretion (2,3), but it now appears that TI dopamine, which reaches the pituitary via the pituitary-portal circulation, may directly inhibit prolactin release from the pituitary (22,23). A major theory of the etiology of schizophrenia is that it is due to a functional increase in dopamine activity in at least some brain dopamine tracts, possibly the mesolimbic (ML) and/or recently discovered cortical dopamine tracts (24–27). In rats stimulating ML neurons increases the activity of TI neurons (28). Conceivably, then, there might be a hyperdopaminergic state in the ML and TI tracts in some schizophrenics, especially in the acute state, which might lead to decreased serum prolactin levels since increased dopamine activity in the TI tract is associated with decreased serum prolactin (20,21,28). Thus if serum prolactin levels were depressed in untreated acute schizophrenics, it would tend to confirm the dopamine theory of schizophrenia.

Meltzer, Sachar, and Frantz studied 8 A.M. serum prolactin in 16 newly admitted, unmedicated acute schizophrenics and five chronic schizophrenics, and found normal serum prolactin levels in all (6). We have now studied serum prolactin levels in an additional 31 unmedicated newly admitted schizophrenic patients. For most of the patients, we studied two or three samples during the first week of hospitalization and in almost all instances found less than a 25% variation in samples collected several days apart. Although it is preferable to study multiple samples from the same subject obtained at frequent intervals on a given day or days, the fact that samples obtained over several days showed so little variation indicates that these values may accurately reflect the mean level of prolactin secretion at that time of day. There is a variation in prolactin secretion at different times of day in relation to the sleep cycle, such that the peak prolactin level occurs generally just before awakening (29). Since many of our patients had marked sleep disturbance, this could very well affect the levels at 8 A.M. Future studies should certainly control for this variable. The results are given in Table 1 and Fig. 1. One of the 31 schizophrenic patients had a serum prolactin level below the 95% confidence limit. Four of the 31 schizophrenic patients (12.9%) had elevated serum prolactin levels. The significance of this is discussed subsequently.

We also studied serum prolactin levels in 8 A.M. samples from three manicdepressive patients, manic phase; nine psychotically depressed patients; and five nonpsychotic patients with mixed anxiety and depressive symptomatology. All 17 subjects had normal serum prolactin levels (Table 1). Meltzer et al. had previously found slightly elevated 8 A.M. serum prolactin levels in both of two manic patients and in one patient with severe anxiety neurosis (6).

Meltzer and Sachar discussed the results of their first study of serum prolactin in newly admitted psychiatric patients as they related to both the dopamine theory of schizophrenia and the effect of stress on serum prolactin in man and rat. *Acute* stresses elevate serum prolactin levels markedly in both man and laboratory animals (30–32). Many newly admitted psychi-

TABLE 1. *Serum prolactin levels in newly admitted psychiatric patients*

Subjects	No.	Prolactin (ng/ml)	
		Mean ± SD	Range
Normal controls			
Males	60	22 ± 6.5	7.6–37
Females	48	31.8 ± 13.7	6.5–51
Schizophrenics			
Males	15	22.0 ± 10.2	5–45
Females	16	29.0 ± 13.1	12.6–54
Manic-depressive, manic			
Females	3	28.2 ± 7.4	20.8–35.5
Psychotic depression			
Males	3	14.6 ± 5.4	7.4–20.5
Females	6	19.8 ± 4.5	15.3–26
Nonpsychotic			
Males	5	21.4 ± 7.2	15.7–35.5

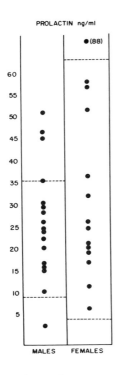

FIG. 1. Serum prolactin levels in unmedicated, newly admitted male and female schizophrenic patients.

atric patients, especially those with acute psychoses, severe depressions, and anxiety states, appear to be severely stressed, give self-reports of experiencing severe stress, and have elevated serum cortisol levels or cortisol production rates (33,34). Conceivably then, newly admitted psychiatric patients might have elevated serum prolactin levels on the basis of stress. The results of the Meltzer and Sachar study (6) and this study do indeed indicate elevated serum prolactin levels in some newly admitted psychiatric patients of all diagnostic types, but it appears to be a small minority of such patients. We studied serum cortisol levels in the admission samples of 11 schizophrenic patients of this sample in an effort to determine if this might serve as a biological indicator of the severity of stress. Three of the subjects chosen were those with elevated prolactin. The normal upper limit for cortisol in our laboratory is 25 ng/100 ml for samples collected at 8:00 A.M. Ten of the 11 schizophrenic patients had normal serum cortisol including all three patients with elevated prolactin. Four of the 11 patients appeared to be experiencing extreme turmoil as part of their acute psychoses, but only one of these had a slightly elevated serum cortisol level (28.6 ng/100 ml). The mean ± 1 SD serum cortisol for the 11 patients was 17.2 ± 5.3 ng/ml. From these limited data it appears that neither serum prolactin nor cortisol in single 8 A.M. samples reflect the psychological turmoil of psychotic patients. Conceivably, the effects of acute stress on cortisol and prolactin are not present in the relatively prolonged stress of even an acute psychosis. However, because cortisol is secreted in bursts (35) and the circadian rhythms of schizophrenics must be markedly disturbed during the acute phase, it is quite possible that single cortisol levels are misleading.

The possibility that the normal levels of prolactin in newly admitted psychiatric patients were the results of a balance between the increase in prolactin release due to stress and the decrease which would occur from increased dopamine was considered by Meltzer et al. (6). Little is known about the mechanisms which underlie the effect of stress on prolactin secretion. Meltzer, Fang, and Daniels found that increased brain dopaminergic activity produced by L-DOPA (dopamine precursor), apomorphine (a dopamine agonist), or pargyline (an inhibitor of monoamine oxidase) prevents the increase in serum prolactin due to restraint stress in male rats (*unpublished data*). Thus increased dopaminergic activity in the TI pituitary tract could suppress any increase in prolactin due to stress in psychiatric patients. In the five subjects in this study who had low or high levels of prolactin, there might be more or less dopamine activity, respectively, relative to stress. If this hypothesis of a stress-induced increase and a dopamine-induced decrease in serum prolactin balancing each other is correct, it would be consistent with the hypothesis of increased dopaminergic activity in schizophrenia. However, it is now clear that almost all newly admitted psychiatric patients, not just schizophrenics, have normal serum prolactin. Increased dopaminergic activity in the TI tract might be present in all chronically stressed patients, not just schizophrenics, since stress is known to increase dopamine turnover and tyrosine hydroxylase activity, the rate-limiting step in dopamine synthesis (36,37). Hyperdopaminergic activity of the ML or cortical tracts may be what is unique to schizophrenia, if indeed that theory is correct. Conceivably, studying (a) the circadian rhythms in secretion of prolactin in unmedicated newly admitted psychiatric patients including acute schizophrenic patients, (b) the effects of additional stresses, or (c) the effect of drugs which affect dopaminergic activity might still demonstrate there is something unique about dopamine control of prolactin in schizophrenia. Therefore the status of the TI dopaminergic tract in unmedicated schizophrenics requires further investigation.

EFFECT OF PROLONGED ADMINISTRATION OF PHENOTHIAZINES ON SERUM PROLACTIN LEVELS

There have been numerous reports of increased levels of serum prolactin in normal subjects receiving single doses of phenothiazine and in psychiatric patients who have received phenothiazine acutely or chronically (7,38–40). In almost all subjects the phenothiazines increase serum prolactin, most likely due to their capacity to block TI-pituitary dopamine receptors but also possibly due to inhibition of impulse flow dopamine release (24,41).

There have been no reports of longitudinal changes in serum prolactin levels in psychiatric patients given phenothiazines. Serum prolactin levels and global ratings of psychotic behavior for two schizophrenic patients are given in Figs. 2 and 3. It is apparent that serum prolactin levels, 12 hours after the last previous dose of neuroleptic drugs, are elevated within the first 1 to 2 days of drug administration. (Lack of elevation in serum prolactin in newly admitted patients has proved to be of value as further confirmation that they have not received neuroleptic drugs for at least 24 to 48 hours.) Thereafter they rise to a level which is relatively constant for a given individual at a given dosage. The coefficient of variation (SD/mean × 100) was generally 15% to 30%. We have observed this in 29 schizophrenic patients treated with thioridazine, trifluoperazine, chlorpromazine, or haloperidol. We now wish to discuss these levels as a function of dosage, drug, sex, duration of treatment, and relationship to clinical response.

We obtained two to 18 serum prolactin values (mean, seven) for a given patient. In 16 patients prolactin was studied at several dosage levels. Mean prolactin levels generally but not always increased as the dosage was raised. In four of 16 patients prolactin levels remained stable or actually declined slightly as the dosage was raised by the equivalent of 200 mg chlorpromazine.

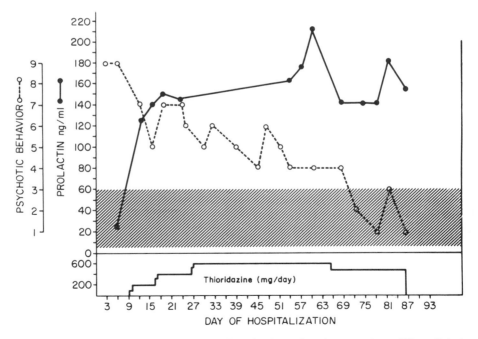

FIG. 2. Serum prolactin levels and psychotic behavior ratings in an acute undifferentiated schizophrenic female. Note that prolactin levels are elevated prior to marked improvement and remain elevated for at least 9 weeks.

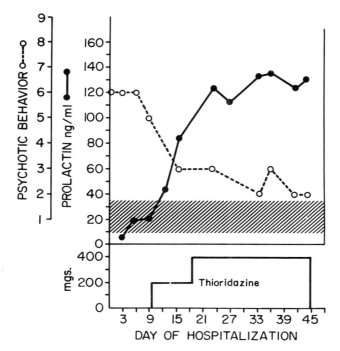

FIG. 3. Serum prolactin levels and psychotic behavior ratings in an acute paranoid schizophrenic male. Normal prolactin level is stippled. Note slightly decreased serum prolactin prior to treatment. Prolactin levels parallel the clinical improvement.

The magnitude of the increase in serum prolactin varied markedly as dosage was increased. In the patient illustrated in Fig. 1, serum prolactin at 200 mg thioridazine was 132 ± 8 ng/ml; at 400 mg, 147 ± 2.5 ng/ml; at 600 mg, 164 ± 22 ng/ml. Another female patient has a serum prolactin of 89 ± 18 ng/ml after 200 mg thioridazine, 294 ± 57 ng/ml after 400 mg thioridazine, and 310 ± 40 ng/ml after 1,000 mg thioridazine.

Thioridazine (400 mg p.o. q.i.d.) produced larger increases in mean serum prolactin in both males and females compared to chlorpromazine (400 mg) and trifluoperazine (20 mg) plus trihexyphenidyl (10 mg/kg), which produced similar increases (Table 2). The range in mean prolactin levels for males at this dosage level, including all drugs, was 49 to 136 ng/ml (2.6-fold increase). The range in mean prolactin levels for females at this dosage level was 79 to 294 ng/ml (2.7-fold increase). Mean levels in females at this dosage were approximately twice as high as in males (Table 2). It is not possible to determine the effect of the anticholinergic trihexyphenidyl on serum prolactin in man from these studies.

The rapid increase in serum prolactin after beginning treatment with neuroleptics in most patients (7), with further increases with increasing dosage, indicates that blockade of pituitary dopamine receptors begins very

TABLE 2. *Serum prolactin levels in males and females after thioridazine, chlorpromazine, or trifluoperazine*

Drug	No. of subjects	Dose (mg)	Prolactin (ng/ml)	
			Means ± SD	Range
Thioridazine		400		
Males	5		95 ± 26	65–136
Females	5		237 ± 64	147–294
Chlorpromazine		400		
Males	2		57 ± 1	56–57
Females	2		120 ± 10	79–141
Trifluoperazine +		20		
trihexyphenidyl		10		
Males	2		63 ± 14	49–76
Females	5		137 ± 58	64–224

rapidly after initiating treatment but is generally only partially complete with low dosages. It is of interest to contrast this with the time course of clinical response, which is generally not rapid (Fig. 1) but usually varies as a function of dosage. If it is assumed that the cortical or mesolimbic dopamine receptors are blocked as readily as the pituitary dopamine receptors, then it is not clear why significant clinical response to the neuroleptic drugs frequently requires weeks to months even in acute-onset, good-premorbid schizophrenic patients in whom it is not necessary to overcome the effect of chronically disturbed character traits and deficits in socialization. This could suggest that simple dopamine blockade is but one facet of the mechanism of action of the neuroleptic drugs.

The variation in the magnitude of prolactin increase in different patients for a given dosage and drug could be the result of differences in: (a) actual drug intake; (b) absorption and metabolism of the phenothiazines; (c) sensitivity of dopamine receptors to dopamine blockade; (d) other possible prolactin secretion control mechanisms, such as decreased cholinergic activity; (e) clearance of prolactin from blood. It is also conceivable that biological factors related to the psychosis itself affect the magnitude of the serum prolactin increase. For example, the increase in serum prolactin produced by the phenothiazines can be blocked by prior L-DOPA administration (38). Conceivably, if some psychotic patients were producing large amounts of dopamine that reached the pituitary by the pituitary-portal circulation, this would tend to antagonize the effects of phenothiazines and thus lead to smaller increases in prolactin. As previously indicated, a similar effect may explain the lack of increase in prolactin levels in newly admitted, severely stressed psychotic patients. Spinal fluid levels of homovanillic acid (HVA), the major metabolite of dopamine, are not generally elevated in psychotic patients (42,43), but this does not rule out the possibility of selective increases in dopamine turnover in small dopamine tracts such as the TI, ML, or cortical tracts.

If the increase in serum prolactin levels after phenothiazines is the result of dopaminergic blockade or interference with dopamine release, then the magnitude of the increase should reflect the extent to which this process occurs. Conceivably this might also be an index of interference with dopaminergic transmission in the dopaminergic areas of the brain that are most relevant to schizophrenia, e.g., the ML or cortical tracts. If this is so, serum prolactin levels could provide some index of clinical responsiveness to phenothiazine. To test this possibility, we rank-ordered patients on the basis of rate and extent of decline in daily ratings of psychotic behavior. The rankings were done blindly by two psychometricians. We next rank-ordered prolactin levels at the equivalent of 400 mg chlorpromazine. The Spearman correlation (r) for psychosis ratings and prolactin levels for males was .532 ($p < 0.05$); for females it was .150 ($p > 0.25$). A positive correlation indicates a tendency for the good responders to have higher serum prolactin at this dose level. Many more patients of both sexes will need to be studied before there can be more confidence about the relationship between serum prolactin and clinical response. Studying serum prolactin levels after a standardized intramuscular injection in schizophrenics may be a superior method of investigating this issue.

We also calculated the correlation between the maximum dose of medication prescribed to induce remission and mean serum prolactin levels at the dosage of 400 mg chlorpromazine or its equivalent. The maximum dose prescribed may be taken as a crude index of absorption, metabolism, and rapidity and extent of responsiveness to antipsychotic effects of the drug. Many of the people who received higher doses of medication ultimately responded as well as the patients who responded quickly to small doses. We hypothesized that the more responsive patients would show higher prolactin levels than the less responsive patients at a given dosage. For 12 males the product-moment r was .0726 ($p > 0.5$); for 11 females it was .299 ($p > 0.5$). Thus there were no significant relationships between peak prescribed dose and prolactin. This means 8 A.M. prolactin levels may not be of value as an index of ultimate drug needs. Again, the rate and magnitude of increase in prolactin after a standardized intramuscular dose may be more useful as a predictor of ultimate drug needs.

DECREASE IN PROLACTIN AFTER DISCONTINUING NEUROLEPTICS

Medication was temporarily discontinued in eight patients who had been receiving neuroleptics at doses of 200 to 800 mg chlorpromazine or its equivalent for periods of 2 to 3 months. Four male patients had prolactin levels of 60 to 95 ng/ml and four female patients levels of 150 to 230 ng/ml when medication was ceased. In all patients prolactin levels were decreased by 30% to 70% 24 hours after the last dose and were within normal limits by 48 to 72 hours.

This indicates that blockade of TI pituitary dopamine receptors is readily reversible, just as its onset is rapid. One of the eight patients showed a marked intensification of psychotic symptoms as active medication was withdrawn temporarily. The seven others showed no change. Thus the correlation between decrease in TI pituitary dopamine receptor blockade and intensification of psychotic symptoms is not very strong. The rapidity with which serum prolactin levels return to normal after discontinuing neuroleptic drugs, in comparison with the prolonged time in which neuroleptic drugs or their metabolites can be found in urine and the usual lack of rapid clinical deterioration after discontinuing neuroleptic medication, is surprising. If the rapid restoration of TI pituitary dopamine receptors to normal is not paralleled by restoration of ML cortical dopaminergic function, this raises serious questions as to the utility of the TI pituitary system as a model for the ML cortical dopamine system. If dopamine blockade is rapidly terminated in both systems, the question arises as to why clinical deterioration is not more rapid.

It is important to point out that our results differ from those of previous investigators. Prolactin levels did not return to normal for 2 to 3 weeks in some schizophrenic patients following termination of neuroleptic treatment (44). Since increased dopaminergic activity can antagonize the effects of phenothiazines (38), it is possible that the rapid decline in serum prolactin in some patients after discontinuing neuroleptic drugs is an indication that there is hyperdopaminergic activity in some schizophrenics but not in all. Beaumont et al. reported that prolactin levels remained elevated for 1 to 3 weeks when neuroleptic drugs were discontinued in a small group of patients who had been treated for at least 10 years (40). Conceivably the very long duration of treatment in these patients accounts for the difference between them and our patients. In the patients of Beaumont et al., as in ours, there was no relationship between clinical change and prolactin levels (40).

The difference between the time of onset and disappearance of TI pituitary dopamine receptor blockade by the neuroleptic drugs and their effect on clinical state is an important area for further investigation. Conceivably it is due to the fact that there are major differences between the TI pituitary dopamine receptor and the ML cortical dopamine receptor. However, there is considerable evidence that they are similar. For example, in both man and laboratory animals, the capacity of phenothiazines to produce an increase in serum prolactin parallels their clinical efficacy (45; see also Sachar et al., *this volume*). There is evidence that both the TI pituitary dopamine receptor, the ML, and the striatal dopamine receptor are dopamine-stimulated adenyl cyclases (46–49), although they might be isozymes with different properties. If there are major differences between the TI pituitary and ML cortical dopamine receptors, this could indicate there is limited utility in studying serum prolactin levels in psychotic patients as an index of dopaminergic function in more rostral brain areas.

If the TI pituitary dopamine receptors and the ML cortical dopamine receptors respond similarly to neuroleptics, the usual marked differences in time of onset and decay of the serum prolactin effect and the clinical effect of the neuroleptics has implications for the dopamine hypothesis of schizophrenia. It could suggest that dopaminergic blockade – if it is important at all to the clinical effect of the antipsychotics – is only the initial event in a much more complex sequence of neuronal events. Conceivably events with a comparatively long half-life (e.g., slow axonal transport, induction of synthesis of new enzymes, or axonal sprouting) which might be dependent on dopaminergic neuronal activity must be altered before the antipsychotic effect is manifest. For example, on the basis of finding considerable sprouting of subterminal motor nerve endings in psychotic patients, one of us has speculated that comparable processes might occur in the central nervous system and play an important if not decisive role in recovery from a psychotic episode, which might be conceived of as a type of acute brain disease (50). It is conceivable that the neuroleptic drugs via their capacity to block dopaminergic receptors or to interfere with impulse flow dopamine release might either facilitate sprouting that is restorative or inhibit disruptive sprouting. Either effect would presumably take days to weeks to become apparent in terms of behavior. It must also be recognized that any or all of these longer-term events which might be influenced by neuroleptics may have nothing to do with dopamine neurons; but in light of all the evidence for a role of dopamine in the etiology of schizophrenia, this seems unlikely.

In any event, the rapid decay of the effect of neuroleptic drugs on serum prolactin suggests that for some pharmacological studies in man it may be appropriate to wait only brief intervals after discontinuing medication, particularly if the subject has been receiving medication for only a brief time. This might be especially true for studies involving dopamine receptors. Other effects of neuroleptics might be much more persistent.

EFFECT OF IMIPRAMINE AND LITHIUM CARBONATE ON SERUM PROLACTIN LEVELS

Turkington measured serum prolactin levels using a bioassay in 20 psychiatric patients who had been treated with imipramine or amitryptyline (dose not specified) for at least 2 weeks (44). All 20 patients had elevated serum prolactin levels; in five the level was > 100 ng/ml. No lithium-treated patients were studied.

We studied serum prolactin in six patients treated with 150 to 300 mg imipramine for 4 to 8 weeks, and in five patients treated with lithium carbonate we used 1,200 mg/day for up to 2 months. In all but one patient serum prolactin levels remained at or near the upper limits of normal. In one man with bipolar illness who was first studied during a severe depressive phase, serum prolactin levels ranged from 16 to 27 ng/ml. Following an excellent response to electroconvulsive treatment (ECT), he was treated with lithium carbonate. During the next 4 weeks serum prolactin levels were measured six times and ranged from 18 to 68.5 ng/ml (mean 45.4 ± 12.6 ng/ml). No prolactin levels were determined following ECT and prior to lithium carbonate, so it is not clear if the elevated prolactin was related to ECT, lithium, or neither. In view of the blockade of serotonin reuptake by imipramine (51), and of the numerous effects of lithium on brain norepinephrine and serotonin metabolism, which are believed to produce a decrease in the functional availability of both amines (52–55), these results suggest that norepinephrine and serotonin have little influence on the basal secretion of prolactin in man. Furthermore, if it is correct that dopamine is the main determinant of prolactin secretion, this suggests that imipramine and lithium have little effect on TI pituitary dopaminergic function in man. Effects on other dopaminergic tracts are not precluded.

SUMMARY

Serum prolactin levels in 8 A.M. samples from 31 unmedicated, newly admitted schizophrenic patients were generally within normal limits. Similar results were observed in 17 nonschizophrenic newly admitted psychiatric patients. Serum cortisol levels were also normal in 10 of 11 schizophrenics. These results are discussed in reference to the dopamine theory of schizophrenia. Increased brain dopaminergic activity in psychiatric patients under stress could inhibit the increase in serum prolactin which usually develops during acute stress.

Phenothiazine administration promptly elevated serum prolactin levels in patients. Levels reached a relatively steady plateau, which usually increased as dosage was increased. When medication was stopped, levels returned to normal within 48 to 72 hours. In male (but not female) schizophrenic patients, there was a significant correlation between clinical response and magnitude of prolactin levels. Imipramine and lithium carbonate did not increase serum prolactin levels.

ACKNOWLEDGMENTS

This work was supported in part by United States Public Health Service (25,116) and a grant from the Department of Mental Health, State of Illinois (431-13-RD).

Dr. Meltzer is the recipient of USPHS Career Development Award MH 47,808.

REFERENCES

1. Brown, G. M., and Martin, J. B. (1973): Neuroendocrine relationships. *Prog. Neurol. Psychiatry,* 28:193–240.
2. McCann, S. M., Kalra, P. S., Donoso, A. P., Bishop, W., Schneider, H. P. G., Fawcett, C. P., and Krulich, L. (1972): The role of monoamines in the control of gonadotropin and prolactin secretion. In: *Median Eminence: Structure and Function,* edited by K. M. Knugge, D. F. Scott, and A. Weindl, pp. 224–235. Karger, Basel.
3. Meites, J., Lu, K.-H., Wuttke, W., Welsch, C. W., Nagasawa, H., and Quadri, S. K. (1972): Recent studies on function and control of prolactin secretion in rats. *Recent Prog. Horm. Res.,* 28:471–526.
4. Sachar, E. J., Mushrush, G., Perlow, M., Wertzman, E., and Sassin, J. (1972): Growth hormone responses to L-dopa in depressed patients. *Science,* 178:1304–1305.
5. Sachar, E. J., Frantz, A., Altman, N., and Sassin, J. (1963): Growth hormone and prolactin in depressed and manic-depressive patients: Responses to L-dopa and hypoglycemia. *Am. J. Psychiatry,* 130:1362–1967.
6. Meltzer, H. Y., Sachar, E. J., and Frantz, A. G. (1974): Serum prolactin levels in unmedicated schizophrenic patients. *Arch. Gen. Psychiatry,* 31:564–569.
7. Meltzer, H. Y., Sachar, E. J., and Frantz, A. G. (1975): Dopamine antagonism by thioridazine in schizophrenia. *Biol. Psychiatry,* 10:53–57.
8. Meltzer, H. Y. (1975): Neuromuscular abnormalities in the major mental illnesses. I. Serum enzyme studies. In: *The Biology of the Major Psychoses: A Comparative Analysis,* edited by D. X. Freedman. Raven Press, New York (*in press*).
9. Hwang, P., Guyda, H., and Friesen, H. (1971): A radioimmunoassay for human prolactin. *Proc. Natl. Acad. Sci. U.S.A.,* 68:1902–1906.
10. Kalo, Y., Nakai, Y., Imura, H., Chichara, K., and Ohgo, S. (1974): Effect of 5-hydroxytryptophane (5-HTP) on plasma prolactin levels in man. *J. Clin. Endocrinol. Metab.,* 38:695–697.
11. Libertun, C., and McCann, S. M. (1973): Blockade of the release of gonadotropins and prolactin by subcutaneous or intraventricular injection of atropine in male and female rats. *Endocrinology,* 92:1714–1724.
12. Grandison, L., Gelato, M., and Meites, J. (1974): Inhibition of prolactin secretion by cholinergic drugs. *Proc. Soc. Exp. Biol. Med.,* 145:1236–1239.
13. Lawson, D. M., and Gala, R. R. (1975): The influence of adrenergic, dopaminergic, cholinergic and serotoninergic drugs on plasma prolactin levels in ovariectomized, estrogen-treated rats. *Endocrinology,* 96:313–318.
14. Harms, P. G., Ojeda, S. R., and McCann, S. M. (1973): Prostaglandin involvement in hypothalamic control of gonadotropin and prolactin release. *Science,* 181:760–761.
15. Jacobs, L. S., Snyder, P. J., Wilber, J. F., Utiger, R. D., and Daughaday, W. H. (1971): Increased serum prolactin after administration of synthetic thyrotropin releasing hormone (TRH) in man. *J. Clin. Endocrinol. Metab.,* 33:996–998.
16. Kalra, P. S., Fawcett, C. P., Krulich, L., and McCann, S. M. (1973): The effects of gonadal steroids on plasma gonadotropins and prolactin in the rat. *Endocrinology,* 92:1256–1268.
17. Davis, S. L. (1972): Plasma levels of prolactin, growth hormone, and insulin in sheep following the infusion of arginine, leucine and phenylalanine. *Endocrinology,* 92:1256–1268.
18. Kamberi, I. A., Mical, R. S., and Porter, J. C. (1971): Effect of anterior pituitary perfusion and intraventricular injection of catecholamines on prolactin release. *Endocrinology,* 88:1012–1020.
19. Meites, J., and Clemens, J. (1972): Hypothalamic control of prolactin secretion. *Vitamin Horm.,* 30:165–221.
20. Donoso, A. O., Bishop, W., Fawcett, C. P., Krulich, L., and McCann, S. M. (1971): Effects of drugs that modify brain monoamine concentrations on plasma gonadotropin and prolactin levels in the rat. *Endocrinology,* 89:774–784.
21. Lu, K.-H., Amenomori, Y., Chen, C.-L., and Meites, J. (1970): Effects of central acting drugs on serum and pituitary prolactin levels in rats. *Endocrinology,* 87:667–672.
22. Macleod, R. M., and Lehmeyer, J. E. (1974): Studies on the mechanism of the dopamine-mediated inhibition of prolactin secretion. *Endocrinology,* 94:1077–1085.

23. Sachar, C. J., and Clemens, J. A. (1974): The role of catecholamines in the release of anterior pituitary prolactin in vitro. *Endocrinology,* 95:1202–1212.
24. Carlsson, A., and Linquist, M. (1963): Effect of chlorpromazine or haloperidol on the formation of 3-methoxytryramine and normetanephrine in mouse brain. *Acta Pharmacol. Toxicol.,* 20:140–144.
25. Matthysse, S. (1973): Antipsychotic drug actions: A clue to the neuropathology of schizophrenia. *Fed. Proc.,* 32:200–205.
26. Snyder, S. H., Banerjee, S. P., Yamamura, H. I., and Greenberg, D. (1974): Drugs, neurotransmitters and schizophrenia. *Science,* 184:1243–1253.
27. Thierry, A. M., Blanc, G., Sobel, A., Stinus, L., and Glowinski, J. (1973): Dopaminergic terminals in the rat cortex. *Science,* 182:499–501.
28. Lichtensteiger, W., and Keller, P. J. (1974): Tubero-infundibular dopamine neurons and the secretion of luteinizing hormone and prolactin extrahypothalamic influences, interaction with cholinergic systems and the effect of urethane anesthesia. *Brain Res.,* 74:279–303.
29. Sassin, D. F., Frantz, A., Kapen, S., and Weitzman, E. (1973): The nocturnal rise of human prolactin is dependent on sleep. *J. Clin. Endocrinol. Metab.,* 37:436–440.
30. Neill, J. D. (1970): Effect of "stress" on serum prolactin and luteinizing hormone levels during the estrous cycle of the rat. *Endocrinology,* 87:1192–1197.
31. Ajika, K., Kalra, S. P., Fawcett, CP., Krulich, L., and McCann, S. M. (1972): The effect of stress and nembutal on plasma levels of gonadotropins and prolactin in ovariectomized rats. *Endocrinology,* 90:707–715.
32. Noel, G. L., Suh, H. K., Stone, G. J., and Frantz, A. F. (1972): Human prolactin and growth hormone release during surgery and other conditions of stress. *J. Clin. Endocrinol. Metab.,* 35:840–851.
33. Sachar, E. J., Kanter, S. S., Buie, D., Engle, R., and Mehlamn, R. (1970): Psychoendocrinology of ego disintegration. *Am. J. Psychiatry,* 126:1067–1078.
34. Sachar, E. J., Hellman, L., Fukushima, D., and Gallagher, T. (1970): Cortisol production in depressive illness: A biochemical and clinical clarification. *Arch. Gen. Psychiatry,* 23:289–298.
35. Hellman, L., Nakada, F., Curti, J., Weitzman, E. D., Kream, J., Roffwarg, H., Ellman, S., Fukushima, D. K., and Gallagher, T. F. (1970): Cortisol is secreted episodically by normal man. *J. Clin. Endocrinol. Metab.,* 30:411–422.
36. Bliss, E. L., Ailion, J., and Swanziger, J. (1968): Metabolism of norepinephrine, serotonin and dopamine inrat brain with stress. *J. Pharmacol. Exp. Ther.,* 164:122–134.
37. Thoenen, H., Otten, U., and Oesch, F. (1973): Trans-synaptic regulation of tyrosine hydroxylase. In: *Frontiers in Catecholamine Research,* edited by E. Usdin and S. Snyder, pp. 170–185. Pergamon Press, New York.
38. Kleimberg, D. L., Noel, G. L., and Frantz, A. G. (1971): Chlorpromazine stimulation and L-dopa suppression of plasma prolactin in man. *J. Clin. Endocrinol. Metab.,* 33:873–876.
39. Friesen, H., Guyda, H., Hwang, P., Tyson, J. E., and Barbeau, A. (1972): Functional evaluation of prolactin secretion: A guide to therapy. *J. Clin. Invest.,* 51:706–709.
40. Beumont, P. J. V., Corker, C. S., Friesen, H. G., Kolakowska, T., Mandelbrote, B. M., Marshall, J., Murray, M. A. F., and Wiles, D. H. (1974): The effects of phenothiazines on endocrine function. II. Effects in men and post-menopausal women. *Br. J. Psychiatry,* 124:420–430.
41. Seeman, P. (1972): The membrane action of anesthetics and tranquilizers. *Pharmacol. Rev.,* 24:583–656.
42. Rimon, R., Roos, B.-E., Rakkolainen, V., and Alanen, Y. (1973): The content of 5-HIAA and HVA in the CSF of patients with acute schizophrenia. *J. Psychosom. Res.,* 15:375–378.
43. Bowers, M. B., Jr. (1973): 5-Hydroxyindoleacetic acid (5-HIAA) and homovanillic acid (HVA) following probenecid in acute psychotic patients treated with phenothiazines. *Psychopharmacologia,* 28:309–318.
44. Turkington, R. W. (1972): The clinical endocrinology of prolactin. *Adv. Intern. Med.,* 18:363–387.
45. Clemens, J. A., Smalstig, E. B., and Sawyer, B. D. (1974): Antipsychotic drugs stimulate prolactin release. *Psychopharmacologia,* 40:123–127.
46. Ojeda, S. R., Harms, P. G., and McCann, S. M. (1974): Possible role of cyclic AMP and

prostaglandin E on the dopaminergic control of prolactin release. *Endocrinology*, 94:1650–1657.

47. Karobath, M., and Leitch, H. (1974): Antipsychotic drugs and dopamine-stimulated adenylate kinase prepared from corpus striatum of rat brain. *Proc. Natl. Acad. Sci. U.S.A.*, 71:2915–2918.

48. Miller, R. J., Horn, A. S., and Iversen, L. J. (1974): The action of neuroleptic drugs on dopamine-stimulated adenosine cyclic 3'-5'-monophosphate production in rat neostriatum and limbic forebrain. *Mol. Pharmacol.*, 10:759–766.

49. Clement-Cormier, Y. C., Kebabian, J. W., Petzold, G. L., and Greengard, P. (1974): Dopamine-sensitive adenylate cyclase in mammalian brain: A possible site of action of antipsychotic drugs. *Proc. Natl. Acad. Sci. U.S.A.*, 71:1113–1117.

50. Meltzer, H. Y., and Crayton, J. W. (1975): Neuromuscular abnormalities in the major mental illnesses. II. Muscle fiber and subterminal motor nerve abnormalities. In: *The Biology of the Major Psychoses: A Comparative Analysis,* edited by D. X. Freedman. Raven Press, New York (*in press*).

51. Corrodi, H., and Fuxe, K. (1968): The effect of imipramine on central monoamine neurons. *J. Pharm. Pharmacol.*, 20:230–232.

52. Colburn, R. W., Goodwin, F. K., Bunney, W. E., Jr., and Davis, J. M. (1967): Effect of lithium on the uptake of noradrenaline by synaptosomes. *Nature (Lond.)*, 215:1395–1397.

53. Katz, R. I., Chase, T. N., and Kopin, T. J. (1968): Evoked release of norepinephrine and serotonin from brain slices: Inhibition by lithium. *Science,* 162:466–467.

54. Ho, A.K.S., Loh, H. H., Graves, F., Hitzemann, R. J., and Gershon, S. (1970): The effect of prolonged lithium treatment on the synthesis rate and turnover of monoamines in brain regions of rats. *Eur. J. Pharmacol.*, 10:72–78.

55. Segaiva, T., and Nakano, M. (1974): Brain serotonin metabolism in lithium treated rats. *Jap. J. Pharmacol.*, 24:319–324.

Hormones, Behavior, and Psychopathology, edited by
Edward J. Sachar. Raven Press, New York © 1976.

Neuroendocrine Regulation in Affective Disorders

Bernard J. Carroll and *Joseph Mendels

*Department of Psychiatry, University of Michigan, Ann Arbor, Michigan 48104; and
*Department of Psychiatry, University of Pennsylvania, and Depression Research Unit,
Veterans Administration Hospital, Philadelphia, Pennsylvania 19104*

This chapter is a review of studies describing neuroendocrine dysfunction in the affective disorders, with particular emphasis on work reported since 1970. Before that date, few investigators were carrying out dynamic, physiologic studies of neuroendocrine control mechanisms, and we realize now that significant advances will be made only by the use of such an approach (1). Two aspects of endocrine function in depression are dealt with: the hypothalamic-pituitary-adrenal cortical (HPA) axis, and the hypothalamic-pituitary-growth hormone system.

HPA SYSTEM

Many clinicians have noted associations between affective disturbances and diseases of the HPA axis, especially Addison's and Cushing's diseases (2–4). Partly as a result of this association and partly for other reasons (e.g., the early development of steroid assays) the HPA system in depression is by far the most intensively studied area in clinical psychoneuroendocrinology. The many studies in this area may be summarized under three categories (Table 1).

Most early studies were concerned with the first and, to a lesser extent, with the second of these general questions. Investigations carried out before 1970 concerning cortisol metabolite excretion and plasma 17-hydroxycorticosteroid (17-OHCS) levels in depression have been reviewed several times (2,3,5,6).

TABLE 1. *General areas of HPA function*

1. Cortisol metabolism: normal or abnormal?
2. Tissue exposure to cortisol: elevated or reduced?
3. Regulatory mechanisms: limbic system-hypothalamus-pituitary-adrenal cortex

CORTISOL METABOLISM

The aspects of cortisol metabolism that have been studied over many years are listed in Table 2. Only the first of these functions is discussed

TABLE 2. *Aspects of cortisol metabolism that have been studied*

Production rate
Metabolic clearance rate
Biologic half-life
17-OHCS excretion
17-Ketogenic steroid excretion
Other metabolites

here. The production (or secretion) rate of cortisol is elevated consistently in most if not all patients suffering from depression. Gibbons (7) reported from England in 1964 that the cortisol production rate was high in depression and that return to more normal values was associated with recovery. Two years later he repeated his original observation and included the finding that the production rate of corticosterone also was elevated in depression (8). In the small series of patients Gibbons studied, cortisol production rates as high as 30 mg/day were recorded, whereas in normals this value rarely exceeds 20 mg/day with the method employed. In 1971 Sachar, working with the Montefiore Hospital endocrine group in New York, also reported high cortisol production rates in depressed patients (9). Nine of the 16 patients in this study produced more than the normal upper limit of 25 mg/day. Similar findings were reported by Carpenter and Bunney in 1971, using two different methods of estimating the production rate of cortisol (5). More recently Sachar's group has again found high production rates, measured from 24-hour venous catheter studies in which cortisol secretory pulses were determined (10). In this study six depressed patients had a mean cortisol production rate of 30.1 mg/day, which fell to a mean of 19.7 mg/day on recovery, whereas eight control subjects had a mean production rate of 17.2 mg/day.

In one study Sachar et al. (9) described a strong relationship between the cortisol production rate on the one hand and the level of anxiety and psychotic disorganization evident in the patients on the other hand. This finding was less evident in the subsequent report from Sachar's group (10) and was not described by Gibbons or by Carpenter and Bunney (5,7,8) although it is not clear that these workers looked specifically for such an association.

TISSUE EXPOSURE TO CORTISOL

The studies of cortisol production rate described above provide an indirect indication of the activity of the "central" part of the HPA axis, i.e., an indication of the rate of ACTH release from the anterior pituitary. Another large body of studies in the endocrinology of depression is directed toward answering a different question: Is the tissue exposure to cortisol normal in depressed patients? The various approaches to this general question are given in Table 3.

TABLE 3. *Approaches to studying tissue exposure to cortisol*

Plasma cortisol (and ACTH)
Plasma CBG capacity
Plasma free cortisol
CSF cortisol
Urinary free cortisol
Brain tissue cortisol

Plasma cortisol levels in depression are usually elevated over control or recovery values. The many studies of this parameter were reviewed recently (6). In general the differences between depressed and control or recovered values are greatest when blood samples are obtained during the evening or night hours, rather than in the early morning (6 to 9 A.M.). The conclusion to be drawn from these studies is that depressed patients have elevated plasma cortisol levels at most sampling times, and especially during the night. The significance of these plasma cortisol elevations and their circadian variations are discussed in a later section.

Plasma ACTH levels have not been studied very systematically in depressed patients. Berson and Yalow (11) included some data on patients suffering from "psychodepressive disorders" in their 1968 report on ACTH measurements. Fifteen patients, presumably with primary affective disorders, but possibly including some schizoaffective patients, had blood taken for ACTH estimation immediately before electroconvulsive therapy (ECT). The mean plasma ACTH value before ECT was 74 pg/ml; comparison values for 70 normal subjects and 223 general hospital inpatients were 22 and 45 pg/ml, respectively.

Eleven depressed patients studied by Endo et al. (12) in Japan had a mean morning plasma ACTH level of 73 pg/ml when ill and 83 pg/ml on recovery. Allen et al. (13) studied two depressed women and reported baseline values immediately before ECT within the normal range (20 to 120 pg/ml). Obviously additional studies of plasma ACTH levels in depressed patients are needed, and attention to circadian variations will be important.

The elevation of plasma cortisol levels in depression was noted above. As measured in all of the studies published to date, "plasma cortisol" refers to the *total* circulating plasma glucocorticoid level (cortisol plus corticosterone) when fluorometric or competitive protein-binding assays are used, or to the total 17-OHCS level (mainly cortisol alone) when colorimetric assays are employed (see ref. 6 for review). It is well known, however, that less than 10% of the total "plasma cortisol" is free in the circulation, and that most of it is bound to a specific α-globulin called "transcortin" (14), or corticosteroid-binding globulin (CBG) (15). The small fraction of plasma cortisol that is not protein-bound is considered to be responsible for the tissue effects of circulating glucocorticoids (16). One way in which the plasma free cortisol level could be increased would be by a reduction of the plasma CBG binding capacity, as is known to occur in some conditions of medical and surgical stress (17,18).

Plasma CBG levels in depressed patients have been measured by only one investigator (19), who reported essentially normal values in most cases. A small number of males with unipolar depressive illness had levels that were reduced slightly by comparison with bipolar males (27.2 and 33.5 mg/liter, respectively). Further studies of this variable would be in order, although it is not likely that major abnormalities will be found.

Plasma free cortisol levels in depression have not been published. Carroll and Curtis (*unpublished observations*) have begun to study this issue and have found that the free cortisol levels of depressed patients are higher than those of control psychiatric inpatients. Nine depressed patients had a mean plasma free cortisol level of 7.7 ng/ml, and five control patients had a mean level of 3.1 ng/ml.

Cortisol levels in the cerebrospinal fluid (CSF) bear a close relationship to plasma free cortisol levels (6) and also indicate the exposure of the central nervous system (CNS) to active cortisol. Rather conflicting results for CSF cortisol values have been reported by the three groups who examined this issue. McClure and Cleghorn (20) studied five depressed patients and five controls on a neurology service who were later found to have no organic lesion. The depressed patients had a mean CSF cortisol value of 7.9 ng/ml, compared with a mean value of 14.4 ng/ml for the controls. On the basis of these results, the authors suggested that depressed patients could have a deficiency of brain cortisol, a proposal which they and others have made elsewhere (5,21).

In another study of CSF cortisol levels, Coppen et al. (22) reported values which were rather low but not different from control values. Like McClure and Cleghorn, this group used a competitive protein binding (CPB) technique for quantitation of CSF cortisol. In Coppen's study, however, an inappropriate version of this technique was used, and their results are subject to serious methodologic criticism. Briefly, the average cortisol mass in the samples was 1.2 ng, and the mass of the lowest standard was 5 ng; and with the microscale assay that was used, the error near zero is large. Those readers familiar with CPB techniques will recognize that an ultramicro scale is required to quantitate accurately such low amounts of cortisol (23). The mean CSF cortisol values given by Coppen et al. (22) were 3.3 ng/ml in 23 depressed patients, 2.98 ng/ml in 18 control patients, and 2.94 ng/ml in 16 manic patients. These are low values and, as indicated above, not reliable.

We obtained CSF cortisol values in depression which are significantly higher than control, normal, or recovered values (6,24). Forty depressed patients studied in Melbourne, Philadelphia, and Ann Arbor had a mean CSF cortisol level of 11.3 ng/ml, whereas 13 control psychiatric inpatients had a mean level of 5.0 ng/ml ($p < 0.0001$). Normal CSF cortisol values by the same technique are given by Murphy et al. (25) as 4 ng/ml. Thirteen depressed patients were found to have a mean CSF cortisol level of 13.3 ng/ml when ill, compared to a mean value of 8.3 ng/ml after recovery ($p < 0.001$) (24).

Thus the largest study of this issue using reliable methods indicates that CSF cortisol values are considerably elevated during depression, in keeping with the elevation of plasma free cortisol levels already noted. In nine depressed patients the mean plasma free cortisol level was 7.7 ng/ml, and the mean value of cortisol in CSF drawn simultaneously was 8.1 ng/ml. The corresponding values for five nondepressed psychiatric inpatients were 3.1 ng/ml (plasma) and 3.3 ng/ml (CSF).

Another method of estimating the tissue exposure to active cortisol is to measure the urinary free cortisol (UFC) excretion (6). This parameter reflects the effective level of circulating plasma cortisol over time (26–30), and excreted free cortisol increases in direct proportion to the plasma free cortisol level (30).

Ferguson et al. measured UFC excretion in a small group of depressed women and found significant elevations, with a return to more normal values after treatment (31). Carroll (32) studied UFC excretion in 21 depressed patients and a contrast group of 10 acute schizophrenics who manifested comparable degrees of anxiety and psychotic disorganization as measured by Sachar's criteria (9). The depressed patients had a mean UFC excretion of 104 μg/day compared with a mean value of 39 μg/day for the schizophrenics. In 11 depressed patients studied before and after treatment (ECT), the mean values were 134 and 79 μg/day, respectively. The mean UFC excretion of normal subjects is 40 μg/day by the method used (33).

By all of these measures (elevated plasma free and total cortisol, normal plasma CBG capacity, elevated CSF cortisol, and elevated urine free cortisol excretion) the evidence points strongly to a significantly increased exposure of the tissues and CNS of depressed patients to active cortisol. The extent of this increase is considerable but not as great as that seen in Cushing's disease.

It comes as a surprise, therefore, that the available data on suicide brains indicate the cortisol concentration to be lower than normal in the CNS of depressives. The brain tissue levels of glucocorticoids are known to parallel the circulating plasma levels rather closely (34), so that one would expect to find elevated brain cortisol values in depression. Brooksbank and associates, working in Coppen's unit in England, reported on two series of brain tissue studies (35,36). In the first series six brains of depressive suicides contained

71.7 ng of cortisol plus dihydrocortisols per gram, and seven control patients who died suddenly had a level of 94.3 ng/g in their brains (mean values). In a second series of eight control and six depressive suicide brains, a reduction of both cortisol and corticosterone by approximately 40% was noted in the suicide brains. Similar results were obtained in a study of this issue carried out in Melbourne (Carroll, *unpublished observations*). Brooksbank and associates (36) also noted that the corticosterone level in human brain was 40% to 50% of the total corticosteroid level, whereas in human plasma this ratio is less than 10%. A similar finding was reported by Fazekas and Fazekas (37). Suicide brain studies are fraught with difficulties of interpretation, and the findings noted above cannot be evaluated properly until more is known about the rate of postmortem decay of cortisol in nervous tissue. Notwithstanding these reservations, the low tissue cortisol level is a remarkable finding and could possibly indicate a steroid binding deficiency in the brains of depressed patients.

HPA REGULATORY MECHANISMS

In recent years it has become obvious that a study of the central regulatory mechanisms of HPA function is necessary to provide an adequate understanding of pituitary-adrenal cortex activity in depression or, indeed, in any clinical psychoendocrine disorder. For many years it was assumed implicitly, if not explicitly, that increased activity of the HPA system in psychopathological states could be "explained" as the result of psychological stress. The interaction between stress and the psychological defense mechanisms of the patient was regarded as critical in determining whether HPA activation would occur (9,38,39).

With the identification of important brain mechanisms which regulate HPA function by clinical endocrinologists, and with the application of these new procedures by psychoendocrinologists, this early explanatory system for HPA overactivity in depression is now becoming less tenable. The recent evidence is beginning to point less toward ego states and more toward limbic system abnormalities as major determinants of the observed HPA disturbances in depression.

The central regulatory mechanisms for HPA function were identified by basic and clinical endocrinologists when attention shifted from the pituitary to the hypothalamus and then to the extrahypothalamic brain regions such as septum, hippocampus, amygdala, and midbrain (see ref. 6 for review). New insights into the complexity of neuroendocrine organization and pathology have resulted from these basic studies; in addition, the clear involvement of the CNS in the regulation of anterior pituitary function immediately suggested the possibility of neuropharmacologic strategies, so that a neuroendocrine approach to testing the biogenic amine theories of depression became feasible.

TABLE 4. *Tests of HPA regulatory mechanisms*

Corticotropin (ACTH)

Lysine vasopressin
Pyrogen
Metyrapone
Insulin hypoglycemia

Circadian rhythm
Circadian profile

Dexamethasone suppression

As a word of caution, however, it should be noted that most of the *clinical* neuroendocrine test procedures which are now commonplace were developed by clinicians to answer pragmatic questions. (For example: Does a given patient have a functional anterior pituitary reserve capacity? Has a particular attempt at pituitary ablation been successful? Is autonomous pituitary or adrenal cortical function present?) These tests (e.g., insulin hypoglycemia stress and dexamethasone suppression tests) were not developed primarily with the needs of psychoneuroendocrinologists in mind, and they are relatively blunt, crude instruments for determining subtle alterations of neuroendocrine function. The regulatory mechanisms of the HPA axis in depressed patients have been examined in several different ways, as outlined in Table 4.

ACTH

Since most clinical studies rely on the measurement of plasma cortisol levels to judge changes in HPA function, it becomes important to know if the adrenal cortex can respond normally to ACTH. Otherwise an impaired response (e.g., to insulin hypoglycemia) cannot be interpreted necessarily to indicate dysfunction of the hypothalamic-pituitary portion of the axis.

In two studies where maximal adrenal cortical stimulation was employed, depressed patients were shown to have normal plasma cortisol responses, with no change on recovery from the depressed state (5,6). Three later studies have been reported in which synthetic ACTH (Synacthen; β-1–24 corticotropin) was employed at more physiological dose levels. Sclare and Grant (40) used 250 μg Synacthen by intramuscular injection at 9 A.M. in 12 patients before and after treatment for their depression. The plasma cortisol response 30 min later was very similar in both phases. Bridges (41) gave the same dose intravenously to 15 severely depressed patients at 10 A.M. following attempted suppression by dexamethasone. In five cases exaggerated plasma cortisol responses were observed. Each of these patients had also failed to suppress normally with the dexamethasone regimen

(2 mg at 10 P.M. on the previous day and 2 mg at 7 A.M. on the day of the Synacthen test). Endo et al. (12) tested 11 depressed patients before and after treatment and obtained slightly increased responses during the illness, compared with the values after recovery. In all cases, however, these responses (which, as in the previous study, were also obtained after dexamethasone suppression) were within normal limits.

The sum of this evidence indicates that in most depressed patients the adrenal cortex responds normally to ACTH. The increased responses seen in Bridges' patients (41) who had failed to suppress with dexamethasone may well reflect the additive effect of endogenous ACTH release occurring at the time of the test, as indicated by their nonsuppressed plasma cortisol levels. These results with ACTH stimulation thus make it most probable that the increased cortisol secretion rates in depression described earlier do reflect increased hypothalamic-pituitary activity and not simply exaggerated adrenal cortical responses to normal amounts of ACTH.

Lysine Vasopressin

Lysine-8-vasopressin (LVP) is a short peptide which causes a prompt increase of plasma cortisol levels in normal subjects, most probably by acting at or above the level of the hypothalamus, since the response can be blocked by dexamethasone, morphine, and chlorpromazine (see ref. 6 for review). The cortisol response is secondary to the release of corticotropin-releasing factor (CRF) from the hypothalamus, which in turn causes ACTH release from the anterior pituitary.

Jakobson and associates (42) tested 16 depressed patients with this procedure and reported that 12 failed to respond normally. In another study (6) it was found that three of 14 patients had no plasma cortisol response to LVP; in six patients tested both before and after treatment similar mean responses to LVP were obtained on each occasion.

From these results it seems that some depressed patients may have an impairment of the central mechanisms which mediate the HPA response to LVP. Since little is known of the mechanism by which LVP causes HPA stimulation, however, or of the neurotransmitter pathways involved, further comment is not possible at the present time.

Pyrogen

Injections of purified bacterial polysaccharide pyrogens can cause HPA activation, independently of the fever which also occurs. The pyrogen test is sometimes used by clinical endocrinologists to evaluate HPA function (see ref. 6 for review). The site of action of pyrogen is probably at or above the level of the hypothalamus.

In one reported investigation of eight depressed patients studied before and after treatment, normal plasma cortisol responses were obtained on each occasion (6). It is not likely that this test will be applied very usefully in future studies, since the precise sites and mechanism of action of pyrogen are not understood.

Metyrapone

Metyrapone (SU-4885) is a drug which can block the 11β-hydroxylase step of cortisol synthesis in the adrenal cortex, leading to a fall in circulating cortisol levels followed by a rise in ACTH production. The increased ACTH levels then cause a large rise in 11-desoxycortisol, which is measured as the index of response to the test (see ref. 6 for review). The metyrapone test is considered to be a method of evaluating the HPA *feedback control* mechanism. A circadian influence on the response to metyrapone is also documented (6).

Two studies in depressed patients using the traditional extended-dosage metyrapone regimen have been carried out (5,43), and essentially normal results were obtained by both groups. More recently Endo et al. (12) used the newer single midnight dose metyrapone test (44) and measured the plasma ACTH response directly. Eleven patients studied before and after treatment had normal and equivalent plasma ACTH increments on each occasion. Thus there appears to be no significant impairment of the HPA response to metyrapone in depressed patients.

Insulin Hypoglycemia

The insulin hypoglycemia test is the best known of the HPA stimulation tests used in clinical endocrinology. A fall in blood glucose levels is produced rapidly by intravenous injection of insulin, followed by a large HPA activation, provided the degree and duration of the hypoglycemia are adequate (6). HPA responses to the insulin tolerance test (ITT) are reproducible when the test is repeated in normal subjects (45), but it can be difficult to be certain that truly comparable degrees of hypoglycemic stress have been obtained when comparisons between patients are made. The wide "normal" range of plasma cortisol responses to the ITT reflects this inherent difficulty with the test procedure.

Perez-Reyes (46) studied 25 normal subjects, 25 neurotic depressives, and 25 psychotic depressives with the ITT. The minimum recorded blood sugar levels were very similar in each group. The psychotic depressives had a much reduced plasma cortisol response (mean rise 12.4 μg/dl) compared with the normal subjects (mean rise 20.2 μg/dl), whereas the neurotic depressives had exaggerated responses (mean 33.2 μg/dl). Similar differences were observed also in the urinary catecholamine responses of these patients to the hypoglycemia.

Carroll (6,47) administered the ITT to 16 depressed patients before and after treatment with ECT. The mean fall in blood glucose was similar on each occasion; the minimum blood glucose level obtained was lower on recovery (28 mg/dl) than during illness (37 mg/dl). The mean plasma cortisol response during illness (9.3 μg/dl) was significantly lower than after treatment (15.9 μg/dl). An association was noted between impaired responses to the ITT and impaired responses of the same patients to the dexamethasone suppression test.

At variance with these results are the findings of two other groups (12,48) who found equal plasma cortisol responses in depressed and recovered patients, despite the production of slightly more profound hypoglycemia during the recovered phase. Three of the four studies with the ITT (12,47, 48) revealed evidence of significant resistance to the hypoglycemic action of insulin in the depressed phase, thus confirming the earlier observation of Mueller et al. (49).

From these four studies it seems that some depressed patients may have impaired HPA responses to hypoglycemia, but that this finding is by no means a consistent one. If an impairment is present it may occur only in those patients with greatly increased cortisol production rates and resistance to dexamethasone suppression (6,47). If this finding should be replicated it would be of much theoretical interest, since a similar constellation of HPA disturbances occurs also in diencephalic Cushing's disease (50).

Circadian Rhythm

The HPA axis of man exhibits a definite circadian variation in spontaneous activity (51–54) and in response to some tests of the regulatory mechanisms, e.g., vasopressin (55), pyrogen (56), metyrapone (57), and dexamethasone suppression (58,59). In clinical practice this circadian variation factor was evaluated initially by comparing morning plasma cortisol levels with those obtained in the afternoon or evening. In normal subjects the later value is the lower, and the absence of such a "diurnal variation" was suggested as a simple screening criterion for Cushing's disease. These aspects have been well reviewed by Krieger et al. (60).

In depressed patients the circadian rhythm of plasma cortisol levels is generally preserved. However, the elevation of evening plasma cortisol over control values is usually much greater than the elevation of morning values (5,6,61–64). Whereas morning plasma cortisol levels in depressed patients are generally within the normal range, the levels found during the night more often lie clearly in the pathologic range (61,63–65). The same is true of Cushing's disease (60,63). The explanation of these findings was not apparent until more detailed knowledge about circadian cortisol secretory patterns was obtained (60,66). Doig et al. (61) recognized the abnormality as well as was possible in 1966 when they observed that a "shift to the left" in the circadian rhythm of plasma cortisol was occurring, so that the peak plasma levels in some depressed patients were seen at 3 A.M. rather than the normal time of 6 A.M. This abnormally high level of HPA activity during the night in depression was confirmed also by other groups (62,63) using both urinary and plasma measures.

Thus most depressed patients do maintain a circadian rhythm of HPA activity (where "rhythm" is defined as morning values being higher than night values). There is, however, a circadian variation in the extent of HPA overactivity in depression: Night values are more abnormal than are morning values.

Circadian Profile of Cortisol Secretion

In recent years there has been a change in our understanding of the normal regulation of HPA activity. Early theories described a negative feedback system for ACTH-cortisol release, with a variable "resetting" of the reference input to the feedback control loop to account for responses to stress (67). Improved versions of this "reset hypothesis" were developed later to take note of the experimental findings that pretreatment with glucocorticoids did not always prevent the HPA activation caused by stress (68,69). The negative feedback concept was maintained in these theories, at least as far as basal, nonstressed HPA activity was concerned.

As a result of more recent studies we know now that under basal conditions in man the HPA axis does *not* operate continuously with minute-to-minute feedback control. Instead, the system is activated in an episodic manner, with intervening periods of total adrenocortical quiescence (60,66). Hellman's group (66) reported that cortisol secretion occurs for a total period of only about 6 hours each day in normal subjects, and that more than half the daily cortisol production takes place at approximately the time of waking. During the late evening and night hours cortisol secretory episodes are rarely seen, and plasma cortisol values below 5 μg/dl are maintained. Thus there appears to be a "CNS program" for HPA activation throughout the circadian cycle; during most of the day the brain exerts an *inhibitory* influence on the HPA axis, with this inhibition being relaxed chiefly during the morning peak of cortisol secretory episodes (6).

Sachar et al. (10) studied depressed patients with the venous catheter technique (60,66) used to characterize the normal circadian secretory profile of cortisol. The results in depression provided clear detail of the circadian HPA disturbance which previous studies using infrequent blood samples had revealed only in outline. Patients with depression were found to have an increased number of secretory episodes, increased total cortisol secretion, and increased time per day during which cortisol secretion was occurring. In particular, active cortisol secretion was observed during the late evening and early morning hours when cortisol secretion is minimal in normal subjects. This active nighttime cortisol secretion took place whether the depressed patients were sleeping or awake. There was no relationship between the nocturnal secretory episodes and epochs of rapid eye movement (REM) sleep. The depressed patients maintained high plasma cortisol levels and an increased number of secretory episodes during the daytime as well. These results were interpreted as indicating a failure of the normal CNS circadian inhibitory influence on HPA activity. Sachar and associates went on to suggest that "the hypersecretion of cortisol in certain depressive illnesses may not be simply a stress response, as such responses are usually envisioned, and as described during the course of neurotic depressive reactions [70], but rather another reflection of apparent limbic system dysfunction, along with disturbances in mood, affect, appetite, sleep, aggressive and sexual drives, and autonomic nervous system activity." Abnormal circadian profiles of cortisol secretion similar to those reported by Sachar's group have been found by the present authors in other depressed patients (71). *The essential neuroendocrine lesion in depression therefore appears to be a disinhibition of the HPA axis,* i.e., a failure of the normal inhibitory influence on the brain on ACTH-cortisol release. In fact, in some depressed patients the HPA circadian profiles cannot be distinguished from those seen in diencephalic Cushing's disease (10,60,71,72).

Dexamethasone Suppression

The final test of HPA regulation is distinct from the stimulation tests and baseline profiles already described. It involves the attempted suppression of HPA activity by exogenous glucocorticoids such as dexamethasone. Following the introduction of this test by Liddle in 1960 (73), a diurnal variation in the response to both orally and intravenously administered dexamethasone was identified (58,59). This led to the development of single-dose dexamethasone suppression tests in which the steroid is given during the critical phase for CNS programming of circadian HPA activity, i.e., at approximately midnight (59,74–78). Following a midnight dose of, for example, 2 mg dexamethasone, normal subjects show maximal suppression of plasma cortisol levels for at least the next 24 hours (76).

Krieger et al. (60) studied this response in detail by the catheter technique, which revealed some interesting and important interactions between dexamethasone dosage and circadian factors. A dose of 0.5 mg dexamethasone given orally at midnight to a normal subject prevented cortisol secretory episodes for 28 hours, i.e., until the next day's major secretory phase was expected. The same dose given at 8 A.M. prevented cortisol secretion for a shorter period (20 hours); secretion of cortisol was re-established at the time of the next expected major circadian increase (approximately 4 A.M.). By contrast, a 1-mg dose of dexamethasone given at 8 A.M. prevented cortisol secretory episodes for 44 hours, i.e., for 24 hours longer than the 0.5-mg dose did. In discussing these results Krieger and associates (60) suggested that "there may be only one 'critical period' of ACTH release within the 24 hour cycle. If the concentration of steroid within the nervous system at this critical time is sufficient to block ACTH release, one can see how circadian variation may be suppressed by small doses of steroid just prior to this time, and by larger doses at times well in advance of this 'critical period.'"

Dexamethasone has a plasma half-life of 3 to 6 hours in man (79–82). In rats the dexamethasone half-life is 6 hours in plasma and anterior pituitary, while in the hypothalamus it is approximately 8 hours (83). By contrast, the half-life of cortisol in human plasma is 60 to 90 min (60,66). These results help to explain why large doses of dexamethasone (1 to 2 mg) can have prolonged effects on the HPA axis. We have confirmed by catheter studies that a single midnight dose of 2 mg dexamethasone suppresses secretory episodes for at least 24 hours in nondepressed patients (71).

Several groups have reported abnormalities of dexamethasone suppression in *depressed patients,* but not all workers have been able to confirm these findings. In the light of Krieger's study (60) and new results of our own to be described below, it is now possible to re-examine these reports and to resolve the apparent differences. In addition, it can be shown that abnormalities of HPA suppression by dexamethasone are closely related to the abnormal circadian secretory profiles which Sachar (10) described in depressed patients.

Gibbons and Fahy (84) were the first to report on this subject. They examined depressed patients and control subjects with a 2-mg dose of dexamethasone given intramuscularly at 2 P.M. Over the following 3 hours comparable decrements of plasma cortisol were observed in both groups. They concluded that the acute suppressive effect of dexamethasone was not affected in depressed patients. That same year Stokes (85) presented results of single-dose midnight dexamethasone tests; some depressed patients had impaired suppression as indicated by high plasma 17-OHCS levels 8 hours after the steroid was administered. Similar results were presented by Stokes in 1970 (86), and he published his findings in detail in 1972 (87). Even with doses of dexamethasone as high as 8 mg, many depressed patients failed to show normal suppression of plasma 17-OHCS levels.

Fawcett and Bunney (88) mentioned in 1967 that they had observed some depressed patients who did not respond to dexamethasone, but they did not provide details. In a later study from NIMH, Carpenter and Bunney (5) used a multiple-dose version of the dexamethasone suppression test. By giving 1 mg twice daily for 2 days, then 2 mg twice daily for 2 more days, they suppressed urinary 17-OHCS excretion in 12 depressed patients to less than 40% of baseline values. These results satisfy the criteria of Liddle (73) for excluding adrenocortical hyperplasia. However, as the authors themselves stated, their technique would not detect more subtle abnormalities of HPA suppressibility.

In 1968 Platman and Fieve (89) published the results of 1-mg single-dose dexamethasone tests in 10 depressed patients. The steroid was administered at 11 P.M., and the response was evaluated by measuring the plasma 17-OHCS level 9 hours later. In seven of the 10 cases, the plasma 17-OHCS levels were greater than 7 μg/dl. This value is considered the upper limit for the range of normal responses to the test used (75). The same patients with impaired suppression also had abnormally high plasma 17-OHCS levels at 11 P.M. when the dexamethasone was given. The authors did not attach significance to these results and did not compare them with established criteria for normal suppression. They stated only that the dexamethasone suppressed the plasma 17-OHCS levels by comparison with the patients' baseline values.

During the same year Butler and Besser (65) described three depressed patients who failed to show normal suppression of plasma 11-OHCS levels in response to 2, 4, or 8 mg dexamethasone per day given in four divided doses. These patients also showed abnormally high night levels of plasma 11-OHCS ("cortisol"); after treatment of the depression these abnormalities were no longer present.

Also in 1968 Carroll et al. (90) reported on 27 depressed and 22 nondepressed psychiatric inpatients who received the 2-mg midnight single-dose dexamethasone suppression test. As judged by the plasma 11-OHCS level 8 hours later, approximately half of the depressed patients had impaired suppression (plasma 11-OHCS > 7 μg/dl), whereas the control patients suppressed normally in all but three cases. This abnormality was correlated with the severity of the depressive profile (91) and with the physiologic symptoms of depression (6). There was also an association of impaired suppression with poor response to antidepressant drugs (92,93). Abnormal responses to dexamethasone were associated as well with abnormal plasma 11-OHCS responses to insulin hypoglycemia (47). After treatment the depressed patients showed more normal responses to the dexamethasone test (6,90).

McClure and Cleghorn (21) reported in 1968 on three depressed patients treated with dexamethasone 0.75 mg at 11 P.M. daily for 3 weeks. One of these patients did not show adequate suppression, even over this prolonged period. Bridges (41) studied 15 depressed patients who were to be treated by psychosurgery. He gave 2 mg dexamethasone at 10 P.M. and 2 mg again at 7 A.M. the next morning, and measured the plasma cortisol level at 10 A.M. One-third of the patients had abnormally high plasma cortisol levels (>7 μg/dl) at that time. The most recent positive report is from Endo and associates in Japan (12). They gave a 1-mg dose of dexamethasone at midnight and measured plasma cortisol 8 hours later. Three of their 11 patients failed to show normal suppression.

Shopsin and Gershon reported in 1971 that they were unable to replicate these findings (94). Using a 2-mg dose of dexamethasone given at 11 P.M. and measuring the plasma 17-OHCS level 9 hours later, they found normal suppression (<5 μg/dl) in 13 depressed and six schizophrenic subjects. A second mainly negative report comes from Verghese et al. in India (95). Using the same technique as Shopsin and Gershon (94) but with a 1-mg dose of dexamethasone, they found that only three of 24 depressed men had plasma 11-OHCS levels above 7 μg/dl. Eighteen of these patients were noted to have 11 P.M. (predexamethasone) plasma 11-OHCS levels in the abnormal range for that time of night (>7 μg/dl).

In order to clarify this area, further studies were carried out with careful attention to the time of blood sampling and with the addition of urinary measures of HPA suppression. Carroll reported (6,32) that urinary free cortisol (UFC) excretion in 24 hours following a midnight 2-mg dose of dexamethasone differentiated depressed patients very clearly from schizophrenic subjects; in this study postdexamethasone plasma cortisol levels taken at 8 A.M. and 4 P.M. were less effective in distinguishing the two patient groups. In another investigation Carroll (6) found evidence from UFC excretions suggesting that some depressed patients had initial suppression of HPA function followed by escape from the suppression abnormally early.

In exploring these findings further by catheter studies, there was clear evidence for this pattern of initial suppression followed by early escape (71). The postdexamethasone plasma cortisol levels of a depressed patient who was studied before and after recovery are shown in Fig. 1. Blood samples were obtained every 30 min on each occasion. When the patient was depressed he had high levels of plasma cortisol at midnight when dexamethasone 2 mg was administered orally. The plasma cortisol levels were suppressed to the relatively normal value of 3.6 μg/dl at 8 A.M., but then an escape from suppression was observed. When the same patient was studied after recovery, the midnight plasma cortisol level was in the normal range (<5 μg/dl) and there was complete suppression of plasma cortisol levels for 24 hours. If a single blood sample taken at 8 A.M. had been used as the sole criterion of response to dexamethasone, it would have been concluded that there was no significant difference between the two tests in this patient. In the studies detailed earlier this was the case—a single morning plasma cortisol value was the index of suppression (85–87,89–95).

As a result of these findings with the catheter technique, we carried out further studies of depressed and nondepressed patients' responses to dexamethasone 2 mg orally given at midnight. By taking baseline and postdexamethasone urine samples for 24-hour UFC measurement, together with postdexamethasone plasma cortisol estimations at 8 A.M., 4 P.M., and midnight on each day, a more complete assessment of suppression can be obtained. The three blood samples are less ideal than continuous catheter sampling but are more practical for routine use. Abnormal cortisol secretory

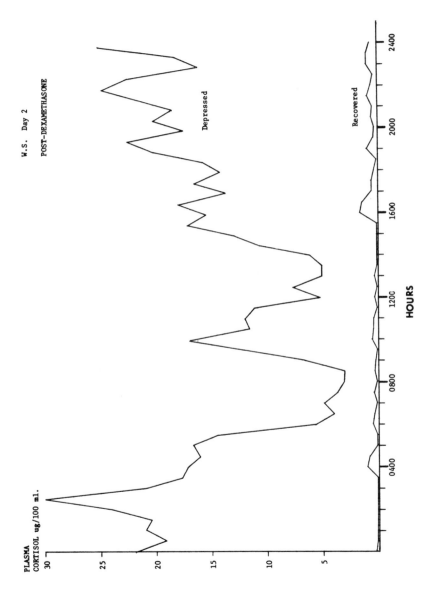

FIG. 1. Plasma cortisol concentrations following dexamethasone 2 mg p.o. in a depressed patient studied before and after treatment. Blood samples were obtained every 30 min through a venous catheter. Dexamethasone was given at midnight on each occasion.

episodes occurring during the 24 hours postdexamethasone may be missed when only three blood samples are taken, but their occurrence is reflected in the 24-hour UFC excretion of that day.

Through the use of this practical test procedure in a large number of patients, it is now obvious that the abnormality previously called "nonsuppression" is really an early escape from suppression, and that it is a graded phenomenon rather than an all-or-none event. Several different observed patterns of response to dexamethasone are illustrated in Fig. 2. The normal pattern is seen in Fig. 2a, where the midnight plasma cortisol level and all three values postdexamethasone are within the range 0 to 6 μg/dl. The pattern in Fig. 2b shows the least severe deviation from normal, with

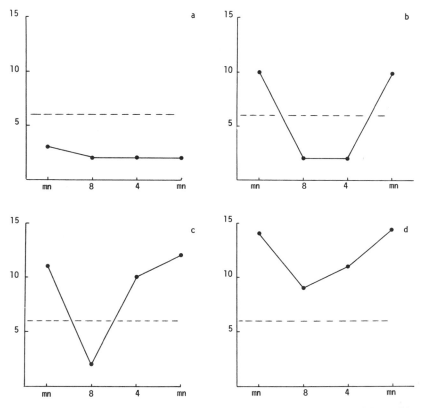

FIG. 2. Patterns of response to dexamethasone 2 mg p.o. given at midnight. Plasma cortisol levels measured at the time of dexamethasone administration and at 8 A.M. (8), 4 P.M. (4), and midnight (mn) on the following day. Horizontal lines indicate the upper limit of normal values. **a:** Normal. **b–d:** Increasing degrees of HPA disinhibition and progressive early escape from suppression.

(usually) minimal elevation of the midnight baseline plasma cortisol level and escape from suppression at midnight only on the next day. More pronounced elevations of the midnight baseline plasma cortisol level are seen in Figs. 2c and 2d, together with escape from suppression earlier on the postdexamethasone day, first by 4 P.M. and then by 8 A.M.

When individual depressed patients are tested in this way at different times during an episode of illness the patterns shown in Fig. 2 occur in a predictable sequence. For example, in one patient pattern 2d was observed 1 week after admission but before treatment commenced, then patterns 2c, 2b, and finally 2a shortly before discharge of the patient. In another patient pattern 2a was found during an episode of mania, then pattern 2b just prior to a switch into depression, and finally pattern 2c when a retarded depressive state had developed. These results will be reported in detail elsewhere (71).

In 60 dexamethasone suppression tests carried out in this way the following four criteria distinguished the 35 depressed patients from 25 inpatients without depression: (a) baseline UFC $> 100\ \mu g/24$ hours ($p < 0.0025$); (b) postdexamethasone UFC $> 25\ \mu g/24$ hours ($p < 0.001$); (c) sum of postdexamethasone plasma cortisol levels $> 12\ \mu g/dl$ ($p < 0.01$); (d) any postdexamethasone plasma cortisol level $> 6\ \mu g/dl$ ($p < 0.0025$). In addition, a high baseline midnight plasma cortisol level strongly predicted escape from suppression on the next day: the sum of the three postdexamethasone plasma cortisol levels correlated $+0.62$ with the baseline midnight plasma cortisol level ($p < 0.0001$). In several patients, however, an elevated baseline midnight plasma cortisol level was found without any plasma or UFC indication of escape from suppression postdexamethasone. The *baseline* plasma cortisol levels at 8 A.M., 4 P.M., and midnight were much less powerful than the postdexamethasone plasma cortisol values in distinguishing depressed from nondepressed patients. In this series the depressed patients who displayed abnormal escape from suppression were those with the endogenous or endogenomorphic (96) symptom profile, while those with neurotic and reactive depressions showed normal HPA suppression. As in the earlier studies, however, not every patient with an "endogenomorphic" symptom profile had an abnormal response to the suppression test. The precise determinants of this disturbance have yet to be identified. In general the more severely depressed patients have abnormal test results. A careful prospective study of clinical features and response to treatment in relation to the dexamethasone test responses is now in progress.

The conclusions which can be stated at the present time are: (a) Depressed patients display a range of abnormal responses to a midnight dose or dexamethasone 2 mg. (b) The abnormalities appear in a predictable sequence, occurring progressively earlier in the circadian cycle with increasing severity of illness. (c) The abnormalities are strongly related to the disinhibition of circadian cortisol secretion described by Sachar et al. (10).

It seems very probable now that the dexamethasone test results reflect the same basic disturbance of HPA function as described by Sachar et al. (10), i.e., a failure of the normal CNS inhibitory mechanism for ACTH-cortisol release. When viewed in the light of Krieger's findings in normal subjects (60), the early escape from suppression is quite abnormal and the circadian factor seen in Fig. 2 is striking. Suppression of episodic release of cortisol is achieved temporarily by giving dexamethasone at the critical circadian period, but escape from suppression is seen as the steroid is metabolized over the next 24 hours. At first this occurs late on the postdexamethasone day when the steroid concentration in the nervous system is lowest. Then as the depression becomes more severe, escape occurs earlier in the day, overriding higher steroid concentrations in the nervous system. This interpretation of the results, if correct, lends support to the proposal made in 1968 that "the steroid-sensitive neurons of the hypothalamus (in depressed patients) are subjected to an abnormal drive from other limbic areas" (90). With increasing severity of depression this abnormal drive increases, and escape from suppression occurs earlier on the postdexamethasone day. Further studies of the rate of dexamethasone absorption and metabolism need to be carried out in the patients tested in order to sustain the interpretation given above.

Returning to the earlier studies in which only the morning postdexamethasone value was obtained, it now appears that the patients who had abnormal suppression by that criterion were those with the earliest abnormal escape. Many patients who escaped later in the day must have been undetected by these early studies. Thus the mainly negative results of Shopsin and Gershon (94) and of Verghese et al. (95) must be re-evaluated with this new information in mind.

The gradations of HPA disinhibition which may now be described are:

Grade 1: Cortisol production rate elevated, mainly during daytime; marginal elevation of nighttime cortisol secretion; normal suppression response.

Grade 2: Cortisol secretion elevated in both daytime and nighttime; normal suppression response.

Grade 3: Cortisol secretion elevated during both daytime and nighttime; early escape from suppression occurs 16 to 24 hours postdexamethasone.

Grade 4: As in grade 3, but escape occurs 8 to 16 hours postdexamethasone.

Grade 5: As in grade 4, but escape occurs within 8 hours postdexamethasone.

From the results of our recent study described above (71), 25 nondepressed psychiatric inpatients had grade 1 or 2 disturbances, while the more severe grades were seen only among the 35 depressed patients. Some depressed patients, however, showed only grade 1 or 2 disturbances.

Summary of HPA Disturbance in Depression

We now understand the nature of HPA disturbance in depression in the following way. Depressed patients secrete abnormally large amounts of cortisol as a result of increased ACTH release rather than from altered adrenal cortical responsiveness to ACTH. This increased activation of the HPA axis occurs throughout the day and especially at inappropriate times of the circadian cycle when HPA inhibition should be present. As a consequence of the increased cortisol production, there is increased exposure of peripheral tissues and the CNS to free cortisol. The basic defect is regarded as a failure of the normal circadian inhibitory influence on the HPA axis. These findings indicate a serious disturbance of HPA regulatory mechanisms in depressed patients.

Other standard clinical tests of HPA regulatory mechanisms also provide evidence of central neuroendocrine dysfunction in depressed patients. In particular, suppression responses to exogenous glucocorticoids such as dexamethasone are impaired in many cases. The abnormal suppression responses correlate strongly with disinhibition of baseline HPA function. Patients with the most severe suppression abnormality also fail to have adequate HPA responses to hypoglycemia. Responses to lysine vasopressin and pyrogen are impaired in some depressed patients.

The HPA disturbances seen in depression are relatively specific. They do not correlate strongly with subjective distress or with breakdown of ego defense mechanisms. They occur in both bipolar and unipolar depressives with an endogenomorphic clinical profile but not characteristically in neurotic, reactive, or secondary depressive states.

The constellation of HPA disturbances in primary depressive illness is remarkably similar to that seen in diencephalic Cushing's disease. In this form of Cushing's disease (50,97) depressive symptoms are common, whereas mood changes are not prominent in Cushing's symdrome resulting from primary adrenal tumors and other causes (98).

Thus the neuroendocrine findings in depression lend additional support to the clinical association between primary HPA disorders and mood change. The frequent occurrence of depression in Addison's disease can also be understood better in light of these new findings. Diencephalic Cushing's disease, Addison's disease, and primary depressive illness all share the common features of mood disorder and disinhibition of central HPA mechanisms (2,3,60,72,99). In addition, the central limbic sites involved in HPA regulation (hippocampus, amygdala, septal region, midbrain) are closely related to the limbic areas which regulate mood and emotions (6).

Research in this field has reached a point, therefore, which confirms the early intuition of Rubin and Mandell (3) that "functional depressive states may be concomitants of a suprahypophyseal brain dysfunction which also causes stimulation of the anterior pituitary." New directions for this field will involve careful study of the specificity of the HPA disturbances by screening other psychiatric patients; clarification of the clinical features of depressed patients which relate to the HPA disturbances; and, most importantly, study of the neurotransmitter basis of the observed HPA disturbances. In this way direct clinical information of relevance to the biogenic amine theories of depression may be obtained.

GROWTH HORMONE

The second major neuroendocrine system to be studied in affective disorders is the hypothalamus-pituitary-growth hormone system. Developments in this area began within the last 10 years, when radioimmunoassay methods for growth hormone (GH) measurement became generally available. Secretion of GH is affected by a large number of metabolic, pharmacologic, and neural factors (100–102). The stimuli used in depressed patients include slow-wave sleep, L-DOPA, L-5-hydroxytryptophan, apomorphine, ACTH 1-24, and insulin hypoglycemia.

Basal GH Levels

Growth hormone is frequently considered to be one of the "stress-responsive" hormones, and elevations of GH are well documented in physical and psychologic stresses (103,104). In depressed patients, despite the distress of their illnesses, basal GH levels are consistently reported as normal. It should be noted that normal fasting plasma GH levels are very low (less than 2 ng/ml). Thus this particular stress-responsive system is not activated in depression, in contrast to the HPA activation described earlier. Patients with anorexia nervosa, however, who have marked loss of body weight, do show elevated plasma GH levels (105).

Sleep

In normal subjects a large increase of plasma GH levels is observed soon after sleep onset, in association with the first episode of slow-wave (stage 3–4) sleep (106,107). An early report suggested that this "physiologic" release of GH was reduced in depressed patients (108). However, it is known that aging is associated with a marked decrease of this nocturnal GH release (109,110), and a subsequent study of age-matched controls revealed that the initial finding was probably related to this age effect (111). Patients with diencephalic Cushing's disease also have impaired nocturnal GH release during both the active disease stage and the remission after treatment, indicating that the defect is not related to hypercortisolemia (97). The same patients had persistent impairment of stage 3–4 sleep, a defect seen also in depressed patients (112). In view of the other neuroendocrine similarities between depression and diencephalic Cushing's disease a re-examination of nocturnal GH release in depression would be in order. The antidepressant drug imipramine is known to block the GH rise associated with slow-wave sleep (106).

L-DOPA

The amino acid precursor of the catecholamines, L-dihydroxyphenylalanine (L-DOPA) causes a rise in plasma GH levels in most normal subjects and in patients with Parkinson's disease (113–115). This response can be prevented by hyperglycemia induced before (116) but not after (117) the L-DOPA is administered, and it is reduced or absent in obese patients (118). An early report (119) from Sachar's group suggested that depressed patients may have impaired GH responses to L-DOPA. In this study a definite influence of age was noted as well, with older normal subjects responding less than young subjects. In a later study (120) patients with unipolar depression were found to have poor GH responses to L-DOPA, whereas bipolar patients had normal responses. Equivalent decrements of plasma prolactin levels were measured in all three patient groups, which suggested that the absorption of L-DOPA and its entry into the brain were not impaired in the unipolar patients. However, as the authors pointed out, there was a preponderance of postmenopausal women in the unipolar patient group. It was then established that postmenopausal women secrete less GH in response to L-DOPA than do age-matched men, and that normal and depressed postmenopausal women had equal responses (121). Thus this group ultimately found no systematic abnormalities in GH response to L-DOPA among depressed patients. A similar conclusion was reached in our own study of GH responses to L-DOPA carried out with male depressed patients (122). In view of the previous discussion of diencephalic Cushing's disease, it is of interest that patients with this condition are known to have absent GH responses to L-DOPA. This finding applied to those with clinical remission as well as to those with active disease (123). This experience with the L-DOPA test, from the initial promising report to the final realization that there was no defect in depressed patients, illustrates very well the problems of applying clinical neuroendocrine test procedures in psychoendocrine studies.

L-5-Hydroxytryptophan

The amino acid L-5-hydroxytryptophan (5HTP) is the immediate precursor of the neurotransmitter 5-hydroxytryptamine (serotonin) and was reported by Imura et al. (124) to cause plasma GH elevations when given to normal subjects in a dose of 150 mg orally. As with L-DOPA, the GH response to this agent is blocked by hyperglycemia. L-Tryptophan, the precursor of 5HTP, has been given intravenously in large doses (5 to 10 g) without any consistent effect on GH levels in plasma (125). In one other study, however, 5 g infused L-tryptophan was reported to raise plasma GH levels (126).

Takahashi and associates (127) tested 14 depressed patients with an oral dose of 200 mg L-5HTP. Each of three bipolar depressed patients and two of three unipolar depressives had no GH response. Six patients with chronic depressive states (four neurotic, two unipolar) also failed to respond. Four manic patients had normal responses. Interpretation of these data is somewhat difficult because some patients had plasma GH elevations before the 5HTP was administered, and age- and sex-matched control responses are not known. In view of the experience with L-DOPA, these control issues must be studied carefully before speculation as to the significance of these results can be entertained.

Apomorphine

Apomorphine is structurally similar to dopamine and is known to act directly on dopamine receptors in the brain (128). When given in subemetic doses to normal subjects it causes GH release (129,130), which can be antagonized by the dopamine receptor-blocking agent chlorpromazine (131). As with L-DOPA, this effect is age-related: Older normal subjects have lesser responses than young adults (132). Other influences on the GH response to apomorphine include sex, progesterone, menstrual cycle, and hyperglycemia (133). In male depressed and manic patients no consistent abnormality of GH response to apomorphine was observed (122).

ACTH 1-24

Synthetic ACTH fragments, both 1-18 (134) and 1-24 (135–137), cause release of GH in normal subjects. The effect is independent of the cortisol response to ACTH, since it is observed also in patients with Addison's disease (138) and in children on long-term steroid therapy (136). The response is not seen in patients with Cushing's syndrome, and it is not prevented by hyperglycemia (137).

The responses of depressed patients to ACTH 1-24 have been studied by Endo and associates (12). Using a dose of 0.25 mg intravenously following overnight dexamethasone suppression, they found GH responses just within the normal range in eight patients. After recovery the same patients had improved responses. Details of the responses of individual patients were not provided. Further exploration of this provocative stimulus for GH release would be in order, although, once again, the mechanism by which the ACTH fragments cause this effect is not known and the variables of age, sex, ovarian status, and obesity have not been systematically studied.

Hypoglycemia

The most widely used clinical procedure for stimulating GH release is insulin-induced hypoglycemia (45,103). Interest in this procedure as a means of evaluating central neurotransmitter mechanisms in depression was stimulated by the discovery that the GH response was apparently mediated by a catecholamine system in the brain. Phentolamine, an α-adrenergic receptor-blocking agent was found to reduce the response, whereas propranolol, a β-adrenergic receptor blocker, potentiated the response in man (139). From these results the concept of an α-adrenergic stimulatory mechanism and a β-adrenergic inhibitory mechanism was developed.

Mueller first reported on GH responses of depressed patients to hypoglycemia (49). The hypoglycemic response to insulin was impaired in the patients, indicative of resistance to the action of insulin. Neurotic depressives (five patients) had normal GH responses. Manic-depressives (six patients) had adequate GH responses; psychotic depressives (seven patients) had the lowest responses, but the mean GH response was within normal limits. Details of the individual responses were not given. Following treatment with amitriptyline, the GH responses of the manic-depressive and psychotic depressive patients improved, in keeping with a more profound degree of hypoglycemia.

Carroll reported on fourteen unipolar depressed patients studied before and after treatment with electroconvulsive therapy (6). Resistance to the hypoglycemic action of insulin was observed in this study also. Four of ten depressed patients in whom the blood glucose level fell below 45 mg/dl had an impaired GH response (less than 7 ng/ml). Three of these patients continued to have impaired responses after treatment. There was no relationship between these GH responses and the suppression responses of the patients to dexamethasone.

Quite similar findings were reported by Sachar et al. in 1971 (48). Five of 13 patients had deficient GH responses (less than 5 ng/ml), and in four of these cases the response was still absent after recovery. Once again, insulin resistance was observed, and correlation was noted between impaired GH responses and high cortisol production rates. The patients with defective responses were primarily postmenopausal women.

In a later study from Sachar's group (120) unipolar depressives (mainly postmenopausal women) had impaired responses, whereas those of bipolar depressives (all male) were normal. Since ovarian status is known to influence the GH response to hypoglycemia (140,141), these results could have been related to the sex difference between the patient groups rather than to their depressive conditions. In order to resolve this issue, a very carefully controlled study was carried out by Gruen, Sachar, and associates (142). Ten postmenopausal women were matched with normal postmenopausal women for age, body weight, baseline GH levels, and blood sugar response to insulin. A clear reduction of the GH response was seen in the depressed women.

An additional study of this matter was reported briefly by Endo and associates (12). Eleven depressed patients had a lower mean GH response to hypoglycemia when ill than after recovery. The group mean responses were not significantly different and individual responses were not given, so that the number of absent responses could not be determined from the report.

Once again, it is noteworthy that in diencephalic Cushing's disease GH responses to insulin hypoglycemia are impaired or absent, regardless of the stage of activity of the disease (123).

Conclusions—Growth Hormone System

The history of the psychoendocrine study of GH in depression, although it goes back only 7 years, illustrates most of the serious problems associated with this kind of clinical investigation. The most obvious problem is that of control issues—age, sex, ovarian status in particular—which were not appreciated when the early positive reports appeared describing defective GH responses of depressed patients to sleep, L-DOPA, and hypoglycemia. The second general problem is that of variance associated with the tests themselves. In the case of hypoglycemic stress, for example, the degree and duration of hypoglycemia are critically important determinants of the GH response (45). The reports of Mueller, Carroll, and Sachar (6,48,49,120) can all be criticized on this ground, since the degree of hypoglycemia in the depressed patients was less than in the recovered or control subjects. The insulin tolerance test is in fact a difficult procedure to standardize rigidly, although Gruen et al. (142) succeeded very well in doing so. When the responses of individual patients are being compared in the depressed and recovered phases, it is important to remember that even normal subjects show a great variance of GH response to hypoglycemia when tested on two occasions (45).

The final general problem is one of interpretation. As psychoendocrinologists we tend to make use of clinical screening procedures (e.g., L-DOPA, L-5HTP, hypoglycemia) soon after their introduction by clinical endocrinologists and before the mechanisms of these provocative tests are understood. From a little information—e.g., about the effects of adrenergic blocking drugs (139)—we are tempted to construct explanatory hypotheses, e.g., about endogenous brain catecholamine defects in depression (142). Such hypotheses may have heuristic value to workers in the field, but they can be misleading to general readers. In the example given, the hypothesis suggested from the hypoglycemia study might appear to give experimental support to the general catecholamine theory of depression (143,144). This impression would be an unfortunate one, since there are several other neuroendocrine variables which need to be considered. The two most notable are the role of serotonin mechanisms in mediating the GH response to hypoglycemia (145) and the role of somatostatin, which antagonizes GH responses to every provocative stimulus (146). The growth hormone system is an extremely complex one (102); it is by no means thoroughly understood, and at this stage clinical findings must be interpreted with caution.

ACKNOWLEDGMENTS

Parts of this work were supported by the National Health and Medical Research Council of Australia, by Research funds from the Veterans Administration, and by the Mental Health Research Institute, University of Michigan.

REFERENCES

1. Carroll, B. J. (1975): Review of clinical research strategies in affective illness. In: *Psychobiology of Depression,* edited by J. Mendels, pp. 143–159. Spectrum, New York.
2. Michael, R. P., and Gibbons, J. L. (1963): Interrelationships between the endocrine system and neuropsychiatry. In: *International Review of Neurobiology,* Vol. 5, pp. 243–292. Academic Press, New York.
3. Rubin, R., and Mandell, A. (1966): Adrenal cortical activity in pathological emotional states: A review. *Am. J. Psychiatry,* 123:4, 387–400.
4. Whybrow, P., and Hurwitz, T. (1976): Psychological disturbances associated with endocrine disease and hormone therapy. *This volume.*
5. Carpenter, W. T., Jr., and Bunney, W. E., Jr. (1971): Adrenal cortical activity in depressive illness. *Am. J. Psychiatry,* 128:1,31–40.
6. Carroll, B. J. (1972): The hypothalamic-pituitary-adrenal axis in depression. In: *Depressive Illness: Some Research Studies,* edited by B. Davies, B. J. Carroll, and R. M. Mowbray, pp. 23–201. Charles C Thomas, Springfield, Ill.
7. Gibbons, J. L. (1964): Cortisol secretion rate in depressive illness. *Arch. Gen. Psychiatry,* 10:572–575.
8. Gibbons, J. L. (1966): The secretion rate of corticosterone in depressive illness. *J. Psychosom. Res.,* 10:263–266.
9. Sachar, E. J., Hellman, L., Fukushima, D. K., and Gallagher, T. F. (1971): Cortisol production in depressive illness: A clinical and biochemical clarification. *Arch. Gen. Psychiatry,* 23:289–298.
10. Sachar, E. J., Hellman, L., Roffwarg, H., Halpern, F., Fukushima, D., and Gallagher, T. (1973): Disrupted 24 hour patterns of cortisol secretion in psychotic depression. *Arch. Gen. Psychiatry,* 28:19–24.
11. Berson, S., and Yalow, R. (1968): Radioimmunoassay of ACTH in plasma. *J. Clin. Invest.,* 47:2725–2751.
12. Endo, M., Endo, J., Nishikubo, M., Yamaguchi, T., and Hatotani, N. (1974): Endocrine studies in depression. In: *Psychoneuroendocrinology,* edited by N. Hatotani, pp. 22–31. Karger, Basel.
13. Allen, J., McGilvra, R., Kendall, J., Denny, D., and Allen, C. (1974): Hormone release during multiple electroconvulsive therapy in man. *World J. Psychosynthesis,* 6:25–29.
14. Slaunwhite, W. R., and Sandberg, A. A. (1959): Transcortin: A corticosteroid-binding protein of plasma. *J. Clin. Invest.,* 38:384–391.
15. Daughaday, W. H. (1958): Binding of corticosteroids by plasma proteins. IV. The electrophoretic demonstration of corticosteroid binding globulin. *J. Clin. Invest.,* 37:519–523.
16. Slaunwhite, W. R., Lockie, G. N., Back, N., and Sandbert, A. A. (1962): Inactivity in vivo of transcortin-bound cortisol. *Science,* 135:1062–1063.

17. Murray, D. (1967): Cortisol binding to plasma proteins in man in health, stress and at death. *J. Endocrinol.,* 39:571–591.
18. Hamanaka, Y., Manabe, H., Tanaka, H., Monden, Y., Uozumi, T., and Matsumoto, K. (1970): Effects of surgery on plasma levels of cortisol, corticosterone and non-protein-bound cortisol. *Acta Endocrinol. (Kbh.),* 64:439–451.
19. King, D. J. (1973): Plasma cortisol-binding capacity in mental illness. *Psychol. Med.,* 3:1,53–65.
20. McClure, D. J., and Cleghorn, R. A. (1970): Hormone deficiency in depression. In: *Science and Psychoanalysis,* Vol. 17, edited by J. H. Masserman, pp. 12–19. Grune & Stratton, New York.
21. McClure, D. J., and Cleghorn, R. A. (1968): Suppression studies in affective disorders. *Can. Psychiatr. Assoc. J.,* 13:477–487.
22. Coppen, A., Brooksbank, R., Noguera, R., and Wilson, D. (1971): Cortisol in the cerebrospinal fluid of patients suffering from affective disorders. *J. Neurol. Neurosurg. Psychiatry,* 34:432–435.
23. Murphy, B. E. P. (1967): Some studies of the protein-binding of steroids and their application to the routine micro and ultramicro measurement of various steroids in body fluids by competitive protein-binding radioassay. *J. Clin. Endocrinol. Metab.,* 27:973–990.
24. Carroll, B. J., Curtis, G., and Mendels, J. (1976): Cerebrospinal fluid and plasma free cortisol levels in depression. *Psychol. Med. (in press).*
25. Murphy, B. E. P., Cosgrove, J. B., McIlquham, M. C., and Pattee, C. J. (1967): Adrenal corticoid levels in human cerebrospinal fluid. *Can. Med. Assoc. J.,* 97:13–17.
26. West, H. F. (1957): Corticosteroid metabolism and rheumatoid arthritis. *Ann. Rheum. Dis.,* 16:173–182.
27. Greaves, M. S., and West, H. F. (1960): Relation of free corticosteroids in urine to steroid dosage. *Lancet,* 1:368.
28. Rosner, J. M., Cos, J. J., Biglieri, E. G., Hane, S., and Forsham, P. H. (1963): Determination of urinary unconjugated cortisol by glass fiber chromatography in the diagnosis of Cushing's syndrome. *J. Clin. Endocrinol. Metab.,* 23:820–827.
29. Schedl, H. P., Chen, P. S., Greene, G., and Redd, D. (1959): The renal clearance of plasma cortisol. *J. Clin. Endocrinol. Metab.,* 19:1223–1229.
30. Beisel, W. R., Cos, J. J., Horton, R., Chao, P. Y., and Forsham, P. H. (1964): Physiology of urinary cortisol excretion. *J. Clin. Endocrinol. Metab.,* 24:887–893.
31. Ferguson, H. C., Bartram, A. C. G., Fowlie, H. C., Cathro, D. M., Birchall, K., and Mitchell, F. L. (1964): A preliminary investigation of steroid excretion in depressed patients before and after electroconvulsive therapy. *Acta Endocrinol. (Kbh.),* 47:58–68.
32. Carroll, B. J. (1976): Limbic system-adrenal cortex regulation in depression and schizophrenia. *Psychosom. Med. (in press).*
33. Murphy, B. E. P. (1968): Clinical evaluation of urinary cortisol determinations by competitive protein-binding radioassay. *J. Clin. Endocrinol. Metab.,* 28:343–348.
34. Carroll, B. J., Heath, B., and Jarrett, D. (1975): Corticosteroids in brain tissue. *Endocrinology,* 97:290–300.
35. Brooksbank, B. W. L., Brammall, M. A., Cunningham, A. E., Shaw, D. M., and Camps, F. E. (1972): Estimation of corticosteroids in human cerebral cortex after death by suicide, accident or disease. *Psychol. Med.,* 2:56–65.
36. Brooksbank, B. W. L., Brammall, M. A., and Shaw, D. M. (1973): Estimation of cortisol, cortisone and corticosterone in cerebral cortex, hypothalamus and other regions of the human brain after natural death and after death by suicide. *Steroids Lipids Res.,* 4:162–183.
37. Fazekas, I. G., and Fazekas, A. T. (1967): Die Corticosteroid-Fraktionen des menschlichen Gehirns. *Endokrinologie,* 51:183–210.
38. Hamburg, D. A. (1962): Plasma and urinary corticosteroid levels in naturally occurring psychologic stresses. *Res. Publ. Assoc. Nerv. Ment. Dis.,* 40:406–413.
39. Mason, J. W. (1968): A review of psychoendocrine research on the pituitary-adrenal cortical system. *Psychosom. Med.,* 30:576–607.
40. Sclare, A. B., and Grant, J. K. (1972): The Synacthen test in depressive illness. *Scott. Med. J.* 17:7,7–8.

41. Bridges, P. K. (1973): Methods of assessing patients for psychosurgery and their outcome after operation. *Psychiatr. Neurol. Neurochir.* 76:335–344.
42. Jakobson, T., Blumenthal, M., Hagman, H., and Heikkinen, E. (1969): The diurnal variation of urinary and plasma 17-hydroxycorticosteroid (17-OHCS) levels and the plasma 17-OHCS response to lysine-8-vasopressin in depressive patients. *J. Psychosom. Res.*, 13:363–375.
43. Jakobson, T., Stenback, A., Strandstrom, L., and Rimon, R. (1966): The excretion of urinary 11-deoxy- and 11-oxy-17-hydroxy-corticosteroids in depressive patients during basal conditions and during the administration of methopyrapone. *J. Psychosom. Res.*, 9:363–374.
44. Jubiz, W., Meikle, A. W., West, C. D., and Tyler, F. H. (1970): Single-dose metyrapone test. *Arch. Intern. Med.*, 125:472–474.
45. Greenwood, F. C., Landon, J., and Stamp, T. C. B. (1966): The plasma sugar free, fatty acid, cortisol and growth hormone response to insulin. 1. In control subjects. *J. Clin. Invest.*, 45:429–436.
46. Perez-Reyes, M. (1972): Differences in the capacity of the sympathetic and endocrine systems of depressed patients to react to a physiological stress. In: *Recent Advances in the Psychobiology of the Depressive Illnesses*, edited by T. Williams, M. Katz, and J. Shield, Jr., pp. 131–135. Government Printing Office, Washington, D.C.
47. Carroll, B. J. (1969): Hypothalamic-pituitary function in depressive illness: Insensitivity to hypoglycemia. *Br. Med. J.*, 3:27–28.
48. Sachar, E. J., Finkelstein, J., and Hellman, L. (1971): Growth hormone responses in depressive illness. I. Response to insulin tolerance test. *Arch. Gen. Psychiatry*, 25:263–269.
49. Mueller, P. S., Heninger, G. R., and McDonald, R. K. (1969): Insulin tolerance test in depression. *Arch. Gen. Psychiatry*, 21:587–594.
50. James, V. H. T., Landon, J., Wynn, V., and Greenwood, F. C. (1968): A fundamental defect of adrenocortical control in Cushing's disease. *J. Endocrinol.*, 40:15–28.
51. Pincus, G. (1943): A diurnal rhythm in the excretion of urinary ketosteroids by young men. *J. Clin. Endocrinol. Metab.*, 3:195–199.
52. Migeon, C. J., Tyler, F. H., Mahoney, J. P., Florentine, A. A., Castle, H., Bliss, E. L., and Samuels, L. T. (1956): The diurnal variation of plasma levels and urinary excretion of 17-hydroxycorticosteroids in normal subjects, night workers and blind subjects. *J. Clin. Endocrinol. Metab.*, 16:622–633.
53. Doe, R. P., Flink, E. B., and Goodsell, M. G. (1956): Relationship of diurnal variation in 17-hydroxycorticosteroid levels in blood and urine to eosinophils and electrolyte excretion. *J. Clin. Endocrinol. Metab.*, 16:196–206.
54. Kobberling, J., and von zur Muhlen, A. (1974): The circadian rhythm of free cortisol determined by urine sampling of two-hour intervals in normal subjects and patients with severe obesity or Cushing's syndrome. *J. Clin. Endocrinol. Metab.*, 38:313–319.
55. Clayton, G. W., Librik, L., Gardner, R. L., and Guillemin, R. (1963): Studies on the circadian rhythm of pituitary adrenocorticotrophic release in man. *J. Clin. Endocrinol. Metab.*, 23:975–980.
56. Takebe, K., Setaishi, C., Hirama, M., Yamamoto, M., and Horiuchi, Y. (1966): Effects of a bacterial pyrogen on the pituitary-adrenal axis at various times in the 24 hours. *J. Clin. Endocrinol. Metab.*, 26:437–442.
57. Jubiz, W., Matsukura, S., Meikle, A. W., Harada, G., West, C. D., and Tyler, F. H. (1970): Plasma metyrapone, adrenocorticotropic hormone, cortisol and deoxycortisol levels: Sequential changes during oral and intravenous metyrapone administration. *Arch. Intern. Med.*, 125:468–471.
58. Nichols, T., Nugent, C. A., and Tyler, F. H. (1965): Diurnal variation in suppression of adrenal function by glucocorticoids. *J. Clin. Endocrinol. Metab.*, 25:343–349.
59. Ceresa, F., Angeli, A., Boccuzzi, G., and Molino, G. (1969): Once-a-day neurally stimulated and basal ACTH secretion phases in man and their response to corticoid inhibition. *J. Clin. Endocrinol. Metab.*, 29:1074–1082.
60. Krieger, D., Allen, W., Rizzo, F., and Krieger, H. (1971): Characterization of the normal temporal pattern of plasma corticosteroid levels. *J. Clin. Endocrinol. Metab.*, 32:266–284.
61. Doig, R. J., Mummery, R. V., Wills, M. R., and Elkes, A. (1966): Plasma cortisol levels in depression. *Br. J. Psychiatry*, 112:1263–1267.

62. Fullerton, D. T., Wenzel, F. J., Lohrenz, F. N., and Fahs, H. (1968): Circadian rhythm of adrenal cortical activity in depression. *Arch. Gen. Psychiatry,* 19:674–681.
63. Knapp, M. S., Keane, P. M., and Wright, J. G. (1967): Circadian rhythm of plasma 11-hydroxycorticosteroids in depressive illness, congestive heart failure and Cushing's syndrome. *Br. Med. J.,* 2:27–30.
64. McClure, D. J. (1966): The diurnal variation of plasma cortisol levels in depression. *J. Psychosom. Res.,* 10:189–195.
65. Butler, P. W. P., and Besser, G. M. (1968): Pituitary-adrenal function in severe depressive illness. *Lancet,* 2:1234–1236.
66. Hellman, L., Nakada, F., Curti, J., Weitzman, E. D., Kream, J., Roffwarg, H., Ellman, S., Fukushima, D. K., and Gallagher, T. F. (1970): Cortisol is secreted episodically by normal man. *J. Clin. Endocrinol. Metab.,* 30:411–422.
67. Yates, F. E., and Urquhart, J. (1962): Control of plasma concentration of adrenocortical hormones. *Physiol. Rev.,* 42:359–443.
68. Dallman, M. F., and Yates, F. E. (1968): Anatomical and functional mapping of central neural input and feedback pathways of the adrenocortical system. *Mem. Soc. Endocrinol.,* 17:39–72.
69. Yates, F. E., Brennan, R. D., and Urquhart, J. (1969): Adrenal glucocorticoid control system. *Fed. Proc.,* 28:71–83.
70. Sachar, E. J., Mackenzie, J., Binstock, W., and Mack, J. (1967): Corticosteroid responses to psychotherapy of depressions. I. Evaluations during confrontation of loss. *Arch. Gen. Psychiatry,* 16:461–470.
71. Carroll, B. J., Curtis, G., Mendels, J., and Sugerman, A. (1976): Neuroendocrine regulation in depression. *Arch. Gen. Psychiatry (in press).*
72. Hellman, L., Weitzman, E., Roffwarg, H., Fukushima, D., Yoshida, K., and Gallagher, T. F. (1970): Cortisol is secreted episodically in Cushing's syndrome. *J. Clin. Endocrinol. Metab.,* 30:686–689.
73. Liddle, G. W. (1960): Tests of pituitary adrenal suppressibility in the diagnosis of Cushing's syndrome. *J. Clin. Endocrinol. Metab.,* 20:1539–1560.
74. Nugent, C. A., Nichols, T., and Tyler, F. H. (1965): Diagnosis of Cushing's syndrome: Single dose dexamethasone suppression test. *Arch. Intern. Med.,* 116:172–176.
75. Pavlatos, F. C., Smilo, R. P., and Forsham, P. H. (1965): A rapid screening test for Cushing's syndrome. *J.A.M.A.,* 193:720–723.
76. McHardy-Young, S., Harris, P. W. R., Lessoff, M. H., and Lyne, C. (1967): Single-dose dexamethasone suppression test for Cushing's syndrome. *Br. Med. J.,* 1:740–744.
77. Connolly, C. K., Gore, M. B. R., Stanley, N., and Wills, M. R. (1968): Single-dose dexamethasone suppression in normal subjects and hospital patients. *Br. Med. J.,* 2:665–667.
78. Asfeldt, V. H. (1969): Simplified dexamethasone suppression test. *Acta Endocrinol. (Kbh.),* 61:219–231.
79. Rabhan, N. B. (1968): Pituitary-adrenal suppression and Cushing's syndrome after intermittent dexamethasone therapy. *Ann. Intern. Med.,* 69:1141–1148.
80. Haque, N., Thrasher, K., Werk, E., Jr., Knowles, H., Jr., and Sholiton, L. (1972): Studies of dexamethasone metabolism in man: Effect of diphenylhydantoin. *J. Clin. Endocrinol. Metab.,* 34:1,44–50.
81. Hichens, M., and Hogans, A. (1974): Radioimmunoassay for dexamethasone in plasma. *Clin. Chem.,* 20:2,266–271.
82. Meikle, A., Lagerquist, L., and Tyler, F. (1973): A plasma dexamethasone radioimmunoassay. *Steroids,* 22:193–202.
83. De Kloet, E., VanDerVies, J., and de Wied, D. (1974): The site of the suppressive action of dexamethasone on pituitary-adrenal activity. *Endocrinology,* 94:61–73.
84. Gibbons, J. L., and Fahy, T. J. (1966): Effect of dexamethasone on plasma corticosteroids in depressive illness. *Neuroendocrinology,* 1:358–363.
85. Stokes, P. E. (1966): Pituitary suppression in psychiatric patients. Presented at The Endocrine Society (USA), 48th Meeting.
86. Stokes, P. E. (1970): Alterations in hypothalamic-pituitary-adrenocortical (HPAC) function in man during depression. In: *The Endocrine Society (USA), 52nd Meeting St. Louis, June 10–12,* Abstract 337.
87. Stokes, P. E. (1972): Studies on the control of adrenocortical function in depression. In:

Psychobiology of Depressive Illness, edited by T. Williams, M. Katz, and J. A. Shield, pp. 199–220. Government Printing Office, Washington, D.C.

88. Fawcett, J. A., and Bunney, W. E. (1967): Pituitary-adrenal function and depression. *Arch. Gen. Psychiatry,* 16:571–535.
89. Platman, S. R., and Fieve, R. R. (1968): Lithium carbonate and plasma cortisol response in the affective disorders. *Arch. Gen. Psychiatry,* 18:591–594.
90. Carroll, B. J., Martin, F. I. R., and Davies, B. M. (1968): Resistance to suppression by dexamethasone of plasma 11-OHCS levels in severe depressive illness. *Br. Med. J.,* 3:285–287.
91. Carroll, B. J., and Davies, B. M. (1970): Clinical associations of 11-hydroxycortico-steroid suppression and non-suppression in severe depressive illness. *Br. Med. J.,* 1: 789–791.
92. McLeod, W. R., Carroll, B. J., and Davies, B. M. (1970): Hypothalamic dysfunction and antidepressant drugs. *Br. Med. J.,* 2:480–481.
93. McLeod, W. R. (1972): Poor response to antidepressants and dexamethasone non-suppression. In: *Depressive Illness: Some Research Studies,* edited by B. M. Davies, B. J. Carroll, and R. M. Mowbray, pp. 202–206. Charles C Thomas, Springfield, Ill.
94. Shopsin, B., and Gershon, S. (1971): Plasma cortisol response to dexamethasone suppression in depressed and control subjects. *Arch. Gen. Psychiatry,* 24:320–326.
95. Verghese, A., Matthew, J., Mathai, G., Kothoor, A., Saxena, B., and Koshy, T. (1973): Plasma cortisol in depressive illness. *Indian J. Psychiatry,* 15:72–79.
96. Klein, D. F. (1974): Endogenomorphic depression. *Arch. Gen. Psychiatry,* 31:447–454.
97. Krieger, D. T., and Glick, S. M. (1972): Growth hormone and cortisol responsiveness in Cushing's syndrome: Relation to a possible central nervous system etiology. *Am. J. Med.,* 52:25–40.
98. Gifford, S., and Gunderson, J. G. (1970): Cushing's disease as a psychosomatic disorder. *Medicine (Balt.),* 49:397–409.
99. Krieger, D. T., and Gewirtz, G. P. (1974): The nature of the circadian periodicity and suppressibility of immunoreactive ACTH levels in Addison's disease. *J. Clin. Endocrinol. Metab.,* 39:46–52.
100. Glick, S. M. (1969): Regulation of growth hormone secretion. In: *Frontiers in Neuro-endocrinology,* edited by W. F. Ganong and L. Martini, pp. 141–182. Oxford University Press, New York.
101. Brown, G. M., and Reichlin, S. (1972): Psychologic and neural regulation of growth hormone secretion. *Psychosom. Med.,* 34:45–61.
102. Martin, J. B. (1973): Neural regulation of growth hormone secretion. *N. Engl. J. Med.,* 288:1384–1393.
103. Roth, J., Glick, S. M., Yalow, R. S., et al. (1963): Hypoglycemia: A potent stimulus for secretion of growth hormone. *Science,* 140:987–988.
104. Brown, G. M., Schalch, D. S., and Reichlin, S. (1971): Hypothalamic mediation of growth hormone and adrenal stress response in the squirrel monkey. *Endocrinology,* 89:694–703.
105. Landon, J., Greenwood, F. C., Stamp, T. C. B., and Wynn, V. (1966): The plasma sugar free, fatty acid, cortisol and growth hormone response to insulin and the comparison of this procedure with other tests of pituitary and adrenal function. II. In patients with hypothalamic or pituitary dysfunction or anorexia nervosa. *J. Clin. Invest.,* 45:437–449.
106. Takahashi, Y., Kipnis, D. M., and Daughaday, W. H. (1968): Growth hormone secretion during sleep. *J. Clin. Invest.,* 47:2079–2090.
107. Honda, Y., Takahashi, K., Takahashi, S., Azumi, K., Irie, M., Sakuma, M., Tsushima, T., and Shizume, K. (1968): Growth hormone secretion during nocturnal sleep in normal subjects. *J. Clin. Endocrinol. Metab.,* 29:20–29.
108. Sachar, E. J. (1973): Endocrine factors in psychopathological states. In: *Biological Psychiatry,* edited by J. Mendels, pp. 175–197. Wiley, New York.
109. Carlson, H. E., Gillin, J. C., Gorden, P., and Snyder, F. (1972): Absence of sleep-related growth hormone peaks in aged normal subjects and in acromegaly. *J. Clin. Endocrinol. Metab.,* 34:1102–1105.
110. Finkelstein, J. W., Roffwarg, H. P., Boyar, R. M., Kream, J., and Hellman L. (1972): Age-related change in the twenty-four-hour spontaneous secretion of growth hormone. *J. Clin. Endocrinol. Metab.,* 35:665–670.

111. Sachar, E. J. (1975): *Personal communication.*
112. Snyder, F. (1968): Electrographic studies of sleep in depression. In: *Computers and Electronic Devices in Psychiatry,* edited by N. S. Kline and E. Laska, pp. 272–303. Grune & Stratton, New York.
113. Boyd, A. E., Lebovitz, H. E., and Pfeiffer, B. (1970): Stimulation of human growth hormone secretion by L-DOPA. *N. Engl. J. Med.,* 283:1425–1429.
114. Cavagnini, F., Peracchi, M., Scotti, G., Raggi, U., Pontiroli, A. E., and Bana, R. (1972): Effect of L-DOPA administration on growth hormone secretion in normal subjects and parkinsonian patients. *J. Endocrinol.,* 54:425–433.
115. Perlow, M. J., Sassin, J. F., Boyar, R., Hellman, L., and Weitzman, E. D. (1972): Release of human growth hormone, follicle stimulating hormone, and lutenizing hormone in response to L-dihydroxyphenylalanine (L-DOPA) in normal man. *Dis. Nerv. Syst.,* 33: 804–810.
116. Mims, R. B., Scott, C. L., Modebe, O. M., and Bethune, J. E. (1973): Prevention of L-dopa-induced growth hormone stimulation by hyperglycemia. *J. Clin. Endocrinol. Metab.,* 37:660–663.
117. Singh, P., McDevitt, D. G., Mackay, J., and Hadden, D. R. (1973): Effects of L-dopa and chlorpromazine on human growth hormone and TSH secretion in normal subjects and acromegalics. *Horm. Res.,* 4:293–301.
118. Fingerhut, M., and Krieger, D. T. (1974): Plasma growth hormone response to L-dopa in obese subjects. *Metabolism,* 23:267–271.
119. Sachar, E. J., Mushrush, G., Perlow, M., Weitzman, E. D., and Sassin, J. (1972): Growth hormone responses to L-dopa in depressed patients. *Science,* 178:1304–1305.
120. Sachar, E. J., Frantz, A. G., Altman, N., and Sassin, J. (1973): Growth hormone and prolactin in unipolar and bipolar depressed patients: Responses to hypoglycemia and L-dopa. *Am. J. Psychiatry,* 130:1362–1367.
121. Sachar, E. J. (1974): Sex differences in growth hormone response. *Am. J. Psychiatry,* 131:608–609.
122. Mendels, J., Carroll, B. J., Frazer, A., and Pandey, G. N.: *In preparation.*
123. Krieger, D. T. (1973): Lack of responsiveness to L-DOPA in Cushing's disease. *J. Clin. Endocrinol. Metab.,* 36:277.
124. Imura, H., Nakai, Y., and Yoshimi, T. (1973): Effect of 5-hydroxytryptophan (5-HTP) on growth hormone and ACTH release in man. *J. Clin. Endocrinol. Metab.,* 36:204–206.
125. Mac Indoe, J. H., and Turkington, R. W. (1973): Stimulation of human prolactin secretion by intravenous infusion of L-tryptophan. *J. Clin. Invest.,* 52:1972–1978.
126. Nakai, Y., Yoshimi, T., Imura, H., Kato, Y., Yoshimoto, S., and Moridera, K. (1973): Effect of L-tryptophan on anterior pituitary hormone release in man. Cited in ref. 127.
127. Takahashi, S., Kondo, H., Yoshimura, M., Ochi, Y., and Yoshimi, T. (1973): Growth hormone responses to administration of L-5-hydroxytryptophan (L-5-HTP) in manic-depressive psychoses. *Folia Psychiatr. Neurol. Jap.,* 27:187–206.
128. Ernst, A. M. (1967): Mode of action of apomorphine and dexamphetamine on gnawing compulsion in rats. *Psychopharmacologia,* 10:316–323.
129. Lal, S., de la Vega, C. E., Sourkes, T. L., and Friesen, H. G. (1972): Effect of apomorphine on human growth hormone secretion. *Lancet,* 2:570–572.
130. Brown, W. A., van Woert, M. H., and Ambari, L. M. (1973): Effect of apomorphine on growth hormone release in humans. *J. Clin. Endocrinol. Metab.,* 37:463–465.
131. Lal, S., de la Vega, C. E., Sourkes, T. L., and Friesen, H. G. (1973): Effect of apomorphine on growth hormone, prolactin, luteinizing hormone and follicle-stimulating hormone levels in human serum. *J. Clin. Endocrinol. Metab.,* 37:719–724.
132. Maany, I., Frazer, A., and Mendels, J. (1975): Apomorphine: Effect on growth hormone. *J. Clin. Endocrinol. Metab.,* 40:162–163.
133. Ettigi, P., Lal, S., Martin, J. B., and Friesen, H. G. (1974): Effect of sex, oral contraceptives, menstrual cycle and glucose loading on apomorphine-induced growth hormone secretion. Presented to the Canadian Psychiatric Association, 24th Annual Meeting, Ottawa, Canada, Oct. 2–4.
134. Takahara, J., Asaoka, K., and Ofuji, T. (1973): Effect of synthetic corticotropins on human growth hormone secretion. *Endocrinol. Jap.,* 20:411–416.
135. Zahnd, G. R., Nadeau, A., and von Muhlendahl, K. E. (1969): Effect of corticotrophin on plasma levels of human growth hormone. *Lancet,* 2:1278–1280.

136. Girard, J., Stahl, M., and Buhler, U. (1971): Stimulierung der Wachstumshormonsekretion bei Kindern unter ACTH-Belastung. *Schweiz. Med. Wochenschr.*, 101:930–932.
137. Strauch, G., Pandos, P., Luton, J. P., and Bricaire, H. (1971): Stimulation de la secretion d'hormone somatrotrope par la beta 1–24 corticotrophine. *Ann. Endocrinol. (Paris)*, 32:526–536.
138. Ikkos, D., Pantelakis, S., Katsichtis, P., and Velentzas, D. (1970): Effect of corticotrophin on plasma-growth-hormone. *Lancet*, 1:1401.
139. Blackard, W. G., and Heidingsfelder, S. A. (1968): Adrenergic receptor control mechanism for growth hormone secretion. *J. Clin. Invest.*, 47:1407–1414.
140. Frantz, A. G., and Rabkin, M. T. (1965): Effects of estrogen and sex difference on secretion of human growth hormone. *J. Clin. Endocrinol. Metab.*, 25:1470–1480.
141. Merimee, T. J., and Fineberg, S. E. (1971): Studies on the sex based variation of human growth hormone secretion. *J. Clin. Endocrinol. Metab.*, 33:896–902.
142. Gruen, P. H., Sachar, E. J., Altman, N., and Sassin, J. (1975): Growth hormone responses to hypoglycemia in post-menopausal depressed women. *Arch. Gen. Psychiatry*, 32:31–33.
143. Bunney, W. E., and Davis, J. M. (1965): Norepinephrine in depressive reactions. *Arch. Gen Psychiatry*, 13:483–494.
144. Schildkraut, J. J. (1965): The catecholamine hypothesis of affective disorders: A review of supporting evidence. *Am. J. Psychiatry*, 122:509–522.
145. Bivens, C. H., Lebovitz, H. E., and Feldman, J. M. (1973): Inhibition of hypoglycemia-induced growth hormone secretion by the serotonin antagonists cyproheptadine and methysergide. *N. Engl. J. Med.*, 289:236–239.
146. Besser, G. M. (1974): Hypothalamus as an endocrine organ. *Br. Med. J.*, 3:613–615.

Hormones, Behavior, and Psychopathology, edited by
Edward J. Sachar. Raven Press, New York © 1976.

Diagnosis and Psychopathology in Psychiatric Patients Resistant to Dexamethasone

Peter E. Stokes, Peter M. Stoll, Marlin R. Mattson, and
Robert N. Sollod

*Payne Whitney Clinic, New York Hospital–Cornell Medical Center,
New York, New York 10021*

The abnormalities of hypothalamic-pituitary-adrenocortical (HPAC) function identified by work from this and other laboratories have classically been viewed as reflections or concomitants of depression. It can be hypothesized that neuroendocrine dysfunction may in fact also act as a contributor to the development or persistence of depression in susceptible individuals. The clinical and laboratory data supporting this hypothesis are varied. The frequent disturbance of affect on administration of corticosteroids to humans, the data relating early experience to subsequent adult HPAC function in animals, the identification of personality characteristics in adults and behavioral state in human infants with HPAC activity patterns, and the effect of corticosteroids on brain slice uptake of norepinephrine are all compatible with the idea that HPAC function may act as a modifier of behavior. This argument is greatly enhanced by de Wied's pioneering work on the effect of pituitary adrenocortical function on animal behavior (1–6).

We previously reported that a majority of hospitalized psychiatric patients diagnosed as having primary affective disorder and manifesting moderate to severe depression show resistance to dexamethasone suppression (7). The present study was undertaken to determine how specific this endocrine abnormality is to depression and to attempt to identify personality characteristics associated with dexamethasone resistance. Data analysis has not been fully completed, but the portions available are presented here.

Consecutive psychiatric admissions to the Payne Whitney Clinic, New York Hospital–Cornell Medical Center over a 3-month period were asked to undergo a standard 1-mg overnight dexamethasone suppression test after signed informed consent was obtained. Patients who had possible contraindications to taking dexamethasone were eliminated, and some additional patients refused the dexamethasone or associated blood tests. On admission, a diagnosis (DSM II) was made on the basis of a psychiatric history and examination. At the same time a global rating of affect was made, using the Payne Whitney Clinical Global Rating Scale for depression and mania (PWCG), a previously described 13-point scale: 0 to 6 for depression and mania (8). The predexamethasone 9 A.M. cortisol samples were drawn within 2 days of admission to the hospital; 1 mg dexamethasone was given orally that same night at 11 P.M. and the postdexamethasone blood was drawn the following morning. All patients were asked to complete the Minnesota Multiphasic Personality Inventory (MMPI) as soon after dexamethasone administration as possible. A total of 74 completed MMPIs were obtained.

Postdexamethasone cortisol levels were obtained on 138 patients. Mean pre- and postdexamethasone cortisol levels and mean PWCG scores for the various DSM II diagnostic categories are given in Tables 1 and 2. Resistance to dexamethasone suppression was defined as a postdexamethasone plasma cortisol level > 5.0 μg%. In analyzing the diagnostic data, patients were subgrouped into the four categories indicated in Table 1. The frequency of dexamethasone resistance in the depressed group (46%) was significantly greater than the 17% frequency in the schizophrenic group (Fig. 1; $p < 0.01$) and was significantly higher than the 25% frequency seen in all other (79) patients combined ($p < 0.05$).

We examined the data for other possible differentiations between the groups of 47 dexamethasone-resistant and 91 normally suppressing patients. Mean PWCG scores for the two groups did not differ significantly. MMPIs were obtained on 54 suppressors and 20 suppression-resistant patients. Mean scores on the 10 original MMPI clinical scales on 17 research scales examined did not yield any significant differentiation between the two

TABLE 1. *Dexamethasone response and diagnostic category*

Catetory	No.	No. dex. resistant	Predexamethasone plasma (μg%) cortisol[a]	Postdexamethasone plasma (μg%) cortisol[b]	PWCG
Schizophrenic	29	5	18.7 ± 5.5	1.7 (1.0–2.3)	-2.6 ± 1.9
Depressed[c]	59	27	21.9 ± 7.7	3.8 (3.2–4.4)	-3.4 ± 2.4
Schizoaffective	9	3	20.5 ± 9.1	2.1 (1.0–3.0)	-2.9 ± 3.3
Other	41	12	22.1 ± 8.8	2.5 (1.9–3.2)	-2.5 ± 2.6

[a] Mean $\pm$ SD.
[b] Mean $\pm$ 1 SD of log-converted data.
[c] Depressed group: DSM II diagnosis (Table 2).

TABLE 2. *DSM II diagnosis*

DSM II diagnosis	No.	No. dexamethasone-resistant
300.4 Depressive neurosis	36	17
298.0 Psychotic depressive reaction	2	1
296.3 Manic depressive, circular, depressed	3	1
296.2 Manic depressive, depressed	12	4
296.0 Involutional melancholia	6	4

	Dexamethasone resistant	Normal supression	
Depression	27	32	59
Schizophrenia	5	24	29

FIG. 1. Dexamethasone suppression. Relationship to DSM II diagnosis. Chi square = 6.85, $p < 0.01$. Normal suppression, postdexamethasone plasma 17-OH $< 5 \mu g\%$.

groups. However, scatter plots of several individual MMPI scale scores against postdexamethasone cortisol levels showed the kind of relationship indicated in Fig. 2 for the MMPI depression scale. That is, there was a noticeable positive correlation between postdexamethasone cortisol level

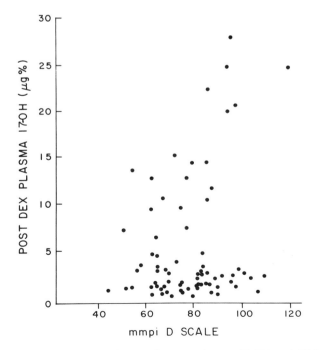

FIG. 2. Relationship between postdexamethasone plasma 17-OH and MMPI D scale in newly hospitalized psychiatric patients.

and MMPI scale score in the group of patients who did not suppress cortisol production completely after dexamethasone and a second group of patients who showed varying MMPI scale score but uniformly low ("normally suppressed") postdexamethasone cortisol level.

Linear correlations between MMPI scale scores and postdexamethasone cortisol levels for dexamethasone-resistant patients (post level >5 $\mu g\%$) were computed. Results are listed in Table 3. Some of the correlations seen in the dexamethasone-resistant group are strikingly large and account for nearly 50% of the variance in postdexamethasone cortisol levels in these patients. Interestingly, none of the 27 scales examined correlated significantly with the predexamethasone cortisol level in these suppression-resistant patients or in the total patient group (Table 3). It appears that in those patients who do not exhibit extreme suppression of HPAC acitivity the morning after dexamethasone the magnitude of postdexamethasone cortisol level is significantly related to a group of MMPI scales. Of the 10 original clinical MMPI scales, those correlating significantly were D, Pt, Hs, Sc, and Hy. Most of these and the six significantly correlating research scales can be characterized as measures of depression, inhibition, anxiety, and/or somatic complaints.

In summary, we draw these conclusions: (a) Resistance to dexamethasone suppression shortly after hospital admission occurs in a wide range of the current DSM II diagnostic classifications but is much more frequent in diagnoses of depression. (b) The degree of resistance of cortisol level to dexamethasone suppression in an HPAC system that is not suppressed below a minimum level of activity appears to be a physiological correlate of several clinically based scales of psychopathology as measured by the MMPI.

TABLE 3. *MMPI scales versus pre- and postdexamethasone plasma cortisol levels in nonsuppressors: significant correlations*

Scales	Postdexamethasone cortisol (N = 20)		Predexamethasone cortisol (N = 20)	
	r	p	r	p
Original MMPI scales				
D, depression	+.721	<0.01	+.234	ns
PT, psychasthenia	+.678	<0.01	+.093	ns
HS, hysteria	+.551	<0.05	+.168	ns
SC, schizophrenia	+.474	<0.05	−.100	ns
HY, hypochondriasis	+.461	<0.05	+.090	ns
Research scales				
AT, manifest anxiety	+.598	<0.01	+.358	ns
LB, lower back pain	+.571	<0.01	+.193	ns
CA, caudality	+.537	<0.05	+.170	ns
MT, maladjustment	+.495	<0.05	+.284	ns
A, anxiety and distress	+.467	<0.05	+.210	ns
R, repression and denial	+.445	<0.05	−.019	ns

ACKNOWLEDGMENT

This work was supported by U.S. Public Health Service grant no. MH-12464.

REFERENCES

1. Clark, L. D., Quarton, Q. C., Cobb, S., and Bauer, W. (1953): Further observations on mental disturbances associated with cortisone and ACTH therapy. *New Engl. J. Med.,* 249:178–183.
2. Ader, R., and Grota, L. J. (1969): Effects of early experience on adrenocortical reactivity. *Physiol. Behav.,* 4:303–305.
3. Poe, R. O., Rose, R. M., and Mason, J. W. (1970): Multiple determinants of 17-hydroxycorticosteroid excretion in recruits during basic training. *Psychosom. Med.,* 32:369–378.
4. Tennes, K., and Carter, D. (1973): Plasma cortisol levels and behavioral states in early infancy. *Psychosom. Med.,* 35:121–128.
5. Maas, J. W., and Mednieks, M. (1971): Hydrocortisone-mediated increase of norepinephrine uptake by brain slices. *Science,* 171:178–179.
6. de Wied, D. (1974): Pituitary-adrenal system hormones and behavior. In: *The Neurosciences, Third Study Program,* edited by F. O. Schmitt and F. G. Worden, pp. 653–666. M.I.T. Press, Cambridge.
7. Stokes, P. E. (1972): Studies on the control of adrenocortical function in depression. In: *Psychobiology of the Depressive Illnesses,* edited by T. Williams and M. Katz, pp. 194–214. Government Printing Office, Washington, D.C.
8. Stokes, P. E., Stoll, P. M., Shamoian, C. A., and Patton, M. J. (1971): Efficacy of lithium as acute treatment of manic-depressive illness. *Lancet,* i:1319–1325.

Hormones, Behavior, and Psychopathology, edited by
Edward J. Sachar. Raven Press, New York © 1976.

Are Nocturnal Cortisol Spikes in Depressed Patients Due to Stress?

George C. Curtis

*Neuropsychiatric Institute, The University of Michigan Medical Center,
Ann Arbor, Michigan 48104*

Should the "night spiking" of plasma cortisol be regarded as a nonspecific "stress" phenomenon? Ader and Friedman (1) and Brown and Martin (2) demonstrated that rats given stressful stimuli during the circadian phase of minimal adrenal secretory activity show prompt rises of plasma corticosterone levels to values similar to those seen during the phase of maximal secretion. This lends indirect support to the concept that night spiking in depressed patients could be a stress phenomenon.

In an ongoing study several University of Michigan psychiatry residents (Randolph Nesse, Martin Buxton, Jesse Wright, and David Lippman) and I have investigated the effects of another type of psychological stress on plasma cortisol in human subjects during the phase of minimal secretory activity. The stress procedure consists of treating patients with severe phobias for physical objects by prolonged, live exposure to the phobic object, the procedure now known as "flooding *in vivo*."

With this method a patient with a phobia—let us say for snakes—is required to approach as rapidly as possible, and eventually to handle, a live (harmless) snake. The pace of the exposure is pushed as rapidly as the patient will tolerate, and anxiety is kept constantly at the highest possible levels. By behavioral and subjective criteria the stress involved is usually substantial and frequently dramatic, with indications of anxiety which include screaming, fleeing, weeping, gooseflesh, chattering teeth, uncontrollable shaking, and startling at slight sounds. Tachycardia is regularly observed. Since the treatment has a predictable effect, and since it occurs at a predetermined time and place, it lends itself to incorporation into research designs for the experimental analysis of anxiety.

Seven subjects with severe object phobias reported to the laboratory for 3-hour sessions on each of five separate evenings. The timing of sessions was standardized so as to occur at the same circadian phase for all subjects on all evenings, and for this phase to occur during minimal adrenal secretory activity. Blood samples were collected at 20-min intervals through an indwelling needle, and self-ratings of anxiety were also made at 20-min intervals on a rating scale of 0 to 100. One-hour treatment periods were held during the second hour of the third and fourth sessions. The remaining observations comprised control data.

Mean anxiety ratings were near zero with the following exceptions: (a) a dramatic and sustained rise during the first treatment session; (b) a smaller and less sustained rise during the second treatment sessions; and (c) rises to approximately 50, apparently anticipatory in nature, at the beginning of the first session, when subjects knew that no treatment would occur, and at the beginning of the third session, when the subjects knew that the first treatment would occur.

Adrenal secretory episodes, by the criteria of Weitzman et al. (3) were small and infrequent, as is characteristic of this circadian phase, equally so during treatment and control sessions. Mean plasma cortisol values showed no significant effect of treatment sessions, although there was an overall decline in plasma cortisol values from the beginning to the end of each session (normal diurnal effect), and from the first to the last sessions in the laboratory, an apparent example of the well-known "novelty effect."

These results show that even a dramatic psychological stress does not necessarily produce night spiking of plasma cortisol values. They do not disprove the hypothesis that night spiking in depressed patients is a stress phenomenon, but they do dictate caution in adopting it.

REFERENCES

1. Ader, R., and Friedman, S. B. (1968): Plasma corticosterone response to environmental stimulation. *Neuroendocrinology,* 3:378–386.
2. Brown, G. M., and Martin, J. B. (1973): Comparison of growth hormone and prolactin responses to handling and novel environment in the rat. *Psychosom. Med.,* 35:447.
3. Weitzman, E. D., Fukushima, D., Nogeire, C., Roffwarg, H., Gallagher, T. F., and Hellman, L. (1971): Twenty-four hour pattern of the episodic secretion of cortisol in normal subjects. *J. Clin. Endocrinol.,* 33:14–22.

Hormones, Behavior, and Psychopathology, edited by
Edward J. Sachar. Raven Press, New York © 1976.

Growth Hormone Responsiveness to Hypoglycemia and Urinary MHPG in Affective Disease Patients: Preliminary Observations

David L. Garver, Ghanshyam N. Pandey, Haroutune Dekirmenjian,
and John M. Davis

Illinois State Psychiatric Institute, Chicago, Illinois 60620

A number of investigators have found that human growth hormone (HGH) response following insulin-induced hypoglycemia (IH) is abnormally low in certain affective disease patients (1–4). Since HGH response to IH is attenuated by the α-adrenergic blocking agent phentolamine in normal subjects (5), there has been speculation that the altered HGH responsiveness in certain affective disease patients may be related to altered norepinephrine (NE) function.

The urinary excretion of a critical NE metabolite, 3-methoxy-4-hydroxyphenylglycol (MHPG), has also been shown to be abnormally low in certain affective disease patients (6–8). Studies in primates have recently shown that 24-hour urinary MHPG assays can be used as an index of central NE turnover (9,10).

We report here preliminary findings relating HGH response to IH and central NE turnover, as measured by 24-hour urinary MHPG levels in a small heterogeneous group of affective disease patients.

METHODS

Twenty-four-hour urine specimens were collected on hospitalized, affective disease patients who were free from medication for at least 2 weeks. The specimens were assayed for MHPG in our laboratories by a modification of the method of Wilk et al. (11).

Following these urine collections, each of the patients underwent neuroendocrine testing by means of induction of hypoglycemia via glucagon-free insulin (0.1 unit/kg). Blood samples were withdrawn through an indwelling catheter every 10 min for 1 hour after the insulin was given for determination of glucose levels. Samples taken at −60, −30, 0, 20, 40, 60, 80, 100, 140, and 180 min surrounding the insulin challenge were assayed for HGH by radioimmunoassay using the double-antibody technique of Odell et al. (12). Patients with one or more preinsulin, baseline samples with HGH

233

above 3 ng/ml were excluded from the data, as were patients on whom full 24-hour urine specimens could not be collected.

RESULTS

Our preliminary data on a heterogeneous group of six affective disease patients (Table 1) demonstrates that those patients who had evidence of lower baseline central NE turnover as reflected in low 24-hour urinary MHPG levels, also subsequently had low peak HGH response to insulin-induced hypoglycemia. Higher HGH responses were found in patients who previously had shown higher levels of MHPG. Patient 1, who would be termed as having inadequate HGH response (i.e., <5 ng/ml) to IH, was the lowest MHPG excretor. In each case serum glucose had fallen to between 35% and 50% of baseline in response to the insulin.

TABLE 1. *HGH response to insulin-induced hypoglycemia and urinary MHPG in affective disease*

Patient No.	Age (yr)	Sex	Glucose[a] (mg/100 ml)	HGH peak[b] (ng/ml)	MHPG[c] (μg/24 hr)
1	60	F	79 → 35	4.8	1,131 (2)
2	37	F	103 → 52	13.4	1,208 (2)
3	22	F	96 → 36	10.1	1,445 (2)
4	36	M	94 → 41	65.0	1,686 (2)
5	25	M	87 → 43	20.4	1,728 (2)
6	26	F	84 → 37	47.0	1,771 (2)

[a] Glucose fall induced by glucagon-free insulin, 0.1 unit/kg i.v.
[b] HGH peak following glucagon-free insulin, 0.1 unit/kg i.v.
[c] Twenty-four-hour urinary MHPG on baseline days preceding neuroendocrine testing.
Figures in parentheses indicate the number of complete collections assayed.

DISCUSSION

The data on a limited number of affective disease patients in this preliminary report suggest that there may be a relationship between baseline NE turnover as measured by urinary MHPG and HGH response to IH in this group of patients. Patient 4 showed excessive psychologic and physiologic stress to both the IH catheter study and to a similar catheter study. No such stress was apparent during the preceding urine collection days. Since psychologic stress of a variety of types has been shown to cause HGH release in normal subjects, even without hypoglycemia (13), psychologic stress and hypoglycemia may be additive in the case of patient 4, resulting in a marked HGH response. The fact that estrogens potentiate HGH response to hypoglycemia contributes another variable; the lowest responder was also the only postmenopausal woman, and three were women who were still having menstrual cycles.

The rank correlation of HGH peaks and MHPG strongly suggests that central NE turnover, as measured by urinary MHPG, plays a major role in modulating the sensitivity of the growth hormone system to the provocative stimulus of insulin-induced hypoglycemia.

REFERENCES

1. Mueller, P. S., Heninger, G. R., and McDonald, R. K. (1969): Insulin tolerance test in depression. *Arch. Gen. Psychiatry,* 21:587–593.
2. Carroll, B. J. (1972): Studies with hypothalamic-pituitary-adrenal stimulation test in de-depression. In: *Depressive Illness,* edited by B. Davies, B. J. Carroll, and R. M. Mowbray, pp. 149–200. Charles C Thomas, Springfield, Ill.
3. Sachar, E. J., Frantz, A. G., Altman, N., and Sassin, J. (1973): Growth hormone and pro-lactin in unipolar and bipolar depressed patients: Responses to hypoglycemia and L-dopa. *Am. J. Psychiatry,* 130:1362–1367.
4. Gruen, P. H., Sachar, E. J., Altman, N., and Sassin, J. (1975): Growth hormone responses to hypoglycemia in postmenopausal depressed women. *Arch. Gen. Psychiatry,* 32:31–33.
5. Blackard, W. G., and Heidingsfelder, S. A. (1968): Adrenergic receptor control mechanism for growth hormone secretion. *J. Clin. Invest.,* 47:1407–1414.
6. Maas, J. W., Fawcett, J. A., and Dekirmenjian, H. (1968): 3-Methoxy-4-hydroxy phenyl-glycol (MHPG) excretion in depressive states. *Arch. Gen. Psychiatry,* 19:129–134.
7. Fawcett, J. A., Maas, J. W., and Dekirmenjian, H. (1972): Depression and MHPG ex-cretion. *Arch. Gen. Psychiatry,* 26:246–251.
8. Beckman, N. H., and Goodwin, F. K. (1975): Antidepressant response to tricyclics and urinary MHPG in unipolar patients. *Arch. Gen. Psychiatry,* 32:17–21.
9. Maas, J. W., Dekirmenjian, H., Garver, D. L., Redmond, D. E., and Landis, D. H. (1972): Catecholamine metabolite excretion following intraventricular injection of 6-OH-dopa-mine. *Brain Res.,* 41:507–511.
10. Maas, J. W., Dekirmenjian, H., Garver, D., Redmond, D. E., and Landis, D. H. (1973): Excretion of catecholamine metabolites following intraventricular injection 6-hydroxy-dopamine in the Macaca speciosa. *Eur. J. Pharmacol.,* 23:121–130.
11. Wilk, S., Gitlow, S. E., Clarke, D. D., and Paley, D. H. (1967): Determination of urinary 3-methoxy-4-hydroxyphenylethylene glycol by gas-liquid chromatography and electron capture detection. *Clin. Chim. Acta,* 16:403–408.
12. Odell, W. D., Rayford, P. L., and Ross, G. T. (1967): Simplified, partially automated method for radioimmunoassay of human thyroid-stimulating, growth, luteinizing and follicle stimulating hormones. *J. Lab. Clin. Med.,* 70:973–979.
13. Brown, G. M., and Reichlin, S. (1972): Psychologic and neural regulation of growth hormone secretion. *Psychosom. Med.,* 34:45–61.

Hormones, Behavior, and Psychopathology, edited by
Edward J. Sachar. Raven Press, New York © 1976.

Responsiveness of Endocrine-Energy Metabolism Systems in Depression and Schizophrenia: Relationships to Symptomatology

*George R. Heninger, †Peter S. Mueller, and *Linda S. Davis

*Department of Psychiatry, Yale University, and Connecticut Mental Health Center, New Haven, Connecticut 06519; and †Department of Psychiatry, Rutgers University, New Brunswick, New Jersey 08902

The review by Carroll and Mendels in this volume clearly indicates the presence of neuroendocrine dysfunction in severe affective disorders. However, important questions remain as to the specific relationships between the type and severity of the affective illness and the magnitude or type of neuroendocrine dysfunction, particularly during symptom changes following various forms of treatment. An additional important question is whether the neuroendocrine dysfunction is specific to the syndrome of affective illness or if it is also observed in other severe mental illness such as schizophrenia.

In our attempts to study these questions, we previously observed (1) that patients with severe depression have a reduced rate of glucose utilization (k) even though they have elevated endogenous serum insulin levels following an intravenous glucose load (GTT). They also demonstrate reduced metabolic responsiveness to exogenous insulin during the insulin tolerance test (ITT) (2). The "psychotic" depressed patients had a significantly smaller human growth hormone (HGH) response in the ITT than the "neurotic" depressed patients. The other GTT and ITT abnormalities were also less obvious in patients with "neurotic" depression compared to patients with "psychotic" depression, and abnormalities went toward normal as the patients improved during amitriptyline treatment. In schizophrenic patients, significant positive correlations were noted between the duration of hospitalization and reduced responsiveness for both glucose and free fatty acids (FFA) in the ITT (3).

The measurements of glucose metabolism and responsiveness to insulin are not highly correlated, and each measurement may reflect quite different physiologic processes. In addition to the possibility that the more symptomatic patients have relatively more abnormalities in glucose metabolism and insulin responsiveness, it is also possible that the pretreatment intensity or magnitude of change during treatment in specific symptoms could relate differently to separate metabolic measurements and that there may be similarities and/or differences between patient groups. If specific relationships can be delineated (e.g., anxiety relating to one metabolic factor and depression to another—or alternatively, anxiety relating to one metabolic factor in depression but not in schizophrenia), then this information would be very useful in clarifying the psychopathophysiologic processes involved.

To evaluate the above findings more fully and to assess whether differential relationships between metabolic measurements and symptoms exist between depression and schizophrenia, 40 patients with a diagnosis of depression and 20 patients with schizophrenia were given a GTT and an ITT when they were symptomatic off medication 14 days. The GTT and ITT were repeated a second time 3 to 8 weeks later when the depressed patients were on amitriptyline and the schizophrenic patients on chlorpromazine or trifluoperazine and most of their symptoms had decreased. Metabolic measurements included serum glucose, insulin, HGH, and FFA. At the time of the GTT and ITT, patients were rated by nursing staff on the short clinical rating scale (SCRS) (4). This allowed: (a) a more detailed evaluation of relationships between symptomatology and metabolic findings, and (b) a comparison of these relationships between the depressed and schizophrenic groups.

RESULTS

The mean symptom ratings between the two patient groups were comparable on many items. The depressives had somewhat more depressed mood and motor agitation, whereas the schizophrenics had more paranoid behavior, thought disorder, and hallucinations. Both patient groups showed improvement in the mean ratings of most items during treatment.

Table 1 lists the comparison between the two patient groups in terms of the statistical significance of the mean changes during treatment of the selected metabolic variables. It can be seen that the depressive patients had a significant decrease in fasting glucose, an increase in the rate of glucose utilization, with a decrease in the insulin level in the GTT and a lower glucose value after insulin in the ITT. In addition there was a mean decrease in the fasting HGH level. In contrast, the schizophrenic group showed no mean change of metabolic variables except a lower glucose level 30 min after the insulin injection. These metabolic changes illustrate the previously reported insulin resistance observed in depression. It is of some interest that a similar phenomenon is not clearly observed in the schizophrenic population.

In Table 2 are the statistically significant correlations between selected rating scale items and the metabolic variables. The coefficients listed under change represent the correlation between the magnitude of the behavioral change during treatment compared to the magnitude of change in the metabolic variable.

TABLE 1. *Change in metabolic variables during psycho-pharmacologic treatment and symptom improvement in depressive and schizophrenic patients*

Metabolic variable	Depressive	Schizophrenic
Fasting glucose	Decrease[a]	ns
Fasting FFA	ns	ns
Fasting insulin	ns	ns
Fasting HGH	Decrease[b]	ns
GTT*k*	Increase[a]	ns
GTT insulin 50 min	Decrease[b]	ns
ITT glucose 30 min	Decrease[a]	decrease[c]
ITT FFA 30 min	ns	ns
ITT HGH maximum	ns	ns

ns, not significant.
[a] $p < 0.001$.
[b] $p < 0.01$
[c] $p < 0.05$.

Increased fasting FFA levels relate positively to high ratings on verbal anxiety in depression but not in schizophrenia. The fasting levels of glucose, insulin, and HGH did not correlate to ratings. In the GTT the low rate of glucose utilization (*k*) is observed in both depressive and schizophrenic patients with relatively more motor retardation. The insulin values do not

TABLE 2. *Significant metabolic-behavioral correlations for depressive and schizophrenic patients*

Metabolic variable	Depressive patients		Schizophrenic patients	
	Pretreatment	Change	Pretreatment	Change
Fasting levels				
FFA vs. verbal anxiety	.43[a]	.37[b]	.11	.01
GTT				
k vs. (motor retardation minus motor agitation)	−.53[c]	−.27	−.50[b]	−.22
ITT				
Glucose 30 min vs. depressed mood	.11	.05	.46[b]	.19
FFA 30 min vs. motor retardation	.38[b]	.65[c]	−.50[b]	−.57[a]
FFA 30 min vs. sum of 13 items	.33[b]	.43[b]	−.52[b]	−.57[a]
HGH maximum vs. depressed mood	−.34[b]	−.37[b]	−.29	.37
HGH maximum vs. verbal anxiety	−.41[b]	−.40[b]	−.18	.15
HGH maximum vs. sum of 13 items	−.42[b]	−.41[b]	−.18	.42

[a] $p < 0.01$.
[b] $p < 0.05$.
[c] $p < 0.001$.

correlate to ratings. In the ITT the glucose at 30 min does not relate significantly to symptoms in depression, but it does in schizophrenia. The FFA at 30 min relates in an opposite manner in depression and in schizophrenia to motor retardation and the sum of 13 items. The maximum HGH response following insulin in depressives correlates significantly to several items, including depressed mood, verbal anxiety, and the sum of the 13 items as listed.

The clearest difference between depressives and schizophrenics on these measurements involved the FFA response to insulin. The retarded depressives had a smaller FFA response and the retarded schizophrenics a larger FFA response following insulin. The depressive and schizophrenic patients also differ, although somewhat less clearly, in that anxiety does not relate to fasting FFA levels in schizophrenics as it does in depressives, the reduced HGH response to insulin observed in the more ill depressives was not found in the more ill schizophrenics, and depressed mood related to increased glucose levels at 30 min in the ITT in schizophrenics but not depressives. The two groups of patients are similar in that both depressed and schizophrenic patients with retardation (retardation minus agitation scale) have the low rates of glucose utilization.

DISCUSSION

The data strongly suggest that a motor activity dimension of symptomatology is important when considering glucose metabolism. There are several studies which indicate that decreased glucose utilization is related to decreased motor activity. However, the low k observed in severe depression cannot be totally explained on the basis of physical activity alone, since the patients had an approximately normal amount of bed rest (8 to 11 hours per day), inactivity was not permitted the patients during the study, and great care was taken to engage the patients in physical and social activity despite psychomotor retardation and emotional withdrawal. Most patients with psychomotor agitation had an increase in their k following recovery from depression despite the fact that their agitation decreased. This indicates that the reduced rate of glucose utilization can be the result of both retardation-type symptoms as well as other factors associated with the illness.

The relationship between fasting FFA levels and anxiety observed in depression is consistent with numerous other reports in the literature linking these two variables. It is of considerable interest that anxiety in schizophrenia does not relate to fasting FFA levels as it does in depression. The lack of fasting FFA-anxiety correlations, the correlations of FFA levels in the ITT to symptomatology that are opposite to those seen in depression, and other studies of FFA metabolism in schizophrenia (5) suggest that the regulation of FFA levels may be abnormal in schizophrenia. Since studies with α- and β-adrenergic blocking agents demonstrate considerable adrenergic regulation of fasting FFA levels (6), the above findings suggest possible abnormalities in catecholamine regulation of FFA release in the schizophrenic population.

The fact that the HGH maximal response to insulin has a general relationship to different depressive symptoms in contrast to the restricted correlations seen with the GTT k or fasting levels is consistent with the review by Carroll and Mendels (*this volume*) suggesting a hypothalamic disorder associated with severe depression. It is of interest that the changes in depressive symptoms following recovery are correlated with the changes in HGH response to insulin. This would be evidence against a chronic hypothalamic disorder in patients with depression and would be in favor of a hypothesis that the hypothalamic disorder reflected in the impaired HGH response to insulin is more related to the depressive episode *per se*.

Even though the correlations in the present study are low, the findings do support the concept that the different symptom components of both depression and schizophrenia have the same and also different physiologic correlates depending on the variables measured. It is important to sort out the similarities and differences between the various syndromes so that in the future accurate diagnoses can be made which are more consistent with the underlying pathophysiologic processes.

SUMMARY

Forty depressed patients and 20 schizophrenic patients were studied with the GTT and ITT and nursing ratings of symptomatology. The findings suggest that motor activity is an important dimension of symptomatology when considering glucose metabolism in both depression and schizophrenia. Anxiety levels in depression relate to fasting FFA levels, but this relationship was not seen in schizophrenia. FFA levels following insulin are very different in schizophrenics and depressives; increased retardation relates to increased levels in depressives and to decreased levels in the schizophrenics. The depressed patients showed several correlations between increased symptomatology and decreased HGH responsiveness in the ITT, but this was not seen in schizophrenia. The type of the correlations suggest that the reduced hypothalamic-pituitary HGH responsiveness in depression is at least partly reversible during amitriptyline treatment. This study demonstrates that the separation of symptomatology into various components when studying metabolic measurements of patients with functional mental illness is a viable and important approach to more specifically describing the altered biologic function associated with different psychopathologic states.

REFERENCES

1. Mueller, P. S., Heninger, G. R., and McDonald, R. K. (1969): Intravenous glucose tolerance test in depression. *Arch. Gen. Psychiatry*, 21:470–477.
2. Mueller, P. S., Heninger, G. R., and McDonald, R. K. (1969): Insulin tolerance test in depression. *Arch. Gen. Psychiatry*, 21:587–594.

3. Schimmelbusch, W. H., Mueller, P. S., and Sheps, J. (1971): The positive correlation between insulin resistance and duration of hospitalization in untreated schizophrenia. *Br. J. Psychiatry,* 118:429–436.
4. Heninger, G. R., French, N. H., Slavinsky, A. T., Davis, L., and Mueller, P. S. (1970): A short clinical rating scale for use by nursing personnel. II. Reliability, validity, and application. *Arch. Gen. Psychiatry,* 23:241–248.
5. Franzen, G. (1972): Plasma free fatty acids, serum cortisol and circulatory response to insulin in acute schizophrenic men. *Psychiatr. Clin.,* 5:201–208.
6. Gagliardino, J. J., Bellone, C. F., Doria, I., Sccenchez, J. J., and Pereyra, V. (1970): Adrenergic regulation of basal serum glucose, NCFA and insulin levels, *Horm. Metab. Res.,* 2:318–322.

Hormones, Behavior, and Psychopathology, edited by
Edward J. Sachar. Raven Press, New York © 1976.

Hormonal and Behavioral Reversals in Hyposomatotropic Dwarfism

John Money, Charles Annecillo, and June Werlwas

Department of Psychiatry and Behavioral Sciences, and Department of Pediatrics, The Johns Hopkins University School of Medicine and Hospital, Baltimore, Maryland 21205

Reversible hyposomatotropic dwarfism is a syndrome characterized by domicile-specific impairment of statural growth and of growth hormone secretion. Both impairments are reversible on change of domicile, as are various behavioral features of the syndrome. Taxonomically, the syndrome has been known as psychosocial dwarfism (1), but reversible hyposomatotropic dwarfism is a more operational term and one which does not imply spurious accuracy concerning etiology. In the past the syndrome has been variously identified with environmental failure to thrive (2), and maternal deprivation (3). It has also been known as deprivation dwarfism (4,5), and as emotional deprivation and growth retardation simulating idiopathic hypopituitarism (6,7).

The syndrome probably occurred in individuals with impaired behavioral development associated with hospitalism or institutionalism, as reported by Spitz (8) and Bowlby (9,10), even though these investigators did not specifically delineate the phenomenon of impaired statural growth as related to impaired somatotropin release. The same applies also to earlier investigations of child abuse, in which associated growth failure is seldom mentioned. Recently, reversible hyposomatotropic dwarfism has been identified in some cases as a concomitant of child abuse (11).

The purpose of this chapter is to review behavioral features of the syndrome of reversible hyposomatotropic dwarfism before and after the blood level of growth hormone becomes elevated and statural growth resumes following a change of domicile. Brown discusses the endocrinological aspects of this syndrome elsewhere in this volume.

SAMPLE

The 40 children and adolescents from whom the data under review were clinically derived were seen in the pediatric endocrine clinic and its psychohormonal research unit at The Johns Hopkins Hospital between 1959 and 1974. No known systematic bias exists with respect to referral for behavioral studies. Probably the process which brings about referral to the hospital from the community at large is subject to bias, although to an unknown degree. The morbidity and mortality of this syndrome are unknown. Affected individuals may or may not survive without receiving professional attention.

The children usually are referred to the hospital because of growth failure. The initial change of domicile is usually from the home of origin to the hospital for an admission of approximately 2 weeks. This admission allows for resumption of growth hormone secretion, and it leads to the onset of catch-up statural growth. These are the essential criteria for establishing the diagnosis. Additional convalescent or foster-care placement permits continued catch-up statural growth.

PROCEDURES

The data for this paper are drawn from published reports based on our pool of patients (11–15) and from information in the actual histories. In the latter, information had been obtained by members of the psychohormonal research unit. Taped and transcribed interviews and notes were obtained from patients, parents or guardians, social workers, and nurses. The interviews always began with open-ended inquiry prior to factual and true-false questioning. Direct observations of the child's behavior in the hospital, visits to their homes, and school reports provided additional information.

FINDINGS

Eating and Drinking

The source of information on the variable "eating and drinking" is from clinical reports and case files, for as yet there has been no published systematic investigation. It is a difficult topic on which to gain accurate information as there is an extreme likelihood of parental prevarication and cover-up concerning restriction of their dwarfed child's nourishment. Some parents may report that the child has a bad reaction to food and therefore must be restricted. The more likely reports, however, are of: eating from unusual places such as garbage cans and drinking from toilet bowls; stealing and hoarding food by day or night (*vide infra*); alleged picky eating; refusal to eat at mealtimes, and ingesting inedible items (*vide infra*). Polyphagia, excessive intake of available food, and polydipsia, excessive drinking, may alternate with vomiting and ostensible self-starvation. Patients may be overweight for their dwarfed height or underweight, but not grossly emaciated as in famine starvation and not suffering from identifiable specific nutritional deficits (11) or malabsorption. In contrast with this syndrome, growth hormone levels are elevated in syndromes of malnutrition including kwashiorkor, marasmus, and anorexia nervosa (13). Eating and drinking symptoms are reversible on change of domicile where food and drink intake is normally timed but no longer restricted.

Sleeping

Money and Wolff (13) reported a systematic investigation of sleep as related to statural growth in hyposomatotropic dwarfism. The characteristics of poor sleep were frequent night crying, inability to fall asleep within 30 min, inability to fall asleep after waking, refusal to nap, duration of sleep shorter than appropriate for age, unpredictable waking times to start the day, and getting up at night to engage in various behaviors including roaming. At night the patients were also known to search for food and drink, go to the bathroom, play, lay awake restless, or disturb other family members' sleep. Especially unusual behavior, such as running out into the street at night, was also reported.

The criteria for poor sleep in reverse became the criteria of good sleep. When elements of both existed, the patient was rated as having mixed sleep. The quality of sleep tended to be related to residence (Figs. 1 and 2). On

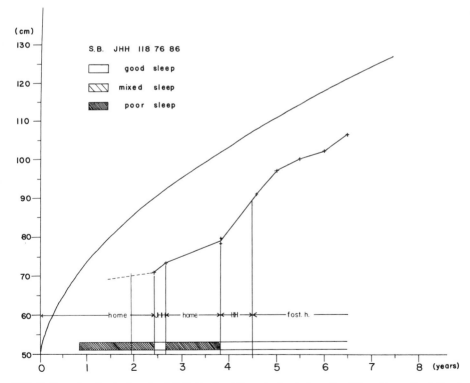

FIG. 1. Growth chart of patient S.B. Vertical lines indicate length of time the patient lived in the respective domicile. The dotted line before the first hospital referral and accurately recorded height measurement represents an estimate of the growth rate between the age of 1 year, the usual age of onset of growth retardation, and the age when the dwarfed stature was first measured. Home, home of origin. JHH, The Johns Hopkins Hospital; HH, Happy Hills Hospital; fost. h., foster home. From Wolff and Money (13), with permission.

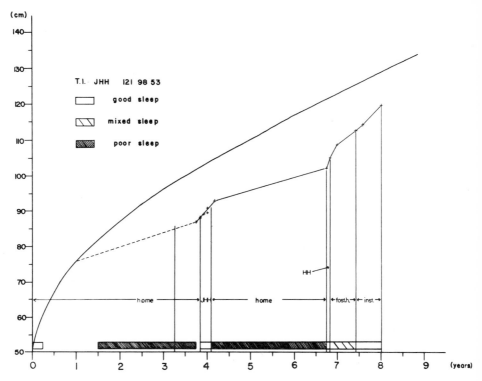

FIG. 2. Growth chart of patient T.I. Home, home of origin. JHH, The Johns Hopkins Hospital. HH, Happy Hills Hospital. fost. h., foster home. inst., residential institution. From Wolff and Money (13), with permission.

removal from the growth-retarding environment, a reversal of sleep behavior from poor to good occurred correlatively with a measured increase in statural growth.

Powell et al. (16) studied growth hormone release during sleep in one patient with the syndrome of reversible hyposomatotropic dwarfism, a 9-year-old girl with a height age of 3.3 years and weight age of 2.5 years. On days 5 and 6 after admission the girl's sleep had already improved—a not uncommon finding. Nonetheless, there was no corresponding improvement in growth hormone release during sleep; the expected release during the initial phase of slow-wave sleep did not occur. By day 40, however, growth hormone function had recovered and there was a marked output during the initial phase of slow-wave sleep.

Enuresis and Encopresis

Enuresis and encopresis are two other symptoms associated with this syndrome for which there is no specific published reference. Although often mentioned in clinical reports, the frequency incidence is not documented. Enuresis and encopresis both refer to failure to regulate elimination either at normal age or, regressively, after toilet training has been achieved.

Wetting and soiling both occurred on some occasions in unusual places possibly in association with roaming. The timing was day or night. They were likely to be associated with periods of solitary isolation in a room or region of the house. For instance, in one case wetting and soiling were known to occur when the patient was locked in a closet or chained to a bedpost. Regardless of previous toilet training history, patients establish or re-establish self-control of elimination after change of domicile.

Pain

The absence of appropriate pain response and/or presence of self-harming activity comprise another reversible aspect of this syndrome. Both types of behavior were the subject matter of a systematic investigation of 32 patients by Money et al. (12). Hyporeactive response to pain was reported in association with physical punishment, as evidenced by bruises, scars, and/or lesions not yet healed; alleged accidents (e.g., a fractured clavicle); infections such as urinary tract and/or inflammatory illnesses; and medical injections, minor surgery, or dental treatment. Parents or guardians of patients reported that at times the children did not complain, cry, or shed tears when punished, and generally did not react or complain when hurt or injured.

The types of self-harming behavior observed or reported were allegedly falling often and/or bumping frequently into furniture, walls, or corners; temper tantrums with head banging and self-biting; head banging without temper tantrums; eating indigestible or harmful objects such as stones, drugs from the medicine cabinet, or soup that was too hot for other family members; twisting and plucking hair; excessive nail-biting and tearing of nailbed skin; and failure to protect with hands when falling.

The evidence of nonrecognition or lack of responsiveness to pain inflicted by self or others is a form of pain agnosia. In an ameliorative environment, pain agnosia was reversible within a short period of time, in some cases 48 hours. Incidences of self-injury sharply decreased, infected lesions healed, and an appropriate negative reaction to injection needles and finger pricks was noted after domicile change.

Social Distancing

On the variable "social distancing" information is drawn from our clinical notes and observations. The term is here used to refer to: (a) not initiating or responding to conversation; (b) turning head away from people; (c) wiggling away from body contact; (d) avoidance of eye contact; (e) a lack of facial expression; (f) a lack of motor response in interpersonal contact.

The patients show a general apathy and inertia. In a quiet and stubborn way they avoid personal contact. In a very short space of time (within 2 weeks) after change of domicile, patients are likely to show a marked increase in interaction with others, which together with an increase in general activity constitutes compensatory hyperactivity (*vide infra*).

Temper Tantrums

Temper tantrums constitute another symptom frequently mentioned in clinical reports and without a published systematic investigation. The inertia and lack of rapport mentioned above is punctuated by an occasional temper tantrum.

These short-lived outbursts correspond to outbursts of excitement in a catatonic patient and may be accompanied by self-harming activity, enuresis or encopresis, and an inability to sleep. Tantrums are not predominantly tearful, nor are they excessively defiant or destructive. They reflect temporary loss of control rather than purposeful revenge and are followed by a return to apathetic withdrawal. After domicile change, temper tantrums disappear, despite the onset of a compensatory increase in overall activity.

Roaming

Another symptom associated with the growth-retarding domicile is roaming. Evidence of its presence comes from clinical notes and observations.

In some instances roaming was reported at night when the family was asleep and by day when the patient was left alone. By night or day, there was a high probability that the patients looked for food or fluid, although the informants did not connect the two activities. In other instances the roaming, as reported, seemed more exploratory, with no specifiable purpose other than to become familiar with the environment or to find playmates or playthings.

Roaming did not include leaving home, hiding, or being unlocated for extended periods. Some parents reported roaming in the form of sleepwalking. There was no available evidence to associate the activity with destructiveness. According to the evidence available, roaming diminishes and disappears after the child moves into a growth-promoting environment.

IQ

IQ elevation by as much as 55 points was documented paralleling change of domicile and catch-up growth in stature. Preliminary data were completed on 16 patients (15). Table 1 shows that 12 of 16 patients have IQ increases, thus supporting the hypothesis of a correlation between improved living environment and IQ elevation. Further support for the hypothesis is evident in Table 2, which demonstrates a trend for the amount of IQ ele-

TABLE 1. *IQ change following change of domicile in reversible hyposomatotropic dwarfism*

No.	Range of change
4	+29 to +55
8	+2 to +14
1	0
3	−1 to −12

TABLE 2. *IQ change and duration of time away from domicile in reversible hyposomatotropic dwarfism*

Group	Duration	IQ change		
		Mean	Range	Median
A (N = 8)	1 mo to 3 yr 7 mo	+8.8	−12 to +37	+5.5
B (N = 8)	4 yr 3 mo to 7 yr 6 mo	+17.5	−1 to +55	+10.5
Total	1 mo to 7 yr 6 mo	+13.1	−12 to +55	+6.5

vation to correlate with the amount of time a child spends away from the growth-retarding domicile. These findings draw attention to the living environment as a powerful determinant of IQ and the change possible when it is altered.

Compensatory Hyperkinesis

According to recorded clinical notes and observations, compensatory hyperkinesis is evidenced only after change of domicile. It appears as a rebound phenomenon, since it is in contrast to an overall deficit of activity prior to the change. The behavior may last for weeks, months, or even longer.

Compensatory hyperkinesis is characterized by: (a) exhaustive talking and questioning with adults and peers; (b) excessive exploratory curiosity; (c) a high rate of motor activity as in running, jumping, climbing, singing, dancing, and cuddling; (d) overreacting to stimuli as in loud crying, boisterous or silly laughing, and emphatic verbalization of anger when frustrated. These activities are performed either playfully or purposefully, but without apparent malicious intent. They eventually decrease, becoming more appropriate and subdued.

Delayed Puberty

There were 12 patients old enough to qualify for a systematic investigation of delayed puberty (14). All of the patients rapidly progressed into

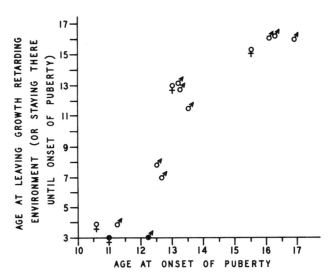

FIG. 3. Relationship between age at onset of puberty and age at leaving the growth-retarding environment (or staying there until the onset of puberty). N = 12. ♀, girl. ♂, boy. Filled symbols represent the norms for girls and boys (17).

puberty following change of domicile after the expected age of normal pubertal onset — based on the criteria of Tanner (17,18).

The oldest patient at the time of domicile change was 16, at which time he had a height age of 8. Signs of puberty began to appear within 1 year. In general, the longer a patient lived in the growth-retarding domicile, the later was the onset of puberty (Fig. 3).

BEHAVIORAL CHARACTERISTICS OF THE FAMILIES IN THE GROWTH-RETARDING ENVIRONMENT

The definitive characteristics of either the growth-retarding or growth-promoting environments constitute a puzzle still unsolved. However, within the growth-retarding environment, evidence of somatic and psychic trauma indicates a type of noxious family influence which needs further investigation and documentation.

Evidence for somatic and psychic trauma ranges widely. It includes identifying the child as different from siblings or peers; rejection of the child from family activities; enforced isolation; questionably authentic accidents and signs of excessive physical trauma including bruises, scars, and infected lesions; restriction of food and water; insufficient warmth in winter or cooling in summer; and punishment at night while asleep.

The validity of parental reporting is often doubtful and requires corroboration from independent sources, which is not easy to obtain. Proof of suspected abuse in the domicile of origin is often extremely difficult to establish because abused children deny, euphemize, and adamantly cover up their plight. The adults in charge, much in the manner of Munchausen's syndrome, are mendacious and deceitful, and are experts at hoodwinking professionals, playing off one against the other (11).

Uncovering the correct biography of suspected abuse is facilitated by the use of auxiliary informants; home visits; precision interviewing of the parents for hour-by-hour, daily, and weekly happenings; and by long-term study until the child loosens up and talks. We are currently investigating these matters further by undertaking an extensive investigation of two families.

SUMMARY

In the syndrome of reversible hyposomatotropic dwarfism, the behavioral symptoms can be classified as disorders of eating, drinking, sleeping, and elimination (enuresis and encopresis), along with pain agnosia, social distancing, temper tantrums, roaming, IQ impairment, and compensatory hyperkinesis. There may also be delayed puberty, behaviorally as well as hormonally. Compensatory hyperkinesis is associated with the early stage of recovery. All the other symptoms, including IQ deficit and delayed puberty, are reversible on change of domicile and onset of catch-up statural growth.

Preliminary data pertaining to intellectual functioning demonstrate a trend for catch-up mental growth and the age of pubertal onset to correlate with age and duration of release from the growth-retarding domicile. While proof of abuse in the domicile of origin is difficult to establish, observations and reports of somatic and psychic trauma point to the likelihood that reversible hyposomatotropic dwarfism is associated with child abuse. Definitive characteristics of both the growth-retarding and the growth-promoting environments, including parental or caretaker influence require further investigation and documentation.

ACKNOWLEDGMENTS

This work was supported by United States Public Health Service grant HD-00325 and by funds from the Grant Foundation, New York.

REFERENCES

1. Reinhart, J. B., and Drash, A. L. (1969): Psychosocial dwarfism: Environmentally induced recovery. *Psychosom. Med.*, 31:165–172.
2. Barbero, G. J., and Shaheen, E. (1967): Environmental failure to thrive: A clinical view. *J. Pediatr.*, 71:639–644.
3. Patton, R. G., and Gardner, L. I. (1969): Short stature associated with maternal deprivation syndrome: disordered family environment as cause of so-called idiopathic hypopituitarism. In: *Endocrine and Genetic Diseases of Childhood*, edited by L. I. Gardner. Saunders, Philadelphia.

4. Silver, H. K., and Finkelstein, M. (1967): Deprivation dwarfism. *J. Pediatr.,* 70:317–324.
5. Gardner, L. (1972): Deprivation dwarfism. *Sci. Am.,* 227:76–83.
6. Powell, G. F., Brasel, J. A., and Blizzard, R. M. (1967): Emotional deprivation and growth retardation simulating idiopathic hypopituitarism. I. Clinical evaluation of the syndrome. *N. Engl. J. Med.,* 276:1271–1278.
7. Powell, G. F., Brasel, J. A., Raiti, S., and Blizzard, R. M. (1967): Emotional deprivation and growth retardation simulating idiopathic hypopituitarism. II. Endocrinologic evaluation of the syndrome. *N. Engl. J. Med.,* 276:1279–1283.
8. Spitz, R. A. (1946): Hospitalism. In: *Psychoanalytic Study of the Child,* Vol. 2, pp. 113–117. International University Press, New York.
9. Bowlby, J. (1969): *Attachment and Loss, Vol. 1: Attachment.* Basic Books, New York.
10. Bowlby, J. (1973): *Attachment and Loss, Vol. 2: Separation.* Basic Books, New York.
11. Money, J. (1975): The syndrome of abuse-dwarfism (psychosocial dwarfism or reversible hyposomatotropinism): behavioral data. *Submitted.*
12. Money, J., Wolff, G., and Annecillo, C. (1972): Pain agnosia and self-injury in the syndrome of reversible somatotropin deficiency (psychosocial dwarfism). *J. Autism Child. Schizo.,* 2:127–139.
13. Wolff, G., and Money, J. (1973): Relationship between sleep and growth in patients with reversible somatotropin deficiency (psychosocial dwarfism). *Psychol. Med.,* 3:18–27.
14. Money, J., and Wolff, G. (1974): Late puberty, retarded growth and reversible hyposomatotropinism (psychosocial dwarfism). *Adolescence,* 9:121–134.
15. Money, J., and Annecillo, C. (1975): IQ change following change of domicile in the syndrome of reversible hyposomatotropinism (psychosocial dwarfism). *Submitted.*
16. Powell, G. F., Hopwood, N. J., and Barratt, E. S. (1973): Growth hormone studies before and during catch-up growth in a child with emotional deprivation and short stature. *J. Clin. Endocrinol. Metab.,* 37:674–679.
17. Tanner, J. M. (1962): *Growth at Adolescence.* Blackwell Scientific Publications, Oxford.
18. Tanner, J. M. (1969): Growth and endocrinology of the adolescent. In: *Endocrine and Genetic Diseases of Childhood,* edited by L. I. Gardner. Saunders, Philadelphia.

Hormones, Behavior, and Psychopathology, edited by
Edward J. Sachar. Raven Press, New York © 1976.

Endocrine Aspects of Psychosocial Dwarfism

Gregory M. Brown

Department of Psychiatry, University of Toronto, and Neuroendocrinology Research Section, Clarke Institute of Psychiatry, Toronto, Ontario, Canada

Prior to availability of suitable endocrine function tests, clinical aspects of the syndrome of dwarfism associated with environmental deprivation were described by many workers including Spitz (1), Widdowson (2), Engel and Reichsman (3), Talbot et al. (4), Bakwin (5), and Patton and Gardner (6). The major endocrine aspects of this syndrome were first described in a now classic paper from Johns Hopkins by Powell et al. in 1967 (7). Since that report there have been a number of studies from several different laboratories which have confirmed and extended these observations. These studies are reviewed in this chapter and the findings subsequently discussed in the light of what is currently known about hypothalamic hormones and neurotransmitters controlling growth hormone release from the pituitary (8–10).

ENDOCRINE FINDINGS IN PSYCHOSOCIAL DWARFISM

Shortness of stature is the most prominent feature of the syndrome known as psychosocial dwarfism. In the 13 patients originally described by Powell et al. (7,11), height age averaged 51% of the chronological age (Table 1). Bone age was also retarded, being more in keeping with height age than with chronological age. Despite retardation in bone age, no delay in dental eruption was found (Table 2). This finding weighs against hypothyroidism as a cause of the delayed bone growth; in fact none had clinical evidence of hypothyroidism, and tests of thyroid function were normal in the majority of these cases (Table 2). Eleven patients had normal protein-bound iodine levels (PBI), one had a low PBI, and another had a high PBI. Thyroxine was normal in all eight patients studied. Radioiodine uptake was normal in nine subjects, low in two, and borderline in one. These findings suggest that disturbance of thyroid function is not a feature of the syndrome.

Investigation of adrenal function showed a low 24-hour urinary secretion of 17-hydroxycorticosteroids (17-OHCS) in nine of 13 and a defective urinary 17-OHCS response to metyrapone in 12 of 13 patients (Table 2). In contrast, the adrenal response to adrenocorticotropic hormone (ACTH)

TABLE 1. *Growth measurements on initial visits in 13 patients with psychosocial dwarfism*

Measurement	Mean
(height age × 100) ÷ (chronological age)	51.4%
Bone age	3.6 years
Height age	2.7 years
Chronological age	5.7 years

Adapted from Powell et al. (11).

TABLE 2. *Endocrine findings on initial visits in 13 patients with psychosocial dwarfism*

Parameter	No. of patients studied	Results
Thyroid — TSH evaluation		
Dental eruption	13	11 normal, 1 low, 1 high
Clinical hypothyroidism	8	8 normal
PBI	13	11 normal, 1 low, 1 high
Thyroxine	8	8 normal
[131]I uptake	12	9 normal, 2 low, 1 borderline low
Adrenal — ACTH evaluation		
17-Hydroxysteroid excretion, 24 hours	13	4 normal, 9 low
17-Hydroxysteroids after metyrapone	13	1 normal, 12 low
17-Hydroxysteroids after ACTH	7	6 normal, 1 low
Growth hormone evaluation		
Response to insulin-induced hypoglycemia	8	2 normal, 6 low
Bone age	13	3 normal, 10 retarded

Adapted from Powell et al. (7).

was normal in six of seven patients. Taken together, these findings suggest that there is a defective release of ACTH in psychosocial dwarfism. Growth hormone (GH) responses to hypoglycemia were defective in six of eight patients tested, and a significant retardation of bone age was seen (Table 2). The latter findings suggest that the endocrine abnormality underlying the decreased growth in psychosocial dwarfism is GH deficiency.

One of the most noteworthy features of the syndrome was reversibility. A move out of the deprived environment led to a remarkable increase in growth rate. Endocrine studies done after regrowth were markedly improved from initial findings (Table 3). Thyroid function remained normal. Resting 24-hour excretion of 17-OHCS and GH responses to hypoglycemia became normal in all patients studied. The only abnormality remaining was a defective response to metyrapone in eight of 10 patients.

To summarize these findings, Powell et al. (7) demonstrated growth

TABLE 3. *Endocrine findings after regrowth in 11 patients with psychosocial dwarfism*

Parameter	No. of patients studied	Results
Thyroid—TSH evaluation		
PBI	11	10 normal, 1 high
Thyroxine	7	7 normal
[131]I uptake	8	8 normal
Adrenal—ACTH evaluation		
17-Hydroxysteroid excretion, 24 hours	10	10 normal
17-Hydroxysteroids after metyrapone	10	2 normal, 8 low
Growth hormone evaluation		
Response to insulin-induced		
hypoglycemia	7	7 normal

Adapted from Powell et al. (7).

retardation together with retardation of bone age. In the majority of patients, this was coupled with reduced 24-hour urinary excretion of 17-OHCS, a reduced 17-hydroxycorticoid response to metyrapone, and a defective GH response to hypoglycemia. Following a period of catch-up growth, significant improvement was seen in these abnormalities with the exception of the response to metyrapone. It was hypothesized that the abnormality in growth was secondary to reduced secretion of GH, which in turn was produced by the environmental deprivation. The major portion of this review is therefore confined to discussing GH.

GH REGULATION IN PSYCHOSOCIAL DWARFISM

Provocative Stimuli

The finding of defective GH response to insulin-induced hypoglycemia in psychosocial dwarfism was subsequently confirmed by several investigators (12–18). The Johns Hopkins series has also been expanded to a total of 22 patients from the original 13 with no change in the original findings (19,20). Because hypoglycemia is only one of many stimuli known to release GH, it is important to know the response to other provocative stimuli. Powell et al. (18) recently examined the response to arginine and found that this was also defective in a single case of psychosocial dwarfism. When examined shortly after admission, no GH elevation occurred after either arginine or insulin. In contrast, after an interval of 35 days in hospital, a clear-cut GH response was found after both arginine and insulin. In the expanded Johns Hopkins series (19,20), responses to arginine were subnormal in eight of 12 subjects. Others have confirmed this finding (14,16). Lacey et al. (17) found reduced GH responses to exercise in two patients with psychosocial dwarfism and to Bovril in one patient.

Sleep-Associated GH Release

One of the most interesting events that relates to GH release is the onset of sleep. This is of interest because, unlike the provocative tests used, it is a naturally occurring event. In normal subjects a rise in GH occurs shortly after onset of sleep usually in association with slow-wave sleep. Wolff and Money (21) examined sleep reports in patients with psychosocial dwarfism and rated sleep on a scale ranging from good to poor depending on the amount of sleep disturbance recorded. In their studies they found that poor sleep correlated with poor growth, and good sleep correlated with good growth. In that study it could not be determined whether the poor growth was secondary to the poor sleep, poor sleep secondary to poor growth, or perhaps both poor growth and sleep were secondary to some other common factor. Powell et al. (18) examined the sleeping pattern of an individual patient with psychosocial dwarfism longitudinally. Unlike the patients described by Wolff and Money (21), this single patient showed a normal pattern of sleep both following admission and later during his hospital stay, with a clear slow-wave sleep episode occurring shortly after initiation of sleep on each occasion when he was tested. Despite the normal sleep pattern, no sleep-associated GH rise occurred during studies done soon after admission. In contrast, after a period of 6 weeks in hospital, GH release occurred after onset of sleep. These findings, together with those cited above, suggest that there is a generalized failure in GH release in psychosocial dwarfism. GH responses to insulin, arginine, exercise, Bovril, and slow-wave sleep may all be diminished.

NUTRITION AND PSYCHOSOCIAL DWARFISM

Let us now turn to the question of nutrition of these patients. Krieger and Good (22) postulated that endocrine findings in patients with deprivation syndrome are related to malnutrition. They selected patients for study on the basis of poor environment together with retardation of psychomotor or language development (selection criteria did *not* include growth retardation). They (15,23) found that GH findings in deprived infants under age three differed from those in children over age three. The latter showed defective GH responses to insulin and did not have elevated resting levels. In contrast, infants had elevated resting GH levels, and none had defective GH responses to hypoglycemia. Elevation of resting GH levels has been reported in association with malnutrition in a variety of conditions, notably protein-calorie malnutrition — kwashiorkor or marasmus (24) — and anorexia nervosa (25–28). Patients with kwashiorkor invariably have elevated resting GH (24,29–34) which is normalized by refeeding with protein but not with carbohydrate (24,29). In marasmus elevated resting GH has been reported by some investigators but not by others. Those reporting elevated GH investigated GH immediately after admission (24,29,30), whereas those who did not find an elevation delayed this investigation (31–34). It appears that elevated GH may be rapidly normalized by refeeding. A subnormal GH response to insulin, however, remains for a prolonged period (32). We recently examined the role of nutrition in producing endocrine abnormalities in anorexia nervosa (28). Elevation of resting GH was found only in patients who had severely reduced food intake, and it was rapidly reversed when the patients began to regain weight. High resting GH levels in young infants from the deprived environments in the study by Krieger and Mellinger (15) therefore suggests that they may have been malnourished. Because of these observations and other evidence (35), Krieger (23) provided home feeding for an additional three cases. In these patients provision of food directly to the home twice daily led to a prompt gain in weight. On the basis of that finding, Krieger concluded that the endocrine abnormalities in psychosocial deprivation could be secondary to malnutrition. Underfeeding due to neglect may not be uncommon in a deprived environment; however, older children would be more capable of resisting underfeeding than would infants. In addition, food intake was carefully assessed in the Johns Hopkins studies, which involved only the older age group (7,20), and malnutrition did not appear to be a significant factor.

ADRENAL FINDINGS

The other major endocrine abnormality in psychosocial dwarfism is a diminished resting 17-OHCS excretion and reduced excretion following metyrapone despite a normal response to ACTH (7,20), suggesting a defective release of ACTH. No diminution in the response to metyrapone was seen by Krieger (15), but a larger dose of metyrapone was used, which may have obscured differences. In contrast to psychosocial dwarfism, patients with marasmus or kwashiorkor are reported to have not only a diminished 24-hour urinary excretion of 17-OHCS (36) but also a reduced response to ACTH (37), suggesting that the adrenal response to ACTH is defective. Findings in marasmus therefore differ from those in psychosocial dwarfism. Adults with protein-calorie malnutrition have an intact 17-OHCS response to both ACTH and metyrapone (38).

REGULATION OF GH

In order to clarify the nature of the GH defect, let us examine some of the factors known to be involved in GH regulation. GH is released from the pituitary as a result of the action of two hypothalamic hormones—one that releases (10) and one that inhibits (39). The latter hormone has been synthesized and is named somatostatin (39). These hormones in turn are controlled by neurotransmitters (see also chapters by Wurtman and Fernstrom and by Sachar et al. in this volume). There is currently convincing evidence for a role of dopamine, norepinephrine, and possibly serotonin in GH regulation. I confine my discussion to the primate because GH regulation in the rat differs in many important respects (10).

ROLE OF NEUROTRANSMITTERS IN GH REGULATION

Administration of L-DOPA (the precursor of dopamine) stimulates GH release after oral ingestion in the human (40) or intravenous infusion in the rhesus monkey (41,42). It has also been shown that α-adrenergic blockade with phentolamine prevents GH release induced by insulin, whereas β-adrenergic blockade can facilitate GH release to stimuli such as arginine or hypoglycemia (8,9,43). These findings suggest that adrenergic mechanisms are intimately involved in GH regulation. Apomorphine (44,45), a specific dopamine agonist, produces a prompt rise in GH. This finding plus collateral evidence has given rise to the concept that there is a central excitatory dopamine mechanism controlling GH. Recently we demonstrated that clonidine, a specific noradrenergic agonist, produces a GH rise in the monkey (41). This effect has been confirmed in the human (46). On the basis of these observations, we postulate the existence of an excitatory noradrenergic mechanism in GH regulation. According to this conceptualization, noradrenergic neurons are capable of stimulating GH release from the pituitary by altering the secretion of GH releasing factor (or of somatostatin). The role of serotonin is less certain. Oral administration of 5-hydroxytryptophan (5-HTP) in the human (47) or intravenous administration in the monkey (41,42) produces a GH rise. Although 5-HTP is the precursor of serotonin, this agent can also displace central norepinephrine and dopamine. Alternatively, it may act via a peripheral mechanism in the same manner as amino acids (42).

There is therefore evidence in the human for a central excitatory dopamine and norepinephrine regulatory mechanism for GH. The role of serotonin is less clear.

CONCLUSION

A number of facts appear to be established: (a) In psychosocial dwarfism, defective GH release appears to be the major factor leading to stunting of growth. (b) Defective GH release is found following provocative stimuli including hypoglycemia, arginine infusion, Bovril administration, and exercise, as well as during sleep. (c) This abnormality is reversible by environmental alterations which produce accelerated growth. Malnutrition may be an important factor, at least in young infants. (d) There is evidence for central excitatory dopamine and norepinephrine mechanisms, and possibly a serotonin mechanism, in GH regulation.

One of the most exciting developments has been the linking of this disorder with a neurotransmitter mechanism. In a single case study done in Japan, Imura et al. (14) reversed the defective GH response to hypoglycemia in a boy with psychosocial dwarfism by treatment with propanolol, a β-adrenergic blocker. Parra at Johns Hopkins (48) studied five patients with both arginine and insulin stimulation. Infusion of epinephrine plus propanolol restored the GH response. These findings suggest that there may be a defective catecholamine mechanism in these patients. It therefore seems logical to suggest that a profitable area for future investigation would be the examination of neurotransmitter regulation of GH in patients with psychosocial dwarfism to determine whether dopamine or norepinephrine mechanisms are impaired. It would also seem appropriate to determine if agents such as propanolol might be useful as therapeutic agents in psychosocial dwarfism.

ACKNOWLEDGMENTS

Investigations in the author's laboratory are supported by the Medical Research Council (grant MT 4749) and the Ontario Mental Health Foundation (grant 324). G. M. Brown is an OMHF research associate.

REFERENCES

1. Spitz, R. A. (1946): Hospitalism: A follow-up report. *Psychoanal. Study Child.,* 2:113–117.
2. Widdowson, E. M. (1951): Mental contentment and physical growth. *Lancet,* 1:1316–1318.
3. Engel, G. L., and Reichsman, F. (1956): Spontaneously and experimentally induced depressions in infant with gastric fistula. *J. Am. Psychoanal. Assoc.,* 4:428–452.
4. Talbot, N. B., Sobel, E. H., Burke, B. S., Lindemann, E., and Kaufman, S. B. (1947): Dwarfism in healthy children: Its possible relation to emotional, nutritional and endocrine disturbances. *N. Engl. J. Med.,* 236:783–793.
5. Bakwin, H. (1949): Emotional deprivation in infants. *J. Pediatr.,* 35:512–521.
6. Patton, R. G., and Gardner, L. I., editors (1962): *Growth Failure in Maternal Deprivation: With an Introduction by Julius B. Richmond.* Charles C Thomas, Springfield, Ill.
7. Powell, G. F., Brasel, J. A., Raiti, S., and Blizzard, R. M. (1967): Emotional deprivation and growth retardation simulating idiopathic hypopituitarism. II. Endocrinologic evaluation of the syndrome. *N. Engl. J. Med.,* 276:1279–1283.
8. Brown, G. M., and Reichlin, S. (1972): Psychologic and neural regulation of growth hormone secretion. *Psychosom. Med.,* 34:45–61.
9. Martin, J. B. (1973): Neural regulation of growth hormone secretion: Medical progress report. *N. Engl. J. Med.,* 288:1384–1393.
10. Brown, G. M., and Martin, J. B. (1973): Neuroendocrine relationships. *Prog. Neurol. Psychiatry,* 28:193–240.
11. Powell, G. F., Brasel, J. A., and Blizzard, R. M. (1967): Emotional deprivation and growth retardation simulating idiopathic hypopituitarism. I. Clinical evaluation of the syndrome. *N. Engl. J. Med.,* 276:1271–1278.
12. Kaplan, S. L., Abrams, C. A. L., Bell, J. J., Conte, F. A., and Grumbach, M. M. (1968): Growth and growth hormone. I. Changes in serum level of growth hormone following hypoglycemia in 134 children with growth retardation. *Pediatr. Res.,* 2:43–63.

13. Hellman, D. A., and Colle, E. (1969): Plasma growth hormone and insulin responses in short children. *Am. J. Dis. Child.*, 117:636–644.
14. Imura, H., Yoshimi, T., and Ikekubo, K. (1971): Growth hormone secretion in a patient with deprivation dwarfism. *Endocrinol. Jap.*, 18:301–304.
15. Krieger, I., and Mellinger, R. C. (1971): Pituitary function in the deprivation syndrome. *J. Pediatr.*, 79:216–225.
16. Illig, R. (1972): Wachstumshormon-untersuchungen bei 225 Kindern mit Minderwuchs. *Schweiz. Med. Wochenschr.*, 102:760–765.
17. Lacey, K. A., Hewison, A., and Parkin, J. M. (1973): Exercise as a screening test for growth hormone deficiency in children. *Arch. Dis. Child.*, 48:508–512.
18. Powell, G. F., Hopwood, N. J., and Barratt, E. S. (1973): Growth hormone studies before and during catch-up growth in a child with emotional deprivation and short stature. *J. Clin. Endocrinol. Metab.*, 37:674–679.
19. Thompson, R. G., Parra, A., Schultz, R. B., and Blizzard, R. M. (1969): Endocrine evaluation in patients with psychosocial dwarfism. *Clin. Res.*, 17:592 (Abstr.).
20. Brasel, J. A. (1973): Review of findings in patients with emotional deprivation. In: *Endocrine Aspects of Malnutrition. Marasmus, Kwashiorkor and Psychosocial Deprivation*, edited by L. I. Gardner and P. Amacher, pp. 115–127. Kroc Foundation, Santa Ynez, California.
21. Wolff, G., and Money, J. (1973): Relationship between sleep and growth in patients with reversible somatotropin deficiency (psychosocial dwarfism). *Psychol. Med.*, 3:18–27.
22. Krieger, I., and Good, M. H. (1970): Adrenocortical and thyroid function in the deprivation syndrome: Comparison with growth failure due to undernutrition, congenital heart disease, or prenatal influences. *Am. J. Dis. Child.*, 120:95–102.
23. Krieger, I. (1973): Endocrines and nutrition in psychosocial deprivation in the U.S.A.: comparison with growth failure due to malnutrition on an organic basis. In: *Endocrine Aspects of Malnutrition. Marasmus, Kwashiorkor and Psychosocial Deprivation*, edited by L. I. Gardner and P. Amacher, pp. 129–162. Kroc Foundation, Santa Ynez, California.
24. Pimstone, B. L., Becker, D. J., and Hansen, J. D. L. (1972): Human growth hormone in protein-calorie malnutrition. In: *Growth and Growth Hormone*, edited by A. Pecile and E. E. Müller, pp. 389–401. Excerpta Medica, Amsterdam.
25. Landon, J., Greenwood, F. C., Stamp, T. C. B., and Wynn, V. (1966): The plasma sugar, free fatty acid, cortisol and growth hormone response to insulin, and the comparison of this procedure with other tests of pituitary and adrenal function. II. In patients with hypothalamic or pituitary dysfunction or anorexia nervosa. *J. Clin. Invest.*, 45:437–449.
26. Giordano, G., Marugio, M., Minuto, F., and Barreca, J. (1971): Somatotropic increment in the course of serious weight loss. *Folia Endocrinol. Jap.*, 24:308–315.
27. Lundberg, P. O., Walinder, J., Werner, I., and Wide, L. (1972): Effects of thyrotrophin-releasing hormone on plasma levels of TSH, FSH, LH and GH in anorexia nervosa. *Eur. J. Clin. Invest.*, 2:150–153.
28. Garfinkel, P. E., Brown, G. M., Moldofsky, H., and Stancer, H. C. (1975): Hypothalamic-pituitary function in anorexia nervosa. *Arch. Gen. Psychiatry*, 32:739–744.
29. Pimstone, B. L., Becker, D. J., and Hansen, J. D. L. (1973): Human growth hormone and sulphation factor in protein calorie malnutrition. In: *Endocrine Aspects of Malnutrition. Marasmus, Kwashiorkor and Psychosocial Deprivation*, edited by L. I. Gardner and P. Amacher, pp. 73–90. Kroc Foundation, Santa Ynez, California.
30. Suskind, R., Amatayakul, K., Leitzmann, C., and Olson, R. E. (1973): Interrelationships between growth hormone and amino acid metabolism in protein-calorie malnutrition. In: *Endocrine Aspects of Malnutrition. Marasmus, Kwashiorkor and Psychosocial Deprivation*, edited by L. I. Gardner and P. Amacher, pp. 99–113. Kroc Foundation, Santa Ynez, California.
31. Beas, F., and Muzzo, S. (1973): Growth hormone and malnutrition: The Chilean experience. In: *Endocrine Aspects of Malnutrition. Marasmus, Kwashiorkor and Emotional Deprivation*, edited by L. I. Gardner and P. Amacher, pp. 1–18. Kroc Foundation, Santa Ynez, California.
32. Godard, C. (1973): Plasma growth hormone levels in severe infantile malnutrition in Bolivia. In: *Endocrine Aspects of Malnutrition. Marasmus, Kwashiorkor and Psychosocial Deprivation*, edited by L. I. Gardner and P. Amacher, pp. 19–30. Kroc Foundation, Santa Ynez, California.

33. Parra, A., Garza, C., Klish, W., Garcia, G., Argote, R., Canseco, L., Cuellar, A., and Nichols, B. L. (1973): Insulin-growth hormone adaptations in marasmus and kwashiorkor as seen in Mexico. In: *Endocrine Aspects of Malnutrition. Marasmus Kwashiorkor and Psychosocial Deprivation,* edited by L. I. Gardner and P. Amacher, pp. 31–44. Kroc Foundation, Santa Ynez, California.
34. Jayarao, K. S., and Raghuramulu, N. (1973): Growth hormone and insulin secretion in protein-calorie malnutrition as seen in India. In: *Endocrine Aspects of Malnutrition. Marasmus, Kwashiorkor and Psychosocial Deprivation,* edited by L. I. Gardner and P. Amacher, pp. 91–98. Kroc Foundation, Santa Ynez, California.
35. Whitten, C. F., Pettit, M. G., and Fischhoff, J. (1969): Evidence that growth failure from maternal deprivation is secondary to undereating. *J.A.M.A.,* 209:1675–1682.
36. Najjar, S. S., and Bitar, J. G. (1967): Adrenal cortex in marasmic children. *Arch. Dis. Child.,* 42:657–658.
37. Beas, F., Ferreira, E., and Rivarola, M. (1973): Adrenal cortical function in infants with malnutrition, as seen in Chile. In: *Endocrine Aspects of Malnutrition. Marasmus, Kwashiorkor and Psychosocial Deprivation,* edited by L. I. Gardner and P. Amacher, pp. 343–353. Kroc Foundation, Santa Ynez, California.
38. Smith, S. R., Bledsoe, J., and Chhetri, M. K. (1975): Cortisol metabolism and the pituitary-adrenal axis in adults with protein calorie malnutrition. *Endocrinology,* 40:43–52.
39. Brazeau, P., Vale, W., Burgus, R., Ling, N., Butcher, M., Rivier, J., and Guillemin, R. (1973): Hypothalamic polypeptide that inhibits the secretion of immunoreactive pituitary growth hormone. *Science,* 179:77–79.
40. Boyd, A. E., Lebovitz, H. E., and Pfeiffer, J. B. (1970): Stimulation of human growth hormone secretion by L-dopa. *N. Engl. J. Med.,* 283:1425–1429.
41. Brown, G. M., Chambers, J. W., and Feldmann, J. (1973): Neurotransmitter regulation of growth hormone release. In: *Program and Abstracts, Society for Neuroscience. Third Annual Meeting,* p. 404 (Abstr.).
42. Jacoby, J. H., Greenstein, M., Sassin, J. F., and Weitzman, E. D. (1974): The effect of monoamine precursors on the release of growth hormone in the rhesus monkey. *Neuroendocrinology,* 14:95–102.
43. Blackard, W. G., and Heidingsfelder, S. A. (1968): Adrenergic receptor control mechanism for growth hormone secretion. *J. Clin. Invest.,* 47:1407–1414.
44. Lal, S., de la Vega, C. E., Sourkes, T. L., and Friesen, H. G. (1972): Effect of apomorphine on human growth hormone secretion. *Lancet,* 2:661.
45. Brown, W. A., Kreiger, D. T., Van Woert, M. H., and Ambani, L. M. (1974): Dissociation of growth hormone and cortisol release following apomorphine. *J. Clin. Endocrinol. Metab.,* 38:1127–1130.
46. Lal, S., Ettigi, P., Martin, J. B., Tolis, G., Brown, G. M., Guyda, H., and Friesen, H. G. (1974): Central catecholamine receptor agonists and anterior pituitary secretion. *Clin. Res.,* 22:732A (Abstr.).
47. Yoshimura, M., Ochi, Y., Miyazaki, T., Shiomi, K., and Hachiya, T. (1973): Effect of 5-HTP on the release of growth hormone, TSH and insulin. *Endocrinol. Jap.,* 20:135–141.
48. Parra, A. (1973): Discussion. In: *Endocrine Aspects of Malnutrition. Marasmus, Kwashiorkor and Psychosocial Deprivation,* edited by L. I. Gardner and P. Amacher, p. 155. Kroc Foundation, Santa Ynez, California.

Hormones, Behavior, and Psychopathology, edited by
Edward J. Sachar. Raven Press, New York © 1976.

Toward An Elucidation of the Psychoendocrinology of Anorexia Nervosa

Jack L. Katz, Robert M. Boyar, Herbert Weiner,
Gregory Gorzynski, Howard Roffwarg, and Leon Hellman

*Departments of Oncology and Psychiatry, and the Institute for Steroid Research, Monte-
fiore Hospital and Medical Center, Bronx, New York 10467*

The dramatic and relentless dieting, capable of producing gross emaciation and even death, and the invariably present amenorrhea have made the syndrome of anorexia nervosa an inviting area of study for both psychiatrists and endocrinologists. A tendency to explain the condition, however, in purely psychological or hormonal terms has been common. Thus psychodynamic formulations have ranged from viewing the illness as an unconscious defense against underlying fears of adult sexuality or oral impregnation to a ritualistic maneuver intended to secure mastery or control in at least one area of the patient's highly dependent existence (1,2). Endocrinologists, on the other hand, suggested for a number of years that anorexia nervosa was merely a variant of Simmonds' disease, i.e., panhypopitutarism (3).

Although dynamic interpretations of anorexia nervosa have remained essentially unchanged, a more precise definition of the endocrinology of this condition has recently begun to evolve. In this chapter we present data supporting the presence of a functional hypothalamic abnormality, whatever its origin, in anorexia nervosa, and suggest some of the psychoendocrine complexities one faces at this point in our knowledge in attempting any single, simple explanation for the total clinical syndrome.

Actually it should be possible to discriminate between anorexia nervosa and panhypopitutarism on clinical grounds alone. Although both conditions are characterized by amenorrhea, Sheehan and Summers showed that Simmond's disease generally is not associated with massive cachexia (4); furthermore, patients with Simmond's disease manifest marked lethargy and asthenia, which contrast dramatically with the robust and actually excessive physical activity characteristically displayed by patients with anorexia nervosa, despite their extraordinary weight loss (5).

The coming of age of more sophisticated endocrine assessment techniques during the past two decades has permitted more definitive confirmation of the absence of panhypopituitarism in anorexia nervosa. Thus Bliss and Migeon (6) and others (7,8) were able to establish that pituitary-adrenal and pituitary-thyroid function were not deficient. On the other hand, low urinary and plasma concentrations of estrogens and gonadotropins have been consistently reported by investigators (6,9,10), and it is to the source for these abnormalities that we have turned our attention.

ONTOGENY OF CIRCADIAN LUTEINIZING HORMONE SECRETORY PROGRAM

The impetus for a specific interest in the ontogeny of the circadian luteinizing hormone (LH) secretory program came from two sets of earlier findings by our Institute for Steroid Research. In the first set it was established that LH is secreted by the pituitary in pulses (11), as is cortisol by the adrenals (12), rather than continuously. In the second it was demonstrated that there exists a developmental or ontogenetic progression in the 24-hour sleep-wake pattern of LH secretion (13). Thus when plasma LH concentrations were determined in blood specimens taken every 20 min throughout a 24-hour period while sleep was being monitored polygraphically, reliably different circadian LH patterns were observed for different stages of sexual development. Figure 1 summarizes the characteristic patterns, as demonstrated by four representative subjects of the 19 studied in that investigation. Stages in females are defined according to Tanner's criteria for breast development (14).

It should be noted that prepubertal children show relatively small pulses and low plasma LH concentrations (mean usually below 7 mIU/ml) during both waking and sleeping; i.e., there is *no* significant circadian variation. Early-mid pubertal adolescents show a significant augmentation of LH secretion, but during sleep only; mid-late pubertal adolescents show a marked increase in pulse amplitude and LH concentration during waking and sleeping, but with nocturnal secretion still significantly greater than diurnal secretion; and, finally, adults show a pattern of relatively large pulses and high concentrations (mean usually above 8 mIU/ml) throughout both sleeping and waking periods. The adult 24-hour pattern of insignificant circadian variation remains the same, incidentally, even during the middle (i.e., ovulatory) part of the monthly cycle in women, plasma concentrations being elevated throughout both sleeping and waking (15).

The immediate seat for the circadian LH secretory program is generally accepted to be within the hypothalamus, its influence over pituitary release of the gonadotropins — LH and follicle-stimulating hormone (FSH) — being mediated by gonadotropin-releasing hormone (GnRH), which it secretes into the pituitary portal vessels (16). Thus normally the existence of a reliably predictable circadian LH secretory pattern provides a new parameter for determining the existence of possible subtle abnormalities of function in the hypothalamic-pituitary-gonadal axis.

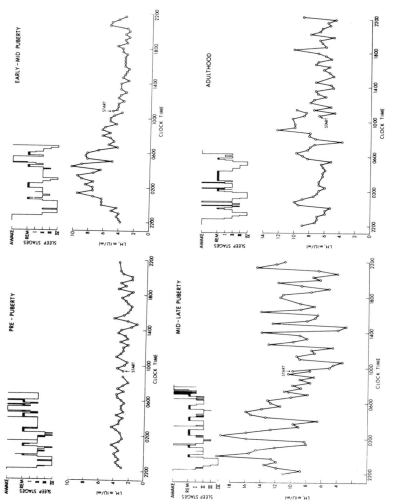

FIG. 1. Twenty-four-hour sleep-wake patterns of LH secretion at different stages of development.

SYNDROME OF ANOREXIA NERVOSA

Because substantial weight loss can occur in a variety of medical and psychiatric illnesses (e.g., as a secondary symptom in schizophrenia or psychotic depression) Bruch emphasized the importance of studying patients who meet the clinical criteria of "true" or "primary" anorexia nervosa, rather than merely accepting a certain minimal percent of weight loss as sufficient for diagnosing a homogeneous group of subjects with this illness (17).

Actually, although Bruch has perhaps contributed the most to defining precisely the clinical syndrome of true anorexia nervosa, this condition has been described with striking consistency since it was first recorded by Gull and then by Laseque approximately a century ago. The following constellation of phenomena are characteristic: (a) a relentless, self-imposed, and consciously enforced dieting, classically initiated by the elimination of carbohydrates, in an obsessive pursuit of thinness and frequently leading to serious emaciation and possibly even to death; (b) a manifest morbid fear of losing control over food intake and weight, with this actually occurring in a substantial number of patients via binge eating, which leads either to excessive cathartic-emetic use or to actual obesity; (c) an incidence highly preponderant in adolescent females, the rare male who manifests the syndrome usually being prepubertal; (d) the inevitable presence of amenorrhea in these women, whether it precedes, accompanies, or follows the dieting and weight loss; (e) a high level of physical activity; (f) the absence of overt (if not covert) psychotic symptomatology, but the presence commonly of an obsessive-compulsive personality style and of manifest conflicts about independence, autonomy, and control; and (g) abnormalities in self-perception, so that obesity versus thinness cannot be discriminated accurately by the patient in herself. The most common background is that of an upper-middle-class family, a domineering and insensitive mother, a distant or affectively uninvolved father, and a home in which food or weight usually has greater than ordinary significance.

SUBJECTS

We report here studies with 10 patients, all of whom met the above criteria for true anorexia nervosa. The 10 women ranged in age from 17 to 23 years, were physically mature, and gave histories of having begun to diet excessively at some point in their teens. All were essentially indifferent about their excessive weight loss, minimizing its significance and expressing concern that if they should attempt to regain the weight lost they would be unable to control their food intake and undoubtedly would overshoot the mark to the point of obesity.

All were white; planning to attend, attending, or graduates of college;

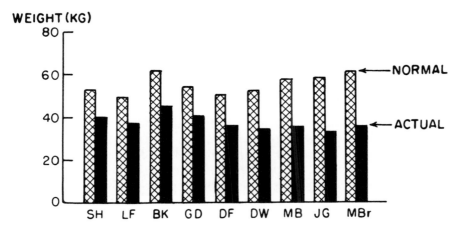

FIG. 2. Actual versus normal weights in nine patients with anorexia nervosa.

and from middle or upper class families in which diet, weight, or appearance was generally a matter of great investment. Manifest precipitating events were described for most episodes of dieting, most commonly a first actual or threatened separation from parents or a first attempt at a sexual relationship. In the seven families that were intact, the mother was described as the domineering parent, while the father was characteristically perceived as more distant or passive. In the other three families the father was absent literally, in two by divorce and in one by death. None of the patients was clinically psychotic, and all were physically active in conjunction with their dieting.

Amenorrhea was present in every patient. Three women had never menstruated; three had stopped menstruating just prior to or concurrently with the start of dieting; and the remaining four became amenorrheic after weight loss had occurred.

As indicated in Fig. 2, weight loss had been substantial. Although thorough medical evaluation failed to uncover any coexistent organic disease, nine patients whose weights are depicted showed deficits below their ideal weights (for their sex, ages, and heights) ranging from approximately 24% to 40%; the tenth patient's deficit was approximately 25%.

METHODS

All patients were admitted to the Clinical Research Center at Montefiore Hospital and Medical Center. No medication was administered during the course of the study, nor had any been recently. The women were adapted to the lightproof, soundproof sleep study room for at least 24 hours prior to the actual 24-hour study.

Shortly before 10 A.M. on the morning of study, a catheter was inserted into an antecubital vein and connected to a longer tube which extended into an adjoining room. The precise details of obtaining and handling blood specimens via this technique in our laboratory have been elaborated elsewhere (12), but its most cogent feature is that it permits frequent blood sampling without disturbance to (actually awareness by) the subject. Beginning at approximately 10 A.M., a blood sample was obtained every 20 min for the next 24 hours. At night the patient had polygraphic monitoring of sleep onset and stages, which were scored according to standardized criteria (18).

Plasma LH was measured by the double-antibody radioimmunoassay method as described by Midgley (19) and modified in our laboratory (20). LH concentrations are expressed in milli-international units per milliliter (mIU/ml). The coefficient of variation for duplicate samples is approximately 10%, and all plasma samples from each 24-hour study were assayed simultaneously. Plasma LH concentration at each 20-min interval was plotted over the 24 hours for each patient, and mean concentrations while asleep versus awake were calculated by averaging all 20-min values that occurred within these respective periods.

In addition, although not providing the principal focus of this report, determinations of cortisol production rate and half-life, and of plasma cortisol, growth hormone (GH), FSH, and prolactin concentrations in the 20-min plasma samples were also carried out for most of these patients.

FINDINGS

Our principal finding was that all 10 women had abnormal circadian patterns of LH secretion when compared to the developmental stage-matched controls of the earlier study. Two groups of patients could be broadly distinguished. Five of the women showed plasma LH concentrations that were significantly low during both sleeping and waking, with generally small pulses throughout the 24 hours, i.e., the characteristic prepubertal pattern. To illustrate the striking similarity, we juxtaposed (Fig. 3) the pattern of one of these women, a 21-year-old patient with secondary amenorrhea, and that of a normal control prepubertal girl. The resemblance speaks for itself.

Figure 4 shows the patterns of three of the remaining four patients with prepubertal patterns. It should be noted that although all are prepubertal a spectrum of developmental maturity can be observed, ranging from the very child-like, virtually flat pattern of patient M.B. to the beginning nocturnal pulsing but otherwise flat pattern of D.E. to the slightly greater but still immature concentrations and pulses of G.D. None of these three patients was younger than 19; the last two had never menstruated, but the first (M.B.), who had the most immature pattern, had had regular periods

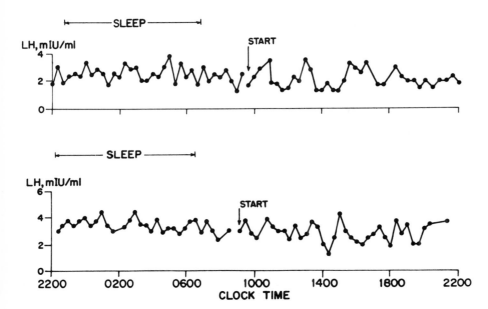

FIG. 3. Comparison of a circadian LH secretory pattern in a 21-year-old woman with anorexia nervosa **(top)** and that in a normal 9-year-old prepubertal girl **(bottom).**

until amenorrhea developed a few months after dieting and weight loss occurred.

The five patients comprising the other group showed circadian LH patterns typical for normal girls in various stages of puberty, not one showing an adult pattern despite three being 20 years of age or more. This is illustrated in Fig. 5 by juxtaposing the pattern of one of these patients, a 17-year-old girl with secondary amenorrhea that actually preceded the dieting and weight loss by approximately 2 months, and that of a normal early-mid pubertal control subject; present in both is the moderate augmentation of LH secretion that characteristically occurs during sleep in early-mid puberty.

Figure 6 incorporates the last patient's pubertal pattern with those of three other such patients. In each instance, as in that of the fifth girl whose graph is not included, mean sleep LH concentrations were significantly greater than awake concentrations, which is the characteristic finding in pubertal rather than in adult patterns. Here too, however, a range is present, one spanning early to late puberty. Thus S.H. shows the slight LH augmentation during sleep seen in early puberty; D.W. the still more significant LH increment and stronger pulsing during sleep seen in early-mid puberty; L.F. the extremely wide LH pulsing during sleep, dramatic circadian variation, and high 24-hour mean concentration characteristic of middle-late puberty;

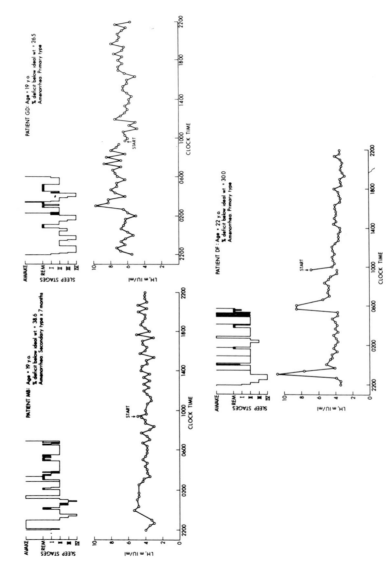

FIG. 4. Prepubertal circadian LH secretory patterns in three women with anorexia nervosa. There is no significant difference between the mean asleep and awake LH concentrations.

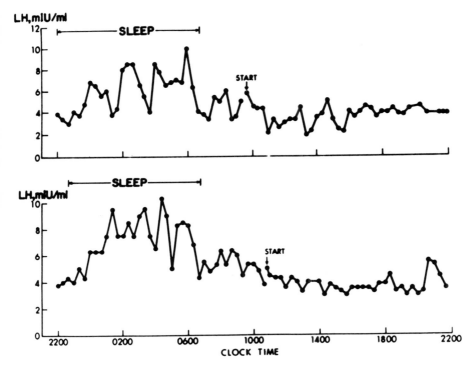

FIG. 5. Comparison of a circadian LH secretory pattern in a 17-year-old woman with anorexia nervosa **(top)** and that in a normal 12-year-old early-mid pubertal girl **(bottom)**.

and B.K. a pattern approaching but not quite attaining the negligible circadian variation, relatively strong pulsing, and more characteristic 24-hour mean concentration of adulthood (her mean sleep LH concentration still being slightly—and significantly—higher than her mean awake concentration).

Age of patients, history of primary versus secondary amenorrhea, or time relationship between secondary amenorrhea and weight loss did not reliably correlate with the degree of immaturity of the circadian pattern. Extent of weight loss also was not correlated with degree of immaturity at a statistically significant level, but a trend in that direction is present ($p < 0.15$).

Although we classified these 10 patients into prepubertal and pubertal groups, it should be stated that we could also regard all 10 patterns as merely falling along one continuum of immature circadian LH secretion, ranging from a virtually child-like pattern at one end to a near-adult pattern at the other.

The data regarding the other endocrine variables are briefly summarized. We have not emphasized FSH because, although this hormone tends to

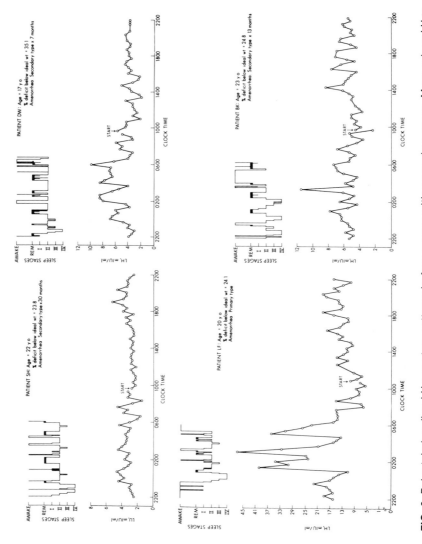

FIG. 6. Pubertal circadian LH secretory patterns in four women with anorexia nervosa. Mean asleep LH > awake LH: $p < 0.05$.

show ontogenetic circadian patterns similar to those of LH in normal developing subjects, the sleep-wake differentiations are not nearly as precise as they are for LH, and this was similarly the case with our patients.

The 24-hour GH and prolactin patterns and values were normal in our patients, again attesting to the absence of a panhypopituitary state in this condition. The data for cortisol, however, proved to be more remarkable.

It should be recalled that the normal 24-hour plasma cortisol secretory pattern shows the following characteristics (12): (a) a fall in plasma cortisol concentration to usually undetectable levels between 12 midnight and 2 A.M.; (b) an onset of episodic secretory activity during the early morning hours, reaching a maximum around the time of rising to a peak concentration of approximately 12 to 15 μg%; and (c) a decrease in the magnitude of secretory episodes during the late afternoon (giving the so-called diurnal variation).

Figure 7 shows the 24-hour plasma cortisol patterns in two representative patients. Note that in both instances the plasma cortisol concentration never falls to zero and that the morning peak values rise to at least 20 μg%. Figure 8 compares the 24-hour mean plasma cortisol concentrations in nine of our patients with those in age-matched normal women. Seven of the nine patients

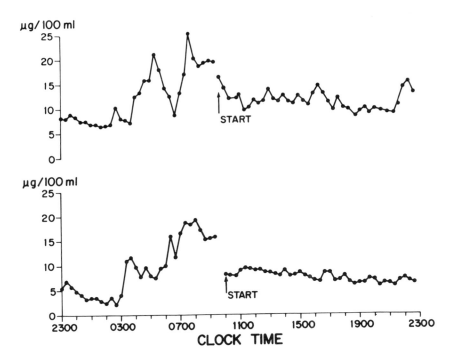

FIG. 7. Circadian cortisol secretory patterns in two women (**top,** 23 years old; **bottom,** 19 years old) with anorexia nervosa.

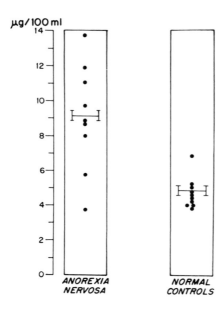

FIG. 8. Comparison of 24-hour mean plasma cortisol concentrations in nine women with anorexia nervosa and 10 normal controls.

had significantly elevated values, the group mean for all nine being 9.2 μg%, compared to 4.8 μg% in the normal controls.

As elevated plasma levels of cortisol or 17-hydroxycorticosteroids (17-OHCS) in anorexia nervosa are often reported as well (21, 22), we attempted to gain further insight into this finding by two additional studies. First, using a radioisotope dilution technique, based on the specific activity of urinary metabolites of ^{14}C-labeled infused cortisol (23), we measured cortisol production rates in our patients. Figure 9 shows these rates in nine of our patients and in six normal adult women. Seven of the nine patients had significantly elevated rates, the patient mean being 28 mg/g creatinine, compared to 12 mg/g creatinine in the controls. Moreover, one patient who had a clinical remission and was restudied 6 months later showed a normalization of her production rate. Since these patients were generally moderately but

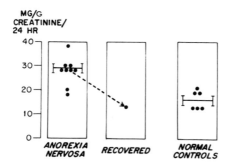

FIG. 9. Comparison of cortisol production rates in nine women with anorexia nervosa (including one during remitted phase) and six normal controls.

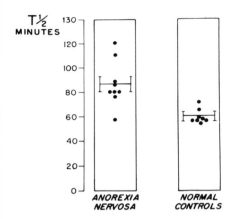

FIG. 10. Comparison of cortisol half-life (using ^{14}C) in nine women with anorexia nervosa and eight normal controls.

not markedly anxious clinically during their hospital stay, we assume that emotional arousal was a contributing factor in their elevated rates of cortisol production but that other variables also may have been operative. Other investigators have even reported a disappearance of the usual diurnal variation in plasma cortisol levels (24).

Second, because urinary corticosteroids are usually relatively low in anorexia nervosa (6,7) despite the elevated plasma levels, we measured the half-life of plasma cortisol in our patients by the isotope dilution technique. Figure 10 shows that the nine patients so studied had a mean cortisol half-life of 87 min, significantly longer than the 60-min mean of our normals. This appears to provide the first proof for the often speculated (6,7) possibility that the patient's hypometabolic state secondary to the malnutrition of anorexia nervosa results in reduced plasma cortisol catabolism and consequently low urinary corticoids in the face of the elevated plasma levels.

DISCUSSION: CONTROL AND LONGITUDINAL STUDIES

To review: we have observed a consistent abnormality (i.e., a developmentally inappropriate program for circadian LH secretion) in each of 10 women with primary anorexia nervosa. Assessment of the significance and implications of this finding is enhanced by the current availability of still other studies.

That hypothalamic function might not be normal in anorexia nervosa had been suggested by the work of Russell, Bell, and co-workers (8,9) who observed some (if not consistent) improvement in the mean daily concentrations of urinary gonadotropins in patients with remitted anorexia nervosa and weight gain, but no return of the normal monthly midcycle peaking of gonadotropins during long periods of follow-up in these women. Marshall and Fraser similarly found low serum LH concentrations in their group of anorexia nervosa patients and low-normal to normal values in their weight-recovered subjects, but with the latter group also failing to show the normal midcycle ovulatory peaking in LH levels (25). Marshall and Fraser also found no effect of clomiphene citrate on serum LH levels in their anorexia nervosa patients, whereas their weight-recovered group showed a response; more recent work by Russell and Wakeling, however, has indicated that even in weight-recovered patients the response to clomiphene is only partially normal — only the immediate, not the delayed (1 week), response being manifested (26). Furthermore, that the pituitary in anorexia nervosa is capable of responding to hypothalamic Gn-RH has now been demonstrated by at least two groups of investigators (27,28).

The possibility that these developmentally inappropriate, hypothalamically mediated, circadian patterns of LH secretion might be found in other types of amenorrhea has been minimized by various studies in our laboratories and elsewhere. Thus women with gonadal dysgenesis (Turner's syndrome) (29), menopause (30), the polycystic ovary (Stein-Leventhal) syndrome (31), or "functional" amenorrhea-galactosemia (32) do not manifest circadian LH patterns that overlap those of our patients, despite the amenorrhea they all manifest in common.

However, because amenorrhea and low urinary gonadotropins also have been observed to occur during starvation (33), we recently initiated studies of circadian LH patterns in emaciated women without true anorexia nervosa. The first such patient for whom complete data are currently available was a clearly schizophrenic 23-year-old woman whose somatic delusions of food "eroding" her stomach and "tasting peculiar" prevented her from eating to the extent that her weight had fallen to 30% below its usual and normal value. She had not wanted to diet and was actually markedly distressed about the weight she had lost. She thus appeared to manifest what Bruch had labeled the "secondary" form of anorexia nervosa. Amenorrhea had developed well after the appearance of significant weight loss.

The patient's 24-hour pattern of plasma LH concentrations is shown in Fig. 11. Her mean LH concentration (17.2 mIU/ml) and the amplitude of the pulses are quite high, and her circadian pattern, although somewhat atypical, is also quite different from that seen in any of the primary anorexia nervosa patients. Unfortunately two aspects of this patient's study make reliability of the data uncertain. For one thing, because of her massive and disorganizing anxiety during the study, she barely slept throughout the 24-hour period; what effect this had on her overall LH pattern and output cannot be specified. For another, the patient could be managed only with medication (chlorpromazine, 50 mg i.m., b.i.d.), which also may have influenced LH secretion; although animal studies have indicated that phenothiazine administration is associated with lowered gonadotropin release (34), which would make this patient's high LH values all the more striking, a

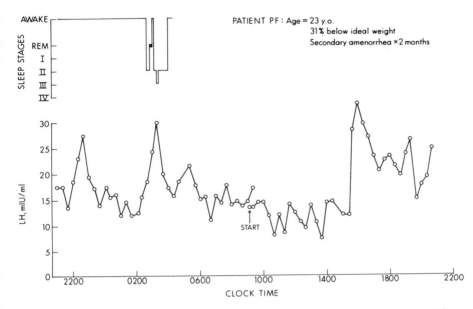

FIG. 11. Circadian LH secretory pattern in a 23-year-old woman with a "secondary" form of anorexia nervosa.

recent report in humans indicated a possible enhancing effect of phenothia-zines on LH secretion (35).

To clarify further the relationships between clinical course, weight, and circadian LH pattern, we have also begun to restudy our patients whenever feasible. To date, four have had follow-up endocrine investigations.

One, markedly underweight and with an extremely immature LH pattern at initial study, was still eating poorly and had gained virtually no weight as of 3 months later, and her LH pattern also remained totally unchanged. A second woman experienced a remission in symptoms after graduation from college several months later, and her weight increased from approxi-mately 76% to 97% of its ideal value. Her initial early pubertal pattern showed a striking "maturization" to a normal adult LH secretory pattern during the remission study approximately 6 months later (Fig. 12). The most likely assumption would be that the change in her circadian LH secre-tory pattern was a reflection of her improved nutritional state, but it is also conceivable that *both* her improved eating and maturer LH pattern were secondary to some more fundamental change in the pathogenetic process.

A third woman also experienced a remission, in this instance during hos-pitalization, and by the end of 2 months had resumed eating reasonably normally and had gained approximately 8 pounds so that rather than being 25% below her ideal weight she had only an 18% deficit. She too showed a

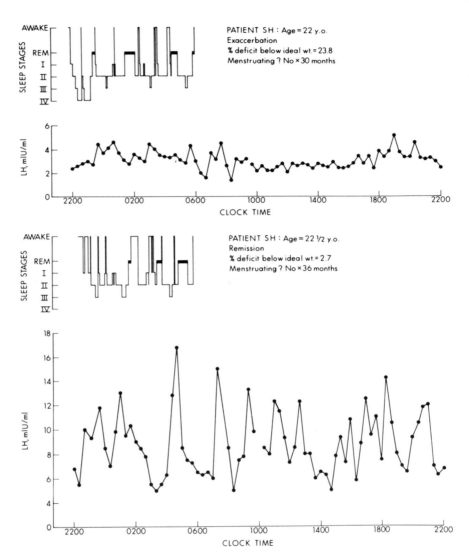

FIG. 12. Six-month follow-up of circadian LH secretory pattern in patient during remission with virtually normal weight.

significant improvement in the maturity of her previously prepubertal 24-hour LH secretory pattern (Fig. 13).

On the other hand, the fourth patient also showed a moderate improvement in her weight, gaining 5 pounds over the next 16 months, but mainly as a consequence of strong pressure from family, friends, and internist, rather than as a reflection of any fundamental inner change in her attitude

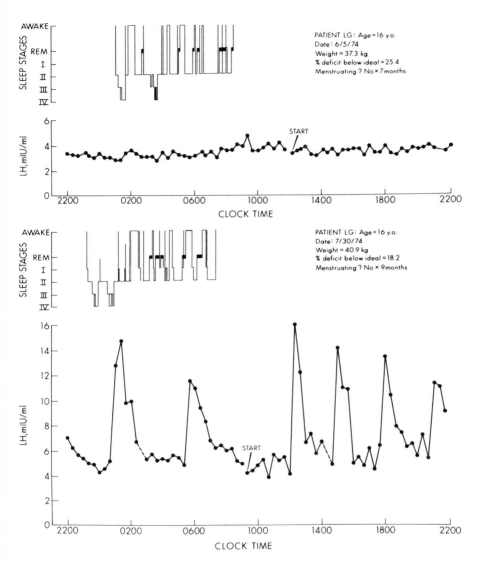

FIG. 13. Two-month follow-up of circadian LH secretory pattern in patient during remission with partial weight recovery.

toward eating and weight. Percent deficit below ideal weight declined approximately 27% to 23%. As shown in Fig. 14, her circadian pattern, despite her weight gain, became even flatter (excepting a solitary strong nocturnal pulse), and the mean 24-hour LH concentration lower (4.5 mIU/ml) than the initial one (6.7 mIU/ml).

The importance of body weight for the onset and maintenance of normal

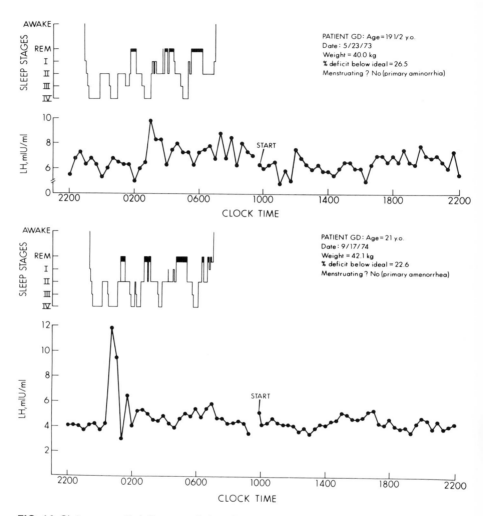

FIG. 14. Sixteen-month follow-up of circadian LH secretory pattern in patient with active anorexia nervosa but slight weight gain.

menses — and by implication for normal circadian and monthly gonadotropin patterns — has received growing attention recently. Frisch and Revelle, for instance, showed that although the age of menarche in white girls in our culture has decreased from 16.5 to 12.5 years during the past 125 years the mean weight at menarche has remained unchanged (36). They suggest that a "critical" body weight is associated with a diminished sensitivity of the hypothalamus to circulating sex steroids and a consequent "feedback" increase in pituitary gonadotropin secretion. More recently, Frisch and

McArthur predicted the minimum weight for height necessary for the onset of menstrual cycles (menarche) and for the restoration of menses in secondary amenorrhea due to undernourishment while also establishing that the latter women, when older than 16 years, require a heavier weight to resume menses than they needed to achieve menarche at puberty (37). These findings are consistent with the observations of Crisp and Stonehill that patients with remitting anorexia nervosa must approach their *ideal* weight, not merely put a little weight back on, if menses are to return (38).

Nevertheless, two facts speak against weight as the sole variable in determining the presence or absence of menses. The first is that the occurrence of amenorrhea prior to or just concurrent with the start of dieting and weight loss is generally observed in 25% to 50% of patients with primary anorexia nervosa (39–41); indeed this occurred in 30% of our 10 patients, while another 30% had never achieved menarche (despite at least two of them clearly having had a fully adequate weight until at least age 13). This observation suggests either that psychological conflict may first evoke "functional" or "psychogenic" amenorrhea in at least some patients, it then being aggravated and maintained by the superimposed bizarre eating behavior and eventual weight loss, or that the amenorrhea and the bizarre eating behavior *both* reflect some hypothalamic abnormality that is being activated.

The second fact is failure of the normal monthly cycling pattern in gonadotropin secretion and of full hypothalamic responsiveness to clomiphene to return despite the return to normal weight in women in remission. On the other hand, this phenomenon could be due merely to a prolonged cessation of function, with prolonged "priming" then being required before normal cycling and responsiveness are restored, a circumstance commonly seen in women who are amenorrheic for several months after going off the "pill" following its prolonged use. It may, alternatively, reflect the lag time required for full restoration of nutritionally dependent precursors for hypothalamic releasing hormones or neurotransmitters. That there are etiologically different subgroups within the anorexia nervosa syndrome of course also remains a possibility.

Thus although the known role of the hypothalamus in eating behavior (42,43) and in endocrine function makes it tempting to postulate a preexisting hypothalamic defect that would account for both the extraordinary dieting and the amenorrhea, such speculation is clearly premature. Although other assessments of hypothalamic function in patients with anorexia nervosa (e.g., water conservation, thermoregulation, carbohydrate metabolism) have also been suggestive of a hypothalamic aberrancy (44,45), we believe that appropriate control and longitudinal psychoendocrine studies now provide the most promising tool for delineating the nature of the hypothalamic defect and for determining whether it is an antecedent, concomitant, or consequence of the illness.

ACKNOWLEDGMENTS

This work was supported in part by grant RR-53 from the General Clinical Research Centers Branch of the National Institutes of Health and by research grants CA 07304, ROH 00331, DH 06209, and HL 14734 from the United States Public Health Service.

REFERENCES

1. Galdston, R. (1974): Mind over matter: Observations on fifty patients hospitalized with anorexia nervosa. *J. Child Psychol. Psychiatry,* 13:246–264.
2. Bruch, H. (1969): Hunger and instinct. *J. Nerv. Ment. Dis.,* 149:91–114.
3. Sheldon, J. H. (1939): Anorexia nervosa. *Proc. R. Soc. Med.,* 32:738–740.
4. Sheehan, H. L., and Summers, V. K. (1948): The syndrome of hypopituitarism. *Q. J. Med.,* 18:319–378.
5. Escamilla, R. F., and Lisser, H. (1942): Simmonds' disease. *J. Clin. Endocrinol. Metab.,* 2:65–96.
6. Bliss, E. L., and Migeon, C. J. (1957): Endocrinology of anorexia nervosa. *J. Clin. Endocrinol. Metab.,* 17:766–776.
7. Marks, V., and Bannister, R. G. (1963): Pituitary and adrenal function in undernutrition with mental illness (including anorexia nervosa). *Br. J. Psychiatry,* 109:480–484.
8. Bell, E. T., Harkness, R. A., Loraine, J. A., and Russell, G. F. M. (1966): Hormone assay studies in patients with anorexia nervosa. *Acta Endocrinol. (Kbh.),* 51:140–148.
9. Russell, G. F. M., Loraine, J. A., Bell, E. T., and Harkness, R. A. (1965): Gonadotrophin and oestrogen excretion in patients with anorexia nervosa. *J. Psychosom. Res.,* 9:79–85.
10. Beumont, P. J. V., Carr, P. J., and Gelder, M. G. (1973): Plasma levels of luteinizing hormone and of immunoactive oestrogens (oestradiol) in anorexia nervosa: Response to clomiphene citrate. *Psychol. Med.,* 3:495–501.
11. Boyar, R., Perlow, M., Hellman, L., et al. (1972): Twenty-four pattern of luteinizing hormone secretion in normal men with sleep stage monitoring. *J. Clin. Endocrinol. Metab.,* 35:73–81.
12. Hellman, L., Nakada, F., Curti, J., et al. (1970): Cortisol is secreted episodically by normal man. *J. Clin. Endocrinol. Metab.,* 30:411–422.
13. Boyar, R., Finkelstein, J., Roffwarg, H., et al. (1972): Synchronization of augmented luteinizing hormone secretion with sleep during puberty. *N. Engl. J. Med.,* 287:582–586.
14. Tanner, J. M. (1962): *Growth at Adolescence, 2nd edition,* p. 32. Blackwell Scientific Publications, Oxford.
15. Kapen, S., Boyar, R., Hellman, L., and Weitzman, E. D. (1973): Episodic release of luteinizing hormone at mid-menstrual cycle in normal adult women. *J. Clin. Endocrinol. Metab.,* 36:724–729.
16. Schally, A. V., Arimura, A., and Kastin, A. J. (1973): Hypothalamic regulatory hormones. *Science,* 179:341–350.
17. Bruch, H. (1965): Anorexia nervosa and its differential diagnosis. *J. Ment. Nerv. Dis.,* 141:555–566.
18. Rechtskaffen, A., and Kales, A., editors (1968): *A Manual of Standardized Terminology, Techniques and Scoring System for Sleep Stages of Human Subjects.* NIH Publication 204. Government Printing Office, Washington, D.C.
19. Midgley, A. R., Jr. (1966): Radioimmunoassay: A method for human chorionic gonadotropin and human luteinizing hormone. *Endocrinology,* 79:10–18.
20. Boyar, R., Perlow, M. Hellman, L., et al. (1972): Twenty-four patterns of luteinizing hormone secretion in normal men with sleep stage monitoring. *J. Clin. Endocrinol. Metab.,* 35:73–81.
21. Marks, V., and Howarth, N. (1965): Plasma growth hormone levels in chronic starvation in man. *Nature (Lond.),* 208:686–687.

22. Landon, J., Greenwood, F. C., Stamp, T. C. B., and Wynn, V. (1966): The plasma sugar, free fatty acid, cortisol, and growth hormone response to insulin, and the comparison of this procedure with other tests of pituitary and adrenal function. II. In patients with hypothalamic or pituitary dysfunction or anorexia nervosa. *J. Clin. Invest.*, 45:437–449.
23. Fukushima, D. K., Bradlow, H. L., Hellman, L., and Gallagher, T. F. (1968): On cortisol production rate. *J. Clin. Endocrinol. Metab.*, 28:1618–1622.
24. Warren, M. P., and Van de Wiele, R. L. (1973): Clinical and metabolic features of anorexia nervosa. *Am. J. Obstet. Gynecol.*, 117:435–449.
25. Marshall, J. C., and Fraser, T. R. (1971): Amenorrhea in anorexia nervosa: Assessment and treatment with clomiphene citrate. *Br. Med. J.*, 4:590–592.
26. Russell, G. F. M., and Wakeling, A. (1974): The endocrine and menstrual response to clomiphene citrate in patients with anorexia nervosa. Presented at the Tenth European Conference on Psychosomatic Research, Edinburgh, Scotland, September 1974.
27. Wiegelmann, W., and Solbach, H. G.: (1972): Effects of LH-RH on plasma levels of LH and FSH in anorexia nervosa. *Horm. Metab. Res.*, 4:404.
28. Yoshimoto, Y., Moridera, K., and Imura, H. (1975): Restoration of normal pituitary gonadotropin reserve by administration of luteinizing-hormone-releasing hormone in patients with hypogonadotropic hypogonadism. *N. Engl. J. Med.*, 292:242–245.
29. Boyar, R. M., Finkelstein, J. W., Roffwarg, H., et al. (1973): Twenty-four hour luteinizing hormone and follicle-stimulating hormone secretory patterns in gonadal dysgenesis. *J. Clin. Endocrinol. Metab.*, 37:521–525.
30. Yen, S. S. C., Tsai, C. C., Naftolin, F., et al. (1972): Pulsatile patterns of gonadotropin release in subjects with and without ovarian failure. *J. Clin. Endocrinol. Metab.*, 34:671–675.
31. Yen, S. S. C., Vela, P., and Rankin, J. (1970): Inappropriate secretion of follicle-stimulating hormone and luteinizing hormone in polycystic ovarian disease. *J. Clin. Endocrinol. Metab.*, 30:435–442.
32. Boyar, R. M., Kapen, S., Finkelstein, J. W., et al. (1974): Hypothalamic-pituitary function in diverse hyperprolactinemic states. *J. Clin. Invest.*, 53:1588–1598.
33. Zubiran, S., and Gomez-Mont, F. (1953): Endocrine disturbances in chronic human malnutrition. *Vitamin Horm.*, 2:97–130.
34. de Wied, D. (1967): Chlorpromazine and endocrine function. *Pharmacol. Res.*, 19:251–288.
35. Santen, R. J., and Bardin, C. W. (1973): Episodic luteinizing hormone secretion in man: Pulse analysis, clinical interpretation, physiologic mechanisms. *J. Clin. Invest.*, 52:2617–2628.
36. Frisch, R. E., and Revelle, R. Height and weight at menarche and a hypothesis of critical body weights and adolescent events. *Science*, 169:397–398.
37. Frisch, R. E., and McArthur, J. (1974): Menstrual cycles: Fatness as determinant of minimum weight for height necessary for their maintenance or onset. *Science*, 185:949–951.
38. Crisp, A. H., and Stonehill, E. (1971): Relation between aspects of nutritional disturbance and menstrual activity in primary anorexia nervosa. *Br. Med. J.*, 3:149–151.
39. Kay, D. W. K., and Leigh, D. (1954): Natural history, treatment, and prognosis of anorexia nervosa, based on a study of 38 patients. *J. Ment. Sci.*, 100:411–431.
40. King, A. (1963): Primary and secondary anorexia nervosa syndromes. *Br. J. Psychiatry*, 109:470–479.
41. Halmi, K. A. (1974): Anorexia nervosa: Demographic and clinical features in 94 cases. *Psychosom. Med.*, 36:18–26.
42. Anand, B. K., and Brobeck, J. R. (1951): Hypothalamic control of food intake in rats and cats. *Yale J. Biol. Med.*, 24:123–140.
43. Balagura, S. (1972): Hypothalamic factors in the control of eating behavior. *Adv. Psychosom. Med.*, 7:25–48.
44. Russell, G. F. M. (1965): Anorexia nervosa. *Proc. R. Soc. Med.*, 58:811–814.
45. Mecklenburg, R. S., Loriaux, D. L., Thompson, R. H., et al. (1974): Hypothalamic dysfunction in patients with anorexia nervosa. *Medicine (Balt.)*, 53:147–159.

Hormones, Behavior, and Psychopathology, edited by
Edward J. Sachar. Raven Press, New York © 1976.

Selective Pituitary Deficiency in Anorexia Nervosa

Katherine A. Halmi

University of Iowa College of Medicine, Iowa City, Iowa 52242

The data presented by Katz's group (1) and reports of other investigators such as Russell et al. (2), Crisp et al. (3), Silverman (4), Danowski et al. (5), and Mecklenberg et al. (6) give evidence to indicate a partial and selective pituitary deficiency in anorexia nervosa. This is most likely secondary

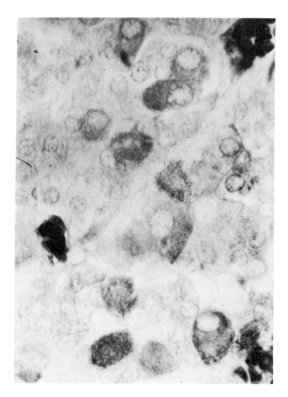

FIG. 1. A: Anterior pituitary gland from patient with anorexia nervosa. Gonadotrops are immunocytochemically stained with antibody against ovine luteinizing hormone (LH) by the unlabeled antibody peroxidase-antiperoxidase (PAP) method of Sternberger et al. (7). Note relatively faint staining (compare with **B**) of ovoid gonadotrops. Intense staining of red blood cells is due to their intrinsic peroxidase activity.

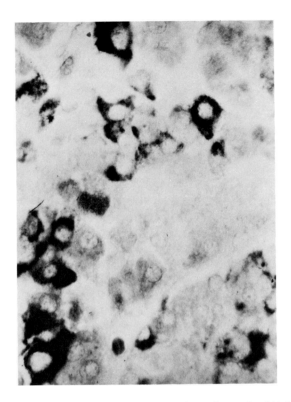

FIG. 1. **B:** Normal gonadotrops stained with antibody against ovine LH, from the pituitary gland of a 39-year-old man who died of aplastic anemia. Note heavy staining of the gonadotrops, which are more irregular in shape than in **A.**

to hypothalamic malfunction. Very few autopsies of patients with this illness have included histological examinations of the hypophysis, and none has utilized modern differential staining and immunocytochemical techniques necessary to distinguish the various adenohypophysial cell types. We applied these advanced methods (Figs. 1 through 4) to the postmortem study of a 30-year-old woman who died of "cardiac arrest" 9 years after the onset of anorexia nervosa. The history of her illness was typical, starting with voluntary dieting at 120 pounds (54.4 kg), onset of amenorrhea at 100 pounds (45.2 kg), and a behavior pattern of bulimia, self-induced

FIG. 2. **A:** Prolactin cells from the pituitary gland of the patient with anorexia nervosa, stained immunocytochemically with antibody against rat prolactin by the unlabeled antibody-PAP method. Pleomorphic, scattered prolactin cells are normal (compare with **B**). For technical data and a detailed description of human prolactin cells, see Halmi et al. (8).
FIG. 2. **B:** Prolactin cells from the hypophysis of the same 39-year-old man who was the source in Fig. 1B. The morphology of these cells is quite similar to that of identically stained cells in Fig. 2A.

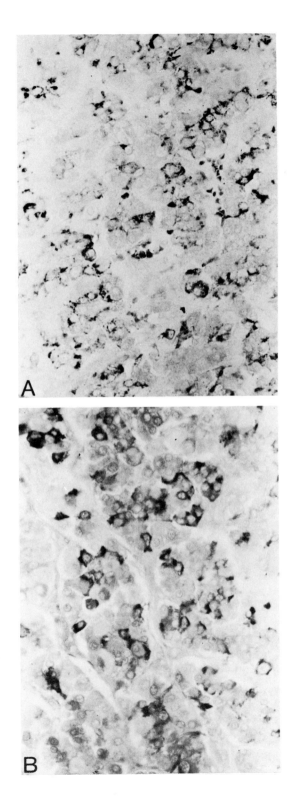

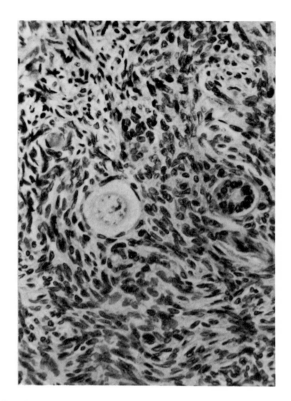

FIG. 3. Ovary of patient with anorexia nervosa. Primordial follicles are surrounded by highly cellular stroma. There are no graafian follicles of corpora lutea present. One corpus albicans in the ovary (not in the field) points to previous occurrence of ovulatory cycles.

vomiting, and vigorous exercising rituals. She resisted a variety of treatment programs offered during multiple hospitalizations at different locations, including her last, when she weighed 31.3 kg. At autopsy most organs were essentially normal. *Staphylococcus aureus* was cultured from the blood.

The atrophic ovaries, which showed hypercellularity, contained only primordial follicles and old corpora albicantia. Profound atrophy of the breast and endometrium attested to estrogen deficiency. The endometrium showed a flat lining epithelium, overall thinness, and short straight glands. The thyroid was normal in size, with flat epithelium. The adrenal glands were normal in size and unremarkable histologically.

The adenohypophysis was examined with a battery of differential stains and immunocytochemically. A section of the anterior pituitary was stained with aldehyde thionine and periodic acid Schiff-orange G. The red-staining

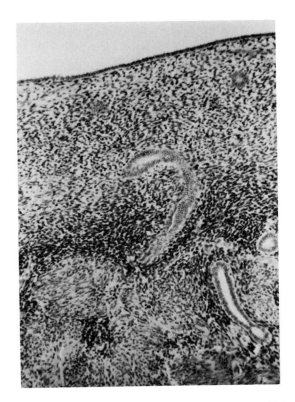

FIG. 4. Atrophic endometrium in patient with anorexia nervosa. Note low lining cells, thinness of endometrium, density of stroma, and short glands. This picture, quite similar to that seen before puberty or after the menopause, suggests estrogen deficiency.

adrenocorticotropic hormone (ACTH) cells and the dark orange-staining growth hormone (GH) cells were normal, as were the angular blue-staining thyroid-stimulating hormone (TSH) cells.

Since gonadotropic cells stain very poorly with aldehyde thionine and prolactin cells are not demonstratable at all with this method, an immunoperoxidase technique was used to demonstrate these cells. The prolactin cells in the pituitary of the anorexia nervosa patient were as numerous and of the same irregular shape as in normal cells. They also stained with the same intensity as those from the normal pituitary controls. The gonadotrops immunostained for luteinizing hormone (LH) stained much more faintly in the anorectic's pituitary than in normal glands.

The morphological findings thus agree with hormone measurements consistently obtained during the acute phase of anorexia nervosa; i.e., plasma LH levels are low and respond poorly to injected synthetic LH-releasing hormone (LH-RH), possibly because of diminished hypophysial LH stores. Plasma estrogen levels are well below the lowest found in women with normal menstrual cycles.

ACKNOWLEDGMENT

I wish to express my grateful appreciation to Dr. Nicholas Halmi for staining the adenohypophysis and to Dr. Sternberger for providing the rabbit perioxidase-antiperoxidase complex for the immunoperoxidase staining.

REFERENCES

1. Boyar, R. M., Katz, J., Finkelstein, J. W., Kapen, S., Weiner, H., Weitzman, E. D., and Hellman, L. (1974): Anorexia nervosa: Immaturity of the 24-hour luteinizing hormone secretory pattern. *N. Engl. J. Med.*, 291:861–865.
2. Russell, G. F. M., Loraine, J. A., Bell, E. T., and Harkness, R. A. (1965): Gonadotrophin and estrogen excretion in patients with anorexia nervosa. *J. Psychosom. Res.*, 9:79–85.
3. Crisp, A. H., Chen, C., MacKinnon, P. C. B. and Corker, C. S. (1973): Observations of gonadotrophic and ovarian hormone activity during recovery from anorexia nervosa. *Postgrad. Med.*, 49:584–590.
4. Silverman, J. A. (1974): Anorexia nervosa: Clinical observations in a successful treatment plan. *J. Pediatr.*, 84:68–73.
5. Danowski, T. S., Livstone, E., Gonzales, A. R., Jung, Y., and Khurana, R. C. (1972): Fractional and partial hypopituitarism in anorexia nervosa. *Hormones*, 3:105–118.
6. Mecklenburg, R. S., Loriaux, D. L., Thompson, R. H., Andersen, A. E., and Lipsett, M. B. (1974): Hypothalamic dysfunction in patients with anorexia nervosa. *Medicine (Balt.)*, 53:147–159.
7. Sternberger, L. A., Hardy, P. H., Jr., Cuculis, J. J., and Meyer, H. G. (1970): The unlabeled antibody enzyme method in immunohistochemistry: Preparation and properties of soluble antigen-antibody complex (horseradish peroxidase-antiperoxidase) and its use in identification of spirochetes. *J. Histochem. Cytochem.*, 18:315–333.
8. Halmi, N. S., Parsons, J. A., Erlandsen, S. L., and Duello, T. (1975): Prolactin and growth hormone cells in the human hypophysis: A study with immunoenzyme histochemistry and differential staining. *Cell Tissue Res.*, 158:497–507.

Hormones, Behavior, and Psychopathology, edited by
Edward J. Sachar. Raven Press, New York © 1976.

The Medical Model in Psychiatry

Charles Shagass

Temple University Medical School and Eastern Pennsylvania Psychiatric Institute,
Philadelphia, Pennsylvania 19129

The medical model in psychiatry has become a much discussed and controversial topic. Articulate proponents of the position that psychiatry should
discard the medical model have excited public interest with book titles like
The Myth of Mental Illness (1) and *The Death of Psychiatry* (2). The
validity of psychiatry's status as a specialty within medicine is being questioned more than ever before in its modern history. In this climate of controversy, it seemed to me that it might be useful to try to answer a few basic
questions: What is the medical model? Is it really responsible for psychiatry's errors and failures? How does our science of psychopathology relate
to the medical model in psychiatry?

CRITIQUES OF THE MEDICAL MODEL IN PSYCHIATRY

Let me first try to state very briefly a few of the main arguments advanced
by two leading proponents of the view that psychiatry should abandon the
medical model. According to Szasz (1), there can be no mental illness since
disease or illness can affect only the body. He views mental illness as a
metaphor and psychiatric diagnoses as stigmatizing labels. He sees no
justification for psychiatric hospitalization or treatment. Torrey (2) asserts
that psychiatry as a medical specialty is dying. He concedes that some major
psychoses may have an organic cause, but states that should an organic
etiology be proved schizophrenia would become a neurological, as distinct
from psychiatric, disease. He proposes an educational model to replace the
medical one for most current psychiatric efforts and argues that behavioral
scientists will do a better job than psychiatric physicians. Torrey counters
objections that nonphysicians will fail to diagnose organic diseases by stating
that anyone with a behavioral science background can be trained within a
few weeks to detect true brain disease. He contends also that the use of
psychoactive drugs could be learned in an equally short time.

Both Szasz and Torrey make the somewhat paradoxical argument that
if it is not organic, it is not disease, and if it should turn out to be organic
it is not psychiatric. Their view of disease seems to be remarkably restricted,
being of "either-or" character. It appears to disregard the generally held
concept of the organism as an integrated system with different levels of

subsystems, in any of which dysfunction can be initiated and toward any of which therapeutic intervention can be directed.

I find myself in sympathy with much that these critics have to say about psychiatry's faults. The province of psychiatry has been overextended by many to fields clearly outside our areas of expertise, and there do seem to have been abuses and misuses of our medical authority. However, I cannot go along with the idea that the medical model is the prime villain responsible for abuses, and that its exorcism from the realm of behavioral disorders will bring about improved mental health care.

THE THREE MEDICAL MODELS

For convenience I have been referring to "the medical model" in the singular; strictly speaking this usage is incorrect, as there is more than one medical model. Siegler and Osmond (3)—to whose book *Models of Madness, Models of Medicine* I owe much of what I say here—describe at least three major medical models. The one most regularly brought to mind is the clinical medical model, which has at its core the reciprocal dyad of doctor and patient. A second major medical model is the public health model; its goal is the prevention of disease and the fostering of health for a particular population. Instead of the patient-doctor dyad, the central roles in the public health model are those of public health official and citizen. The third medical model is the scientific one, in which the doctor's role is that of scientific investigator. When the medical scientist does research with patients his role is quite different from that in the clinical medical model, and so is the role of the patient who is an experimental subject.

Obviously if confusion in discussing the medical model is to be avoided we first need to know which medical model is being considered. Both Szasz and Torrey fail to distinguish between the different models.

CHARACTERISTICS OF THE CLINICAL MEDICAL MODEL

It does seem to be the case that the phrase "medical model" almost inevitably tends to elicit a focus on organic etiology, regardless of the actual status of knowledge about causation. This is a reflection of the impact of advances in scientific medicine, but it ignores the fact that medicine has been around for a very long time and that the clinical medical model was well developed long before most specific causes and treatments of illness were discovered. This brings me to a main point. The clinical medical model does not depend on specific etiological and therapeutic knowledge of a scientifically valid nature. It incorporates and utilizes such knowledge, but the model itself is based on rules of conduct, not specific knowledge. The important contribution of Siegler and Osmond's analysis of the medical model is that they explicitly recognized that rules of conduct in the therapeutic relationship are at its core, rather than scientific knowledge.

In Siegler and Osmond's view, the medical model has two great social roles, both fostering survival of the species. One is that of organizing and conserving knowledge about disease and its treatment, thus healing the ill and saving lives. The other, which is of at least equal importance, is to ensure that no one is blamed for illness. These authors point out that society could hardly survive if every illness or death were perceived as the result of someone's error or malice. Their analysis of the ways in which the medical model goes about performing its twin functions of treating the sick and expunging blame for illness depends on two main concepts: Aesculapian authority and the sick role.

Aesculapian authority, a concept attributed to Paterson (4), refers to the great power conferred on doctors by society. By virtue of this authority, doctors are allowed to probe bodies and minds in ways permitted no one else, and patients, without protest, permit themselves to be subjected to unpleasant, frightening, and even humiliating experiences. Paterson considered Aesculapian authority to derive from a special combination of three kinds of authority, which he designated as: sapiential, moral, and charismatic. Sapiential authority derives from special knowledge, about medicine in this case. Moral authority stems from the fact that the physician acts for the good of the patient, in ways that society expects from him as a doctor. The doctor's charismatic authority is seen as a residual of the original unity of religion and medicine. He still retains some of the priestly role. In western society his priestly attributes are reinforced by the fact that illness raises the possibility of death. Also, since it is impossible to assess the doctor's knowledge fully, faith is required.

The second main factor in the medical model is the function of conferring the sick role. Only someone with Aesculapian authority has the right to confer the sick role. The first step here involves differential diagnosis; the doctor must decide if a potential patient is sick and with what sickness. He then informs the patient clearly about his illness. When the patient accepts that he is ill, the doctor absolves the patient and his family from blame for the illness. This is usually done by implication, rather than directly; it is implied that the sickness results from natural causes, even though its etiology may be unknown. Should the doctor fail to absolve from blame, he runs the risk of not getting essential cooperation from the patient and his family. The doctor then recommends a course of treatment and answers questions about prognosis. He may use a hospital for diagnosis or treatment but never for custodial purposes. His goals are to restore the patient to health if possible, or at least prevent the illness from getting worse. He is dedicated to preservation of life; if the patient's condition involves risk of suicide, he takes steps to minimize the risk.

The sick role appears to be learned rather early in life. It involves both duties and privileges. It is the patient's duty to seek help, accept treatment, and try to get well. He may be required to submit to unpleasant and frightening procedures, prohibited from performing many of his normal functions, and he is expected to follow orders. On the other hand, Aesculapian authority exempts the patient from normal responsibilities. A doctor's certificate has more authority than a court order or an employer's displeasure.

Patients must eventually leave the sick role. They can recover and simply not return for care; they can be pronounced cured by the doctor; or they can die. Aesculapian authority extends into death. A doctor either certifies that the patient died from natural causes, or if he is not satisfied about the cause of death he can order an autopsy to decide the validity of bestowing the sick role retrospectively.

Chronic illness presents difficult problems for the medical model. The crucial question is if anything more can be done. If everything possible has been done and the patient is in a state of permanent impairment but his life is no longer in danger, the doctor's active treatment role should end. It can continue only so long as there is hope of improvement by treatment, or the danger of death. Permanent impairment involves a transition from the sick role to the impaired role, from application of the medical model to use of a model that has been called the "impaired model." Although both sick and impaired persons are expected to see a doctor and report any changes in their condition, the impaired role does not carry the rights and privileges of the sick role. The person in the sick role is encouraged not to engage in normal behavior, not to work, etc.; the person in the impaired role is pressured to act as normally as possible in spite of his handicaps (5).

Siegler and Osmond point out that the deplorable situation that developed in large mental hospitals can be understood, at least in part, by the fact that impaired people were being dealt with by a version of the medical model, rather than by the impaired model. To be treated as sick, rather than impaired, involves discouragement of normal behavior. On the other hand, to be treated as impaired in an institution ostensibly for the sick and organized with the trappings of the medical model certainly involves confusion of role. Furthermore, it results in the patient being deprived of the help of rehabilitation specialists, who are the key personnel in institutions frankly devoted to problems of the impaired.

SCIENCE AND MEDICINE

I hope that I have substantiated the proposition that the clinical medical model rests to a great extent on rules of conduct rather than being based solely on scientific knowledge of disease processes. The doctor's function in society is determined not as much by the state of medical knowledge as by the fact that, in prescribed ways, he does what he can in the realm of his assigned responsibility.

Of course, science is exceedingly important to modern medicine's effectiveness in reducing suffering and prolonging life, but medicine seems to have had a curious set of attitudes about scientific discoveries. On the one hand, major developments, such as understanding bacterial infection and use of anesthesia, generally encountered conservative opposition before they were accepted. On the other hand, the history of medicine indicates that once a discovery meets the pragmatic criterion of being useful it is incorporated without much concern about the school of thought in which it originated. Medicine is pluralistic in its theoretical structure; it accepts multiple causation and diverse views about pathogenesis. In fact, one of the criteria by which medicine can be distinguished from schools of quackery is that the latter tend to have a single theory to account for all illness. I argue that both medical conservatism and its ability to incorporate diverse theoretical views stem from its foundation in rules of conduct for fulfilling a social responsibility.

I have said that medicine's main criterion for accepting new knowledge is pragmatic. Does it work? Does it do any good? How is "good" defined? Without attempting to define the undefinable, it can be said that "good" pertains to clinical medicine's main functions of relieving suffering, ameliorating disability, and saving life, without doing more harm than good. When new knowledge can help to accomplish any of these purposes, it acquires pragmatic relevance. The ideal of scientific medicine is to isolate essential factors, without which disease cannot occur, and to find treatments which remove or control these factors. Successes in the scientific medical enterprise have resulted in a downgrading of merely symptomatic therapies, although they have been the age-old stock in trade of medicine.

In addition to effectiveness, the criterion of pragmatic relevance is strongly governed by such matters as the expenditure of time and money involved in a treatment. The suffering person wants relief as quickly and inexpensively as possible. Thus although the ultimate cause may be an environmental factor that could be removed by social action, as in the case of certain allergies or in defects in the social system itself (as in neuroses associated with racial discrimination), the patient is hardly ready to accept the prescription of reforming society as a cure for his current distress. He wants something that will help now, which is why he turns to the clinical doctor rather than to the public health physician or the politician. Furthermore, medicine responds to his demands, even if only with symptomatic treatment.

The pragmatic basis of the clinical medical model cannot be overemphasized. The case of the major antipsychotic agents is a good example. These drugs do not seem to "cure" psychosis; their mode of action is still poorly understood after two decades of intensive psychopharmacological research; and they are often downgraded as not being "the real" therapy when formal psychotherapy is administered concurrently. Nevertheless, they are widely used because they meet the pragmatic criteria of quickly achieving desirable effects in the realm of relief of suffering, disability, and social disruption.

PSYCHOPATHOLOGY AND THE MEDICAL MODEL

How does psychopathology relate to the medical model? Let me first make clear that I am talking about psychopathology as that science which systematically investigates morbid mental conditions (6) and not as a collective noun used to designate all kinds of behavior that may deviate from some standard. Psychopathology is a broadly based science; it acknowledges virtually no limits to the range of phenomena it accepts as possibly relevant to the understanding of abnormal behavior. The disciplines devoted to the specialized study of these phenomena can extend from molecular biology to social anthropology. Psychopathology as a science is not necessarily tied to psychiatry; potentially its findings can be useful for many other applied disciplines, such as criminal law and special education. The subjects of psychopathology investigations can be animals or humans. However, for studies of mental illness, available animal models are not necessarily convincing, so psychopathology depends heavily on human research and must work closely with psychiatry. As a basic science for psychiatry, psychopathology thus follows the scientific medical model.

Because a great deal of research in psychopathology is concerned with the systematic description of abnormal behavior by means of the methods of clinical observation, there has been some tendency to equate this particular aspect of psychopathological research with the whole field. This obscures the fact that psychopathology's contributions to psychiatry can come from many other levels of investigation. With its multilevel approach, psychopathology hopes to do for psychiatry what pathology has done for medicine in general.

IS MENTAL ILLNESS REAL?

One of the main arguments advanced by critics of the medical model in psychiatry is that psychopathology has failed to prove an organic basis for mental illness. This is hardly a definitive argument when causes of a nonorganic nature are at least equally unproved. Moreover, the critics' argument sets aside a substantial body of evidence which indicates that biological determinants do play an important role in the major "functional" psychoses. The growing literature resulting from biological investigations of the psychoses has provided, among other things, impressive data supporting the role of genetic factors in their causation (7,8). Other lines of study suggest that these conditions involve widespread physiological dysfunction. Among these is evidence of deviant brain activity as reflected in electroencephalograms (EEG) and evoked potentials (9–11), generalized neuromuscular pathology as shown in lesions of muscle and excessive branching of subterminal motor nerves (12,13), and a large number of biochemical aberrations involving particularly neurotransmitters and enzymes (14,15).

Even though there is much evidence for deviant biological mechanisms, definitive causes of the schizophrenic and affective psychoses still remain to be established. Nevertheless, the findings gathered in genetically oriented studies, particularly those involving adoptive rearing, indicate that biological factors acting very early in life play a significant role in causing psychosis. Consequently it seems to me that it is a rather fruitless exercise to argue about whether the psychoses are "real" diseases. In the tradition of the scientific medical model, the important questions concern the specifics of their etiology and pathogenesis, information that could lead to useful psychiatric procedures.

Fortunately for psychiatry, as for medicine in general, we do not have to have complete scientific knowledge of a disease process to help patients. All that is required for effective intervention is one highly relevant fact. Since disease processes involve a complex chain of events, a relevant fact is one which permits altering that chain, with minimal risk of harm, by intervening at some point in it with an easily applied procedure. The main hope for patients suffering from mental illness is that research will lead to development of easily applied, effective treatments, even though psychopathologists might spend the next century explaining in detail how they work.

For at least two reasons, it seems to me that the chances of highly relevant facts being discovered and applied in psychiatry will be greater in the context of the medical model than if it were to be abandoned. One is that the clinical medical model adheres persistently to a value system centered around the benefit of the individual patient. The other is medicine's tradition of applying the test of pragmatic relevance to new ideas and procedures; when they achieve desirable purposes, they are adopted, even though this means changing previous explanations.

If psychiatry were to abandon the medical model, this would almost surely increase its uncertainty about the nature of its responsibilities to the individual patient and his family. Decisions about applying available treatments of known effectiveness would then become more difficult, and the attitudinal climate would favor disregard of new, potentially beneficial facts. Consider suicide. Although it can be argued that a person has a right to take his own life, the clinical medical model places a high positive value on saving life and consequently on preventing suicide. Without such a value, new ways to prevent suicide might more readily be disregarded. Indeed in the existing climate of confusion of values in psychiatry, many seem to have forgotten that the strongest evidence supporting the use of convulsive therapy in severe depressions is that it reduces the death rate (16–18).

WHEN PSYCHIATRY DOES NOT FOLLOW THE MEDICAL MODEL

Critics of the medical model in psychiatry seem to be misplacing their criticisms. They appear to misunderstand the nature of the clinical medical model, with its emphasis on rules of conduct, as well as its pragmatic knowledge base. Although much within psychiatry may merit criticism, I believe that these undesirable events are mainly the consequence of *not* following the medical model. Let me give a few examples.

I have already mentioned one disastrous consequence of "model muddle" in psychiatry, i.e., conditions resulting from the general failure of mental hospitals to distinguish between the impaired and medical models of mental illness. A similar situation seems to prevail outside institutional walls for many chronic patients who have been released into the community. Such patients may be routinely placed in follow-up programs and continued on medication. There seems to be great reluctance to recognize explicitly, however, that a great many are in a state of permanent impairment, and that their best hope lies in the development of specific rehabilitation programs. This may contribute to both the high readmission rate and the tragic picture of large numbers of chronic patients sitting in drugged listlessness in bleak rooming houses. For such patients the medical model is not really being followed since permanent impairment is not explicitly recognized. They need, and do not get, explicit rehabilitation assistance, so they can be helped to act as normally as possible in spite of their handicaps.

The medical model has been blamed for the injustice and cruelty visited on some patients by involuntary commitment practices. Torrey cites loose criteria, cursory psychiatric examinations or no examination at all, casual court proceedings, and absence of legal counsel for the patients. Such practices are wrong, but I submit that when a psychiatrist contributes to injustices resulting from involuntary commitment he usually is deviating from the clinical medical model in one of the following ways: insufficiently careful diagnostic examination; disregard of evidence or failure to acknowledge lack of evidence; making a moral instead of a medical judgment about the patient's conduct; or defining the problem as disposition rather than care, thereby failing to consider more helpful alternatives.

The much publicized psychiatric hospitalization of political dissidents in the Soviet Union hardly requires comment as an example of psychiatrists not following the medical model. Neither does the behavior of those hundreds of psychiatrists who in 1964 accepted publisher Ralph Ginzburg's invitation to diagnose presidential candidate Barry Goldwater from a distance.

The medical model has been blamed for problems of community mental health centers. However, it turns out that the average psychiatrist in such a center spends less than one-fifth of his working time with patients (19). Much of the rest may be spent in activities that seem only remotely related to the clinical medical model. One wonders what the record of the centers might be if they had been organized to provide maximal patient contact by psychiatrists who really behaved in accordance with the medical model (20).

Torrey has given particular attention to long-term intensive psychotherapy. He show that the bulk of psychiatric practice time, at least in urban America, is spent treating the "problems of living" of a relatively small number of minimally ill people. His view that much of this activity is more educational than medical seems to have some validity and is shared by many practitioners. He seems to be accusing the psychiatric psychotherapists both of neglecting the seriously ill and of applying treatment indiscriminately, at least from the standpoint of genuine need. What puzzles me is why he blames the medical model for this situation and why he thinks that it would be remedied if the medical model were to be abandoned. Indiscriminate use of any therapy, somatic as well as psychological, is a self-evident departure from the medical model.

It is easy to discern reasons for partial abandonment of the medical model by many psychiatrists. We live in a pluralistic society that values free enterprise, a fact reflected in the organization of medical services. Psychiatry achieved status within American medicine only when it was able to leave the confines of the mental hospital and engage in private practice. Given this background, it is not surprising that psychiatrists have attempted to meet the public's demands, even when poorly equipped to do so, or that they have engaged in various forms of promotion which promised more benefits than could be delivered. Much of this has been done in good faith, within the context of the optimistic psychiatric atmosphere that prevailed for a number of years after World War II, during which many psychiatrists received their training. One has only to turn to Smith's 1951 presidential address to our American Psychopathological Association (21) to gain some idea of the power that psychiatry's leaders attributed to psychodynamic concepts during that era. He talked about the responsibility of psychiatry, seeing it as contributing significantly to such matters as improving interracial tensions, interpreting the significance of various diseases for internists, art appreciation, educational, and religion.

In short, it seems that psychiatrists have been pressured to abandon the clinical medical model by demands arising both within and without the profession.

CONCLUSION

There are several medical models; the clinical one, which has at its core the doctor-patient dyad, is based on rules of conduct and pragmatically relevant scientific information about disease. Those who would have psychiatry abandon the clinical medical model seem to misunderstand its nature; their criticisms are directed mainly at psychiatric behavior which departs from this model. I submit that psychiatry's allegiance to the medical model needs to be strengthened, not abandoned.

REFERENCES

1. Szasz, T. S. (1974): *The Myth of Mental Illness*. Harper & Row, New York.
2. Torrey, E. F. (1974): *The Death of Psychiatry*. Chilton Book Co., Radnor, Pa.
3. Siegler, M., and Osmond, H. (1974): *Models of Madness, Models of Medicine*. Macmillan, New York.
4. Paterson, T. T. (1974): Cited in Siegler and Osmond (3).
5. Gordon, G. (1966): *Role Theory and Illness: A Sociological Perspective*. College and University Press, New Haven, Conn.
6. English, H. B., and English, A. C. (1958): *A Comprehensive Dictionary of Psychological and Psychoanalytical Terms*. Longmans, Green and Co., New York.
7. Kety, S. S., Rosenthal, D., Wender, P. H., and Schulsinger, K. F. (1968): The types and prevalence of mental illness in the biological and adoptive families of adopted schizophrenics. In: *The Transmission of Schizophrenia*, edited by D. Rosenthal and S. S. Kety. Pergamon Press, New York.
8. Kety, S. S. (1974): From rationalization to reason. *Am. J. Psychiatry*, 131:957–963.
9. Itil, T. M., Saletu, B., and Davis, S. (1972): EEG findings in chronic schizophrenics based on digital computer analysis and analog power spectra. *Biol. Psychiatry* 5:1–13.
10. Shagass, C., Overton, D. A., and Straumanis, J. J. (1974): Evoked potential studies in schizophrenia. In: *Biological Mechanisms of Schizophrenia and Schizophrenia-like Psychoses*, edited by H. Mitsuda and T. Fukuda. Igaku-Shoin Co., Tokyo.
11. Shagass, C., Straumanis, J. J., and Overton, D. A. (1975): Psychiatric diagnosis and EEG-evoked response relationships. *Neuropsychobiology*, 1:1–15.
12. Meltzer, H. Y., and Engel, W. K. (1970): Histochemical abnormalities of skeletal muscle in acutely psychotic patients. *Arch. Gen. Psychiatry*, 23:492–502.
13. Meltzer, H. Y., and Crayton, J. W. (1974): Muscle abnormalities in psychotic patients. II. Serum CPK activity, fiber abnormalities, and branching and sprouting of subterminal nerves. *Biol. Psychiatry*, 8:191–208.
14. Snyder, S. H. (1974): *Madness and the Brain*. McGraw-Hill, New York.
15. Snyder, S. H., Banerjee, S. P., Yamamura, H. I., and Greenberg, D. (1974): Drugs, neurotransmitters, and schizophrenia. *Science*, 184:1243–2153.
16. Huston, P. E., and Locher, L. M. (1948): Manic-depressive psychosis: Course when treated and untreated with electric shock. *Arch. Neurol. Psychiatry*, 60:37–48.
17. Huston, P. E., and Locher, L. M. (1948): Involutional psychosis: Course when untreated and treated with electric shock. *Arch. Neurol. Psychiatry*, 59:385–394.
18. Ziskind, E., Somerfield-Ziskind, E., and Ziskind, L. (1945): Metrazol and electric convulsive therapy of the affective psychoses. *Arch. Neurol. Psychiatry*, 53:212–217.
19. Glascote, R. M., and Gudeman, J. E. (1969): *The Staff of the Mental Health Center*. APA and NIMH Joint Publication, Washington, D.C.
20. Goldberg, D. (1971): The scope and limits of community psychiatry. In: *The Role of Drugs in Community Psychiatry. Modern Problems of Pharmacopsychiatry*, Vol. 6, edited by C. Shagass. Karger, Basel.
21. Smith, L. H. (1953): The responsibility of psychiatry. In: *Current Problems in Psychiatric Diagnosis*, edited by P. H. Hoch and J. Zubin. Grune & Stratton, New York.

Subject Index

Abused children, reversible dwarfism in, 243-251

ACTH
  adrenal response to
    in depression, 199-200
    in reversible hyposomatotropic dwarfism, 253-254, 257
  behavioral responses to, 2-6, 10
  fragments affecting growth hormone levels in depression, 216
  levels in depression, 195
  secretory responses to amphetamines, 162
  therapy with, psychiatric symptoms from, 129-130

Addison's disease, psychiatric symptoms in, 127, 193, 213

Adenyl cyclase activity, thyroid hormones affecting, 53

Adrenal gland
  behavioral responses to hormones of, 1
  disorders with psychiatric symptoms
    Addison's disease, 127, 193, 213
    Cushing's syndrome, 128-129
  function in reversible hyposomatotropic dwarfism, 253-254, 257
  hypothalamic-pituitary system in depression and, 193-213
  neuroendocrine transducer cells in medulla, 146
  psychiatric symptoms from steroid therapy, 129-131
  response to ACTH, in depression, 199-200

Adrenalectomy in females, behavior changes and, 122

Adrenogenital syndrome, 75, 76

Affective disorders, 193-218; see also Depression in endocrine disturbances, 127-141

Aggressive behavior, therapy for sex offenders and, 116-118

Alcohol withdrawal symptoms, effects of TRH and, 20

Alcoholism, imipramine in, with testosterone, 60

Amenorrhea, in anorexia nervosa, 266-267, 276

Amines, biogenic, role in mental disorders, 24-25

Amitryptyline in depression, effects of thyroid hormones and, 47

Amphetamines, neuroendocrine function and, 161-163

Androgens, see also Testosterone
  antiandrogenic therapy for sex offenders, 105-123
  effects in early development, 70-76, 86

Anorexia nervosa, 263-281
  amenorrhea in, 266-267, 276
  cortisol levels in, 273-275
  growth hormone levels in, 244, 256, 273
  hypothalamic function in, 275-281, 286
  luteinizing hormone secretion in, 268-273, 275-279
  pituitary deficiency in, 285-290
  prolactin levels in, 273

Apomorphine
  growth hormone regulation and, 258
  growth hormone response to, in depression, 215

Arcuate nucleus of hypothalamus, lesions of, 153-157

Arginine, growth hormone response to, in reversible hyposomatotropic dwarfism, 255

Behavior, hormones affecting, 1-11
Boeck's sarcoid, mental disorders in, 133
Brain, behaviorally active peptides in, 9-10

Calcium levels, in parathyroid disorders with psychiatric symptoms, 131-134
Castration, male, affecting behavior, 122
Catecholamines, interaction with thyroid hormones, 52-54
Caudate nucleus, thioridazine effects in, 168
Cerebrospinal fluid
  cortisol levels in depression, 196-197
  homovanillic acid levels in psychoses, 184
  prostaglandins in, 159
Chlorpromazine
  effects enhanced by thyroid hormones, 49
  galactorrhea from, 149
  prolactin response to, 164-166, 171
  in schizophrenia, 181-185
Chromosomes
  androgen insensitivity in patients with X, Y chromosomes, 76
  therapy for 46,XY and 47,XYY sex offenders, 105-123

Nervous system, central
cortisol levels in depression, 197-198
gonadal steroids affecting, 100-102
in prenatal development, 70, 73,
85-86
thyroid hormone affecting, 137
Neuroendocrine transducer cells, 145-147
Neuroleptics, *see* Psychotropic drugs
Neurotransmitters
growth hormone regulation and, 258
monoamine, activity of, 147-149
Norepinephrine
action of psychotropic drugs and, 147-149
behavioral effects of, 21
growth hormone regulation and, 258
interaction with thyroid hormones, 52-54
turnover of, growth hormone response
to hypoglycemia and, 233-235
19-Norprogesterone, behavioral responses
to, 5
Nutrition, *see* Eating patterns

Pain responses in reversible hyposomato-
trophic dwarfism, 247
Pancreas, neuroendocrine transducer cells
in, 146
Parafascicular nuclei, lesions in, behavioral
effects of, 10
Paranoid symptoms, from imipramine with
testosterone, 59, 60
Parathyroid disorders, psychiatric symptoms in, 131-134
hyperparathyroidism, 132-134
hypoparathyroidism, 131-132
Parkinsonism
effects of MIF and, 37
effects of TRH and, 20
Pentobarbital
action of SRIF and, 21
antagonism by TRH, 17
Phenothiazines
galactorrhea from, 149, 164
prolactin response to, 166, 181-185
Phentolamine, growth hormone release
and, 258
Phobias, stress procedure in treatment
of, 231-232
Physostigmine, prolactin secretion and, 169
Pimozide
affecting dopamine receptors, 163, 157
prolactin response to, 167
Pineal gland, neuroendocrine transducer
cells in, 146
Piperidine phenothiazines, prolactin response to, 166
Pituitary gland
behavioral responses to hormones of, 1-11

Pituitary gland (*contd.*)
deficiency in anorexia nervosa, 285-290
hypothalamic-adrenal cortical system in
depression and, 193-213, 225
hypothalamic-growth hormone system in
depression and, 213-218
interaction with hypothalamus, 147
Pregnancy, *see* Prenatal hormone effects
Pregnenolone, behavioral responses to, 5
Prenatal hormone effects, 69-88
animal data on, 69-74
California study of, 78-83
controversies in, 83-87
human research in, 74-83
sexual behavior and, 71-73
Pressinamide, behavioral responses to, 2
Progesterone
behavioral responses to, 5, 99-102
postpartum levels of, 122-123
Progestins, and prenatal development, 75,
76-79
Prolactin
behavioral responses to, 10
inhibition of secretion, 147
levels in anorexia nervosa, 273
levels in schizophrenia, 177-188
after discontinuing neuroleptics,
185-187
imipramine affecting, 187-188
lithium affecting, 187-188
in newly admitted patients, 178-181
phenothiazines affecting, 166, 181-185
prostaglandins affecting levels of, 160
psychotropic drugs affecting secretion
of, 149-150, 163-173
chlorpromazine, 163-166, 171
clozapine, 170
fluphenazine, 166, 173
haloperidol, 166, 172
physostigmine, 169
pimozide, 167
promazine, 170
promethazine, 170
thiethylperazine, 171
thioridazine, 168, 171
stress affecting levels of, 179-181
variations in sleep cycle, 179
Promazine, prolactin secretion and, 170
Promethazine, prolactin secretion and, 170
Prostaglandins, 159-160
interaction with norepinephrine, 53
Pseudohypoparathyroidism, mental disorders in, 132
Psychiatric symptoms
in endocrine disorders, 125-142
adrenal, 127-131
parathyroid, 131-134
thyroid, 134-141